BARRON'S

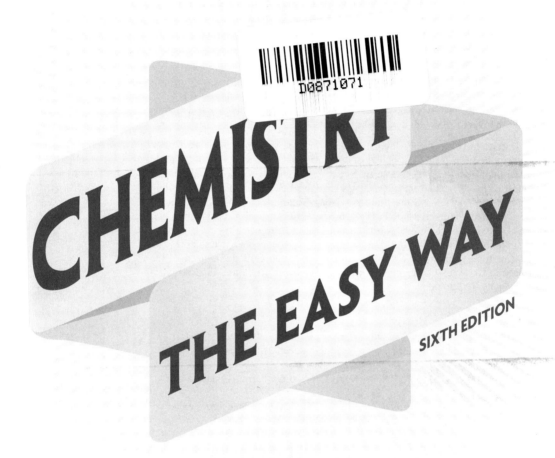

D0871071

CHEMISTRY
THE EASY WAY

SIXTH EDITION

LEARN CHEMISTRY THE EASY WAY!

MARK C. KERNION, M.A.
Chemistry Teacher
Mount Lebanon High School
Pittsburgh, Pennsylvania

JOSEPH A. MASCETTA, M.S., C.A.S.
Former Principal
Chemistry Teacher and Coordinator
Science Department
Mount Lebanon High School
Pittsburgh, Pennsylvania
Science Educational Consultant

About the Authors:

Mark Kernion has been teaching AP and honors chemistry for the past twenty-five years at Mt. Lebanon High School in Pittsburgh, Pennsylvania. In 2006, he was inducted into the Cum Laude Society, which is dedicated to honoring scholastic achievement in secondary schools. He was the recipient of the 2007 Yale Teaching Award. An inventor, he holds a number of patents.

Joe Mascetta has taught high school chemistry for twenty years. He was the science department coordinator and principal of Mt. Lebanon High School in Pittsburgh, Pennsylvania. He also served as a science consultant to the area schools and is a past president of the Western Pennsylvania Association of Supervision and Curriculum Development (ASCD) and the State Advisory Committee of ASCD. He holds degrees from the University of Pittsburgh, the University of Pennsylvania, and Harvard University. He was a participant in Harvard Project Physics, a General Electric Science Fellowship to Union College in Schenectady, New York; the Chemical Bond Approach Curriculum Study at Kenyon College, Ohio; and the Engineering Concepts Curriculum Project and Science Curriculum Supervision at the University of Colorado.

© Copyright 2019 by Kaplan, Inc., d/b/a Barron's Educational Series
© Copyright 2009 by Kaplan, Inc., d/b/a Barron's Educational Series under the title *E-Z Chemistry*.
© Copyright 2003, 1996, 1989, 1983 by Kaplan, Inc., d/b/a Barron's Educational Series under the title *Chemistry the Easy Way*.
© Copyright 1981, 1969 by Kaplan, Inc., d/b/a Barron's Educational Series under the title *How to Prepare for College Board Achievement Tests: Chemistry*.

Published by Kaplan, Inc., d/b/a Barron's Educational Series
750 Third Avenue
New York, NY 10017
www.barronseduc.com

ISBN: 978-1-4380-1210-0

PRINTED IN CANADA

10 9 8 7 6 5 4 3 2 1

Kaplan, Inc., d/b/a Barron's Educational Series print books are available at special quantity discounts to use for sales promotions, employee premiums, or educational purposes. For more information or to purchase books, please call the Simon & Schuster special sales department at 866-506-1949.

CONTENTS

PART II: THE NATURE OF MATTER

2   ATOMIC STRUCTURE AND THE PERIODIC TABLE / 31

PART IV: THE STATES AND PHASES OF MATTER

5 GASES AND THE GAS LAWS / 107

6 CHEMICAL CALCULATIONS (STOICHIOMETRY) AND THE MOLE CONCEPT / 133

PART V: CHEMICAL REACTIONS

8 CHEMICAL REACTIONS AND THERMOCHEMISTRY / 187

9 RATES OF CHEMICAL REACTIONS / 205

10   CHEMICAL EQUILIBRIUM / 213

11   ACIDS, BASES, AND SALTS / 233

PART VIII: TESTING YOUR ACHIEVEMENT IN CHEMISTRY

PRACTICE TESTS IN CHEMISTRY / 385

EQUATIONS AND TABLES FOR REFERENCE / 485

GLOSSARY OF COMMON TERMS / 497

INDEX / 509

PREFACE

Chemistry the Easy Way is designed for students who are presently taking chemistry or need a general review of high school chemistry, for college students who need to get through Chemistry 101, and for anyone who simply wants a refresher in basic chemistry. Four important elements work together in this book to make it easy to learn and review chemistry.

1. **The most up-to-date terminology and usage.** All notations, word usage, units of measure, and terminology reflect the most current practices employed in chemistry. You can be confident you are learning the latest innovations, modern concepts, and language, which will make you more comfortable when moving on to more advanced material. Chapter summaries of the terminology used provide a review and an opportunity to use your computer to do online searches for further expanded explanations.

2. **Detailed explanations, charts, and graphs.** Explanations of major concepts and theories provide a detailed, up-to-date understanding of the basic ideas of chemistry. These presentations are frequently reinforced with additional, concrete examples, and marginal notations highlight important points. More than 150 charts, diagrams, graphs, and illustrations are designed to help you better visualize and understand the material.

3. **Detailed sample problems, end-of-chapter review questions, and explained answers.** The quantitative aspects of chemistry are explained step-by-step and then clarified with sample problems. The dimensional analysis method of keeping track of units is used throughout this book. Numerous sample problems aid in the development of problem-solving skills. "Problem Solving: A Thinking Skill" on page xviii provides a methodology for attacking chemistry problems efficiently. The more than 240 end-of-chapter review questions with explained answers help assess your progress.

4. **Three up-to-date practice tests with answers.** Three complete tests cover the whole range of topics included in a high school chemistry course. The content of all three tests is similar to the College Board's SAT Subject Area Test in Chemistry. Each test is accompanied by a detailed answer section and a self-evaluation section that enables you to identify your strengths and weaknesses. The first two tests are modeled after the SAT Subject Test in Chemistry. The third test is a typical multiple-choice test you might encounter in any full-year chemistry course.

If you are currently enrolled in a chemistry course, you will find this book to be an invaluable supplement to the standard textbook. If you are using this book as a refresher before taking a chemistry class, you will find that your understanding and appreciation of chemistry will be greatly enhanced. *Chemistry the Easy Way* provides you with a proven practical tool to help you achieve an excellent functional knowledge of chemistry.

INTRODUCTION

How to Use This Book

The purpose of this book is to introduce the student to the basic essentials of a good high school chemistry course. These are developed in a simplified manner with the visual reinforcement of charts, graphs, lists, simplified drawings, and chapter review exercises with explained answers.

It is important that the student read and comprehend each section. To check this comprehension, each chapter has a review and chapter summary at the end of the text. These include a variety of questions and a list of terms the student should know. The answers with explanations are given along with the chapter review. A concise glossary of important chemical terms is provided at the end of the book. The complete index assists the student in finding specific topics or information.

Since chemistry is a quantitative science, the mathematical solution of typical problems has been included. This process includes an introduction of the type of problem and a careful development of the solution.

Knowing laboratory setups and the procedures for conducting particular tests is another important aspect of chemistry. The last chapter of this book summarizes typical laboratory techniques and the basic procedures for specific laboratory tests. It also includes a review test with answers. A reference section also gives the student access to the most important tables and charts of chemical properties.

The last portion of the book is devoted to three practice tests. These should be of use to the student in preparing for any chemistry achievement test—be it the Regents test, College Entrance Examination Board test, or one of the privately produced standardized tests. The general advice is the same.

To prepare for a test, the student should allow sufficient time so that last minute cramming will be unnecessary. Start early! Review each of the chapters in this book and then attempt to answer the questions in the review section at the end of the chapter. When you experience difficulty on a particular topic, go back over the information here or in any good high school text until you feel you have mastered that concept. When you have finished this, go on to the general tests in the last section of this book. Attempt to answer the questions as quickly and accurately as possible so that you get some practice in pacing yourself. Omit difficult questions you are not sure of and then go back to them after you have answered the easier ones. Do not attempt questions for which you do not have the background to make a judgment since, on an actual test, a percentage of wrong answers may be automatically subtracted from the number of right answers. When you have finished each test, check your answers to see how well you have done and review material you have missed or omitted.

Diagnosing Your Needs

You can use the practice tests as a means of assessing your achievement in each of the major areas tested and as a guide for studying particular topics. To do this, you should take the practice test simulating the conditions, particularly the length of time allowed, of the actual test. You should give yourself 1 hour to answer the questions in Practice Test 1. Check your answers against the correct answers given immediately after the test. In the section following the answers to this test, directions are given to arrive at subscores for each of the identified areas you need to emphasize by comparing your subscores. Plan to spend more time reviewing the areas in which your subscores were the lowest. However, do not omit any area in your review.

TESTING YOUR ACHIEVEMENT IN A CHEMISTRY COURSE WITH THIS BOOK

1. **Use the Chapter Summaries**

   At the end of each chapter is a **Chapter Summary** section. This section summarizes all the concepts and ideas that were introduced in the chapter. Use these sections to see if you can explain their meaning and how you would use them in chemistry. They appear in boldface type in the chapter to draw attention to them. The boldface type also makes the terms easier for you to look up if you need to. You should also use a search engine on your computer to get a quick and expanded explanation of these terms, laws, and formulas on the Internet.

2. **Take the Review Tests**

   At the end of each chapter is a **Review Test**. Each test has a series of questions that test your knowledge of the material in that chapter. Take the test, and then check your answers in the **Answers and Explanations** section. There are a total of over 240 review questions in these sections.

3. **Take the two tests (Practice Tests 1 and 2) modeled after the SAT Subject Test in Chemistry**
   The College Board Subject Area Test in Chemistry is the most prominent national achievement test in chemistry. It measures the understanding of chemistry independent of a particular textbook or instructional approach and at a level appropriate for college preparation. You are given an hour to complete 85 questions. Details about the test are given in the section "Practice Tests in Chemistry," under General Information (see page 385).

   You should be aware that the ACT Test in Science does not concentrate on chemistry. The test is only a 40-question, 35-minute test that measures the skills required in natural sciences. The test assumes that students are in the process of taking a core science course of study (three years or more) and have completed a course in Earth science and/or physical science and a course in biology. Obviously, it does not attempt to evaluate your achievement in chemistry.

4. **Take the test (Practice Test 3) that uses only a variety of multiple-choice questions**
   This test covers the same areas of chemistry as the previous two practice tests but does not use the true/false questions that are distinctive to the CEEB Subject Area Test in Chemistry. This test uses a variety of multiple-choice type questions that are used on chemistry tests.

Taking a Test

Below are some suggestions for taking actual tests:

1. Get a good night's sleep. If you should become ill while taking the test, report this to the proctor so that he or she may record this information.

2. Read the instructions carefully and be sure you understand them.

3. Skip questions that seem too difficult for you. Go on to the other questions and come back to the omitted ones if time permits. An easy question counts as much as a hard question.

4. Avoid haphazard guessing since this probably will lower your score. If you can eliminate some of the choices to a question, however, it will be to your advantage to answer the question even if you must guess which of the remaining answers is correct. All questions that are answered correctly count the same toward your score.

5. It is important to pace yourself throughout the hour to answer as many questions as possible. Make the best use of your time! Don't be too upset then if you don't finish the test. At least you have done the best you could in the 1-hour time.

Problem Solving: A Thinking Skill*

Chemistry is a subject that deals with many problem situations that you, the student, must be able to solve. Solving problems may seem to be a natural process when the degree of difficulty is not very great. In these cases, you may not need to have a structured method to attack the problem. However, for complex problems you will need an orderly process to solve the problem. The following is such a problem-solving process. Each step is vital to the next step and to the final solution of the problem.

STEP 1. Clarify the problem: to separate the problem into the facts, the conditions, and the questions that need to be answered, and to establish the goal.

STEP 2. Explore: to examine the sufficiency of the data, to organize the data, and to apply previously acquired knowledge, skills, and understanding.

STEP 3. Select a strategy: to choose an appropriate method to solve the problem.

STEP 4. Solve: to apply the skills needed to carry out the strategy chosen.

STEP 5. Review: to examine the reasonableness of the solution and to evaluate the effectiveness of the process.

The steps of the problem-solving process listed above should be followed in sequence. The subskills listed below for each step, however, are not in sequence. The order in which subskill patterns are used will differ with the nature of the problem and/or with the ways in which the individual problem solver thinks. Also, not every subskill need be employed in solving every problem.

CLARIFY THE PROBLEM

Identify the facts. What is known about the problem?

Identify the conditions. What is the current situation?

Identify the questions. What needs to be answered before the problem can be solved?

*Adapted with permission from *Thinking Skills Resource Guide*, a noncopyrighted publication of Mount Lebanon School District, Pittsburgh, PA.

Visualize the problem.

- Make mental images of the problem.

- If desirable or necessary, draw a sketch or diagram, make an outline, or write down symbols or equations that correspond to the mental images.

Establish the goal.   The goal defines the specific result to be accomplished through the problem-solving process. It defines the purpose or function the solution is expected to achieve and serves as the basis for evaluating the solution.

EXPLORE

Review previously acquired knowledge, skills, and understanding.   Determine whether the current problem is similar to a previously seen type of problem.

Estimate the sufficiency of the data.   Does there seem to be enough information to solve the problem?

Organize the data.   There are many ways in which data can be organized. Some examples are outline, written symbols and equations, chart, table, graph, map, diagram, and drawing. Determine whether the data organized in the way(s) you have chosen will enable you to partially or completely solve the problem. The organization of data may suggest what new data need to be collected.

Determine what new data, if any, need to be collected.   What new information may be needed to solve the problem? Can the existing data be reorganized to generate new information? Do other resources need to be consulted? This step may suggest possible strategies to be used to solve the problem.

SELECT A STRATEGY

A strategy is a goal-directed sequence of mental operations. Selecting a strategy is the most important and also the most difficult step in the problem-solving process. Although there may be several strategies that will lead to the solution of a problem, the skilled problem solver uses the most efficient strategy. The choice of the most efficient strategy is based on knowledge and experience as well as a careful application of the clarify-and-explore steps of the problem-solving method. Some problems may require the use of a combination of strategies.

The following search methods may help you to select a strategy. They do not represent all of the possible ways in which this can be done. Other methods of strategy selection are related to specific content areas.

Trial-and-error search.   Such a search either doesn't have or doesn't use information that indicates that one path is more likely to lead to the goal than any other path.

Trial-and-error search comes in two forms, blind and systematic. In *blind search*, the searchers pick paths to explore blindly, without considering whether they have already explored these paths. A preferable method is *systematic search*, in which the searchers keep track of the paths they have already explored and do not duplicate them. Because this method avoids multiple searches, systematic search is usually twice as efficient as blind search.

Reduction method. This involves breaking the problem into a sequence of smaller parts by setting up subgoals. Subgoals make problem solving easier because they reduce the amount of time required to find the solution.

You can set up subgoals by working part way into a problem and then analyzing the partial goal to be achieved. In doing this, you can drop the problem restrictions that do not apply to the subgoal. By adding up all the subgoals, you can solve the "abstracted" problem.

Working backward. When you have trouble solving a problem head-on, it is often useful to try to work backward. Working backward involves a simple change in representation or point of view. Your new starting point is the original goal. Working backward can be helpful because problems are often easier to solve in one direction than in another.

Knowledge-based method. This strategy uses information stored in the problem solver's memory, or newly acquired information, to guide the search for the solution. The problem solver may have solved a similar problem and can use this knowledge in a new situation. In other cases, problem solvers may have to acquire needed knowledge. For example, they may solve an auxiliary problem to learn how to solve the one they are having difficulty with.

Searching for analogous (similar) problems is a very powerful problem-solving technique. When you are having difficulty with a problem, try to pose a related, easier one and hope to learn something that will help you solve the harder problem.

SOLVE

Use the strategy chosen to actually solve the problem. Executing the solution provides you with a very valuable check on the adequacy of your plan. Sometimes students will look at a problem and decide that, since they know how to solve it, they need not bother with the drudgery of actually executing the solution. Sometimes the students are right, but at other times they miss an excellent opportunity to discover that they were wrong.

REVIEW

Evaluation. The critical question in evaluation is, Does the answer I propose meet all of the goals and conditions set by the problem? Thus, after the effort of finding a solution, you must turn back to the problem statement and check carefully to be sure your solution satisfies it.

With easy problems there is a strong temptation to skip evaluation because the probability of error seems small. In some cases, however, this can be costly. Evaluation may prove that errors were present.

Verifying the reasonableness of the answer. It is easy to become so involved with the process and mathematics of a problem that an answer is recorded that is totally illogical. To avoid this mistake, you should simplify the numbers involved and solve for an answer. Having done this, compare your estimated result with your answer to ensure that your answer is feasible.

For example, a problem requires the following operations:

$$5.12 \times 10^5 \times 3.98 \times 10^6 \text{ divided by } 910$$

And doing all the math, you get an answer of

$$0.02239 \times 10^{11} \text{ or } 2.24 \times 10^9$$

To estimate the answer, first simplify the numbers to one significant figure (significant figures are discussed in Chapter 1). This gives

$$5 \times 10^5 \times 4 \times 10^6 \text{ divided by } 9 \times 10^2$$

which is

$$20 \times 10^{11} \text{ divided by } 9 \times 10^2 = 2.2 \times 10^9$$

This is the estimated answer, which validates the answer above.

When you are dealing with test items that provide multiple-choice answers, you can often use estimation to arrive at the answer without doing the more complicated mathematics.

Consolidation. Here the basic question to be answered is, What can I learn from the experience of solving this problem? The following more specific questions may help you to answer this general one:

- Why was this problem difficult?
- Was it difficult to follow a plan?
- Was it difficult to decide on a plan? If so, why?
- Did I take the long way to the answer?
- Can I use this plan again in similar problems?

The important thing is to reflect on the process that you used in order to make future problem solving easier.

PART I

AN INTRODUCTION

INTRODUCTION TO CHEMISTRY

CHAPTER OBJECTIVES

Upon completing this chapter, you will be able to:

- Identify types of matter as elements, compounds, or mixtures

- Identify chemical and physical properties and changes

- Explain how energy is involved in chemical and physical changes

- Identify and use the SI system of measurement

- Do mathematical calculations using scientific notation, dimensional analysis, and significant figures

Matter

DEFINITION OF MATTER

Matter is defined as anything that occupies space and has mass. **Mass** is the quantity of matter that a substance possesses and, depending on the gravitational force acting on it, has a unit of **weight** assigned to it. Although the weight can then vary, the mass of the body is a constant and can be measured by its resistance to a change of position or motion. This property of mass to resist a change of position or motion is called **inertia**. Since matter does occupy space, we can compare the masses of various substances that occupy a particular unit volume. This relationship of mass to a unit volume is called the **density** of the substance. It can be shown in a mathematical formula as $D = m/V$. A common unit of mass (m) in chemistry is the gram (g), and of volume (V) is the cubic centimeter (cm^3) or milliliter (mL).

An example of how density varies can be shown by the difference in the volume occupied by 1 g of a metal, such as gold, and 1 g of styrofoam. Both have the same mass, that is, 1 g, but the volume occupied by the styrofoam is much larger. Therefore, the density of the metal will be much larger than that of the styrofoam. When dealing with gases in chemistry, the standard units for the density of gases are grams per liter at a standard temperature and pressure. This aspect of the density of gases is dealt with in Chapter 6. Basically then, density can be defined as the mass per unit volume.

STATES OF MATTER

Matter occurs in three states: solid, liquid, and gas. A **solid** has both a definite size and shape. A **liquid** has a definite volume but takes the shape of the container, and a **gas** has neither a definite shape nor a definite volume. These states of matter can often be changed by the addition of heat energy. An example of this is ice changing to liquid water and finally to steam.

COMPOSITION OF MATTER

Matter can be subdivided into two general categories: distinct (pure) substances and mixtures. Distinct substances are substances that can be subdivided into the smallest particle that still has the properties of the substance. At that point, if the substance is made up of only one kind of atom, it is called an **element**. Atoms are considered to be the basic building blocks of matter that cannot be easily created or destroyed. The word *atom* comes from the Greek and means the smallest possible piece of something. Today there are approximately 109 different kinds of atoms, each with its own unique composition. These atoms then are the building blocks of elements when only one kind of atom makes up the substance. If, however, there are two or more kinds of atoms joined together in definite grouping, this distinct substance is called a **compound**. Compounds are made by combining elements in a definite proportion (or ratio) by mass and are made up of two or more kinds of atoms. This is called the **Law of Definite Composition (or Proportions)**. The smallest natural occurring unit of a compound is called a *molecule* of that compound. A molecule of a compound has a definite shape that is determined by how the atoms are bonded to or combine with each other. This bonding is described in Chapter 3. An example is the compound water: it always occurs in a two hydrogen atoms to one oxygen atom relationship. **Mixtures**, however, can vary in their composition.

In general, then:

Mixtures	Distinct Substances
	Elements
1. Composition is indefinite (generally heterogeneous).* (Example: marble)	1. Composition is made up of one kind of atom. (Examples: nitrogen, gold, neon)
2. Properties of the constituents are retained.	2. All parts are the same throughout (homogeneous).
3. Parts of the mixture react differently to changed conditions.	
	Compounds
	1. Composition is definite (homogeneous). (Examples: water, carbon dioxide)
	2. All parts react the same.
	3. Properties of the compound are distinct and different from the properties of the individual elements that are combined in its makeup.

*Solutions are mixtures, such as sugar in water, but since the substance, like sugar, is distributed evenly throughout the water, it can be said to be a homogeneous mixture.

The following chart shows a classification scheme for matter.

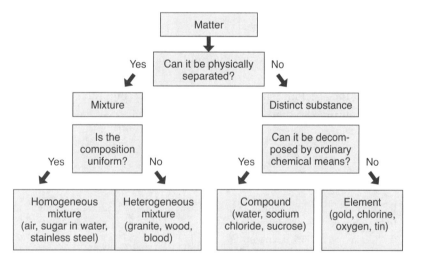

CHEMICAL AND PHYSICAL PROPERTIES

Physical properties of matter are those properties that can usually be observed with our senses. They include everything about a substance that can be noted when no change is occurring in the type of structure that makes up its smallest component. Some common examples are physical state, color, odor, solubility in water, density, melting point, taste, boiling point, and hardness.

Chemical properties are those properties that can be observed with regard to whether or not a substance changes chemically, often as a result of reacting with other substances. For example, iron rusts in moist air, nitrogen does not burn, gold does not rust, sodium reacts with water, silver does not react with water, and water can be decomposed by an electric current.

CHEMICAL AND PHYSICAL CHANGES

The changes matter undergoes are classified as either physical or chemical. In general, a **physical change** alters the physical properties of matter, but the composition remains constant. The most often altered properties are form and state. Some examples are breaking glass, cutting wood, melting ice, and magnetizing a piece of metal. In some cases, the process that caused the change can be easily reversed and the substance regains its original form.

Chemical changes are changes in the composition and structure of a substance. They are always accompanied by energy changes. If the energy released in the formation of a new structure exceeds the chemical energy in the original substances, energy will be given off, usually in the form of heat or light or both. This is called an **exothermic reaction**. If, however, the new structure needs to absorb more energy than is available from the reactants, the result is an **endothermic reaction**. This can be shown graphically.

Notice that in Figures 1 and 2 the term **activation energy** is used. The activation energy is the energy necessary to get the reaction going by increasing the energy of the reactants so that they can combine. You know you have to heat paper before it burns. This heat raises the energy of the reactants so that the burning can begin; then enough energy is given off from the burning so that an external source of energy is no longer necessary.

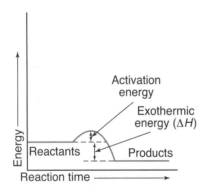

FIGURE 1. An exothermic reaction.

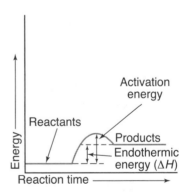

FIGURE 2. An endothermic reaction.

CONSERVATION OF MASS

When ordinary chemical changes occur, the mass of the reactants equals the mass of the products. This can be stated another way: In a chemical change, matter can neither be created nor destroyed, but only changed from one form to another. This is referred to as the **Law of Conservation of Matter** (Lavoisier—1785). This law is extended by the Einstein mass-energy relationship, which states that matter and energy are interchangeable (see page 9).

Energy

DEFINITION OF ENERGY

The concept of energy plays an important role in all of the sciences. In chemistry, all physical and chemical changes have energy considerations associated with them. To understand how and why these changes happen, an understanding of energy is required.

Energy is defined as the capacity to do work. Work is done whenever a force is applied over a distance. Thus anything that can force matter to move, to change speed, or to change direction has energy. The following example will help you understand this definition of energy. When you charge a battery with electricity, you are storing energy in the form of chemical energy. The charged battery has a capacity to do work. If you use the battery to operate a toy car, the energy stored is transformed into mechanical energy which exerts a force on the mechanism that turns the wheels and makes the car move. This continues until the "charge" or stored energy is used up. In its uncharged condition, the battery no longer has the capacity to do work.

Just as work itself is measured in **joules** (J), so is energy. In some problems, it may be expressed in *kilocalories* (kcal). The relationship between these two units is that 4.18×10^3 J is equal to 1 kcal.

FORMS OF ENERGY

Energy may appear in a variety of forms. Most commonly, energy in reactions is evolved as *heat*. Some other forms of energy are *light*, *sound*, *mechanical energy*, *electrical energy*, and *chemical energy*. Energy can be converted from one form to another, as when the heat from burning fuel is used to vaporize water to steam. The energy of the steam is used to turn the wheels of a turbine to produce mechanical energy. The turbine turns the generator armature to produce electricity, which is then available in homes for use as light or heat or for the operation of many modern appliances.

Two general classifications of energy are **potential energy** and **kinetic energy**. Potential energy is stored energy due to overcoming forces in nature. Kinetic energy is energy of motion. The difference can be illustrated by a boulder sitting on the side of a mountain. It has a high potential energy because of its position overcoming gravity above the valley floor. If it falls, however, its potential energy is converted to kinetic energy. This illustration is very similar to the situation of electrons cascading to lower energy levels in the atomic model, where the electrostatic force is at work.

This concept can also be applied at the molecular level. For example, the molecule of N_2 in the diagram below shows three variations of kinetic energy. The N_2 molecule also possesses potential energy because of the chemical bond within the molecule. To break this bond would require energy. The potential energy in each isolated atom of nitrogen would then be greater. This would be similar to raising the boulder to a higher position on the mountainside. Both systems have a higher potential energy and have a tendency to fall back to the lower state, which is more stable.

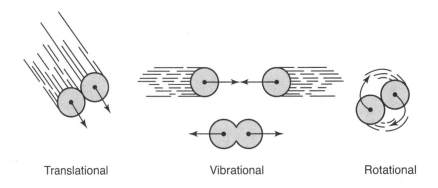

| Translational | Vibrational | Rotational |

TYPES OF REACTIONS (EXOTHERMIC VERSUS ENDOTHERMIC)

When physical or chemical changes occur, energy is involved. If the heat content of the product(s) is higher than that of the reactants, the reaction is **endothermic**. If, on the other hand, the heat content of the product(s) is less than that of the reactants, the reaction is **exothermic**. This change of heat content can be designated as ΔH. The heat content (H) is sometimes referred to as the enthalpy. Every system has a certain amount of heat that changes during the course of a physical or chemical change. The change in heat content (ΔH) is the difference between the heat content of the products and that of the reactants. The equation is

$$\Delta H = H_{products} - H_{reactants}$$

If the heat content of the products is greater than the heat content of the reactants, ΔH is a positive quantity ($\Delta H > 0$) and the reaction is endothermic. If, however, the heat content of the products is less than the heat content of the reactants, ΔH is a negative quantity

($\Delta H < 0$) and the reaction is exothermic. This relationship is shown graphically in Figures 1 and 2 on page 6. This topic is developed in detail in Chapter 8.

CONSERVATION OF ENERGY

Experiments have shown that energy is neither gained nor lost during physical or chemical changes. This principle is known as the **Law of Conservation of Energy** and is often stated as follows: Energy is neither created nor destroyed in ordinary physical and chemical changes. If the system under study loses energy (the reaction is exothermic and ΔH is negative), the surroundings of the system must gain the energy that the system loses so that energy is conserved.

Conservation of Mass and Energy

With the introduction of atomic theory and a more complete understanding of the nature of both mass and energy, it was found that a relationship exists between these two concepts. Einstein formulated the **Law of Conservation of Mass and Energy**, which states that mass and energy are interchangeable under special conditions. The conditions have been created in nuclear reactors and accelerators, and the law has been verified. This relationship can be expressed by Einstein's famous equation:

$$E = mc^2$$

Energy = Mass $\times$ (velocity of light)^2

Measurements and Calculations

THE SCIENTIFIC METHOD

Although some discoveries are made in science by accident, in most cases the scientists involved use an orderly process to work on their projects and discoveries. The process researchers use to carry out their investigations is often called the *scientific method*. It is a logical approach to solving problems by observing and collecting data, formulating a hypothesis, and constructing theories supported by the data. The formulating of a hypothesis consists of carefully studying the data collected and organized to see if a testable statement can be made with regard to the data. The hypothesis takes the form of an "if . . . then" statement. If certain data is true, then a prediction can be made concerning the outcome. The next step is to test the prediction to see if it withstands the experimentation. The hypothesis can go through several revisions as the process continues. If the data from experimentation shows that the predictions of the hypothesis are successful, scientists usually try to explain the phenomena by constructing a *model*.

The model can be a visual, verbal, or mathematical means of explaining how the data is related to the phenomena.

The stages of this process can be illustrated by the diagram shown below:

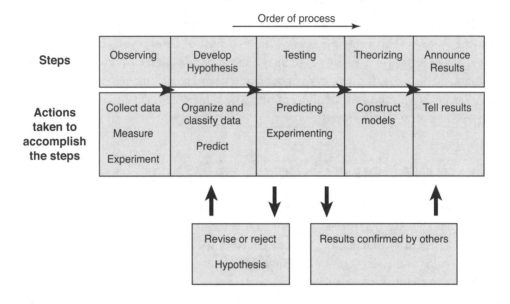

MAKING OBSERVATIONS

Students of chemistry must be able to make good observations. The two types of observations are qualitative and quantitative. Qualitative observations involve descriptions of the nature of the substances under investigation. Quantitative observations involve making measurements to describe the substances under observation. Students must be able to use the correct measurement terms accurately and to solve problems that require mathematical skill, as well as proper terminology, for their correct solution. The following sections review these topics.

SI SYSTEM

It is important that scientists around the world use the same units when communicating information. For this reason, scientists use the modernized metric system, designated in 1960 by the General Conference on Weights and Measures as the International System of Units. This is commonly known as the **SI system**, an abbreviation for the French *Système International d'Unités*. It is the most common system of measurement in the world.

The reason it is so widely accepted is twofold. First, SI uses a decimal system to inflate or deflate units. Second, in many cases units for various quantities are defined in terms of units for simpler quantities.

There are seven basic units that can be used to express the fundamental quantities of measurement. These are called the SI base units and are shown in the following table.

SI BASE UNITS

Quantity	Unit	Abbreviation
mass	kilogram	kg
length	meter	m
time	second	s
electric current	ampere	A
temperature	kelvin	K
amount of substance	mole	mol
luminous intensity	candela*	cd

*The candela is rarely used in chemistry.

Other SI units are derived by combining prefixes with a base unit. The prefixes represent multiples or fractions of 10. The following table gives some prefixes used in the metric system.

PREFIXES USED WITH SI UNITS

Prefix	Symbol	Meaning	Exponential Notation
exa-	E	1,000,000,000,000,000,000	10^{18}
peta-	P	1,000,000,000,000,000	10^{15}
tera-	T	1,000,000,000,000	10^{12}
giga-	G	1,000,000,000	10^{9}
mega-	M	1,000,000	10^{6}
kilo-	k	1,000	10^{3}
hecto-	h	100	10^{2}
deka-	da	10	10^{1}
—	—	1	10^{0}
deci-	d	0.1	10^{-1}
centi-	c	0.01	10^{-2}
milli-	m	0.001	10^{-3}
micro-	μ	0.000 001	10^{-6}
nano-	n	0.000 000 001	10^{-9}
pico-	p	0.000 000 000 001	10^{-12}
femto-	f	0.000 000 000 000 001	10^{-15}
atto-	a	0.000 000 000 000 000 001	10^{-18}

For an example of how the prefix works in conjunction with the base unit, consider the term *kilometer*. The prefix *kilo-* means "multiply the base unit by 1000," and so a kilometer is 1000 m. By the same reasoning, a millimeter is 1/1000 m.

Because of the prefix system, all units and quantities can be easily related by some factor of 10. The following is a brief table of some SI unit equivalents.

LENGTH
10 millimeters (mm) = 1 centimeter (cm)
100 cm = 1 meter (m)
1000 m = 1 kilometer (km)

MASS
1000 milligrams (mg) = 1 gram (g)
1000 g = 1 kilogram (kg)

A unit of length, used especially in expressing the length of light waves, is the nanometer, abbreviated as nm and equal to 10^{-9} meter.

Because measurements are occasionally reported in units of the English system, it is important to be aware of some SI-to-English system equivalents. Some common conversion factors are shown in the following table.

2.54 centimeters = 1 inch (in)	1 kilogram = 2.2 pounds
1 meter = 39.37 inches (10% longer than a yard [yd])	0.946 liter = 1 quart (qt)
28.35 grams = 1 ounce (oz)	1 liter (5% larger than a quart) = 1.06 quarts
454 grams = 1 pound (lb)	

The SI system standards were chosen as natural standards. The meter was first described as the distance marked off on a platinum-iridium bar but is now defined as the length of the path traveled by light in a vacuum during a time interval of $1/2.99792458 \times 10^8$ second.

Volume is an example of a **derived quantity** in the SI system. It is mathematically derived by multiplying the length, width, and height of a regularly shaped object. Since these quantities can all be measured in meters, a derived unit for volume in the SI system is the cubic meter (m^3). Of course, smaller units of distance could be used. This means that cubic centimeters (cm^3) would also represent a derived unit for volume in the SI system. As the figure on page 13 shows, a 10 cm $\times$ 10 cm $\times$ 10 cm cube is also called a liter. Since 1 liter equals 1000 cm^3, then 1 milliliter (mL) equals 1 cm^3. Since 10 cm equals 1 decimeter (dm), 1 liter is also equal to 1 cubic decimeter (dm^3).

There are some interesting relationships between volume and mass units in the SI system. Since water is most dense at 4°C, the gram was intended to be 1 cm^3 of water at this temperature. This means, then, that:

$1000 \text{ cm}^3 = 1 \text{ L}$ of water at 4°C.
1000 cm^3 of water weighs
1000 g at 4°C.

Therefore

1 L of water at 4°C weighs 1 kg

and

1 mL of water at 4°C weighs 1 g.

When 1 L is filled with water at 4°C, it has a mass of 1 kg.

1 mL = 1 cm^3

10 cm

10 cm   10 cm

1 liter
or
1000 cm^3

Most units used by scientists are mathematical manipulations (derived units) of the seven basic units or other derived units. The table below gives examples of commonly used derived units and the special names given to some of them.

DERIVED UNITS

Quantity	Derivation	Unit	Special Name
density	mass/volume	g/mL	—
velocity	distance/time	m/s	—
acceleration	velocity/time	m/s^2	—
force	mass × acceleration	kg · m/s^2	newton
energy	force × distance	kg · m^2/s^2	joule

TEMPERATURE MEASUREMENTS

The most commonly used temperature scale in scientific work is the **Celsius** scale. It gets its name from the Swedish astronomer Anders Celsius and dates back to 1742. For a long time it was called the centigrade scale because it is based on the concept of dividing the distance on a thermometer between the freezing point of water and its boiling point into 100 equal markings or degrees.

Another scale used by the SI system is based on the lowest theoretical temperature (called absolute zero). This temperature has never actually been reached, but scientists in laboratories have recorded temperatures within about a billionth of a unit above absolute zero. William Thomson, also known as Lord Kelvin, proposed this scale. A unit (the kelvin) is the same size as a Celsius degree and is referred to as the **Kelvin** or *absolute* temperature scale. Through experiments and calculations, it has been determined that absolute zero is 273.15° below zero on the Celsius scale. This figure is usually rounded off to −273°C.

The diagram and formulas that follow give the graphic and algebraic relationships between the temperature scales encountered in chemistry.

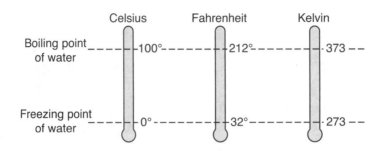

Celsius   Fahrenheit   Kelvin

Boiling point _ _ _ _ $-100°$ _ _ _ _ _ $-212°$ _ _ _ _ _ $\cdot 373 - -$
of water

Freezing point _ _ _ _ $-0°$ _ _ _ _ _ $-32°$ _ _ _ _ _ $\cdot 273 - -$
of water

CONVERSION FORMULAS
$°F = \frac{9}{5}°C + 32°$
$°C = \frac{5}{9}(°F - 32°)$
$K = °C + 273°$
$°C = K - 273$

EXAMPLE 1

$30°C = $ _____ $°F$

SOLUTION

$$°F = \frac{9}{5}(30°) + 32° = 86°$$

EXAMPLE 2

$68°F = $ _____ $°C$

SOLUTION

$$°C = \frac{5}{9}(68° - 32°) = 20°$$

EXAMPLE 3

$10°C = $ _____ K

SOLUTION

$$K = 10 + 273 = 283 \text{ K}$$

Note: In Kelvin notation, the degree sign is omitted: 283 K. The unit is the kelvin, abbreviated K.

EXAMPLE 4

$200 \text{ K} = $ _____ $°C$

SOLUTION

$$°C = 200 - 273 = -73°$$

HEAT MEASUREMENTS

The scales previously discussed are used to measure the degree of heat. A pail of water and a thimble full of water can both be filled with water at 100°C. They both have the same measurement of the degree of heat of the water. However, the pail of water has a greater quantity of heat. This can be easily demonstrated by the amount of ice that can be melted by the water in these two containers. Obviously, the pail of water at 100°C will melt more ice than a thimble full of water at the same temperature. We say that the pail of water contains a greater number of *calories* (cal) of heat. The calorie unit is used to measure the quantity of heat. It is defined as the amount of heat needed to raise the temperature of 1 g of water by 1° on the Celsius scale. This is a rather small unit for the quantities of heat that are involved in most chemical reactions. Therefore, the *kilocalorie* (kcal) is more often used. The kilocalorie is equal to 1000 cal. It is the quantity of heat that will increase the temperature of 1 kg of water by 1° on the Celsius scale. As a unit of heat energy, 1 cal is approximately 4.18 J. As described in an earlier table, the joule is a derived unit in the SI system equal to a kilogram meter squared per second squared ($kg \cdot m^2/s^2$). Since this unit is fairly small, changes in energy associated with chemical reactions are often expressed in kilojoules (kJ).

Problems involving heat transfers in water are called water calorimetry problems and are explained on page 161.

SCIENTIFIC NOTATION

When students must do mathematical operations with numerical figures, the **scientific notation** *system* is very useful. Basically this system uses the exponential means of expressing figures. With large numbers, such as 3,630,000, move the decimal point to the left until only one digit remains to the left (3.630000) and then indicate the number of moves of the decimal point as the exponent of 10 giving you 3.63×10^6. With a very small number such as 0.000000123, move the decimal point to the right until only one digit is to the left 0000001.23 and then express the number of moves as the negative exponent of 10 giving you 1.23×10^{-7}.

With numbers expressed in this exponential form, you can now use your knowledge of exponents in mathematical operations. An important fact to remember is that in multiplication you add the exponents of 10, and in division you subtract the exponents. Addition and subtraction can be performed only if the values have the same exponent.

MULTIPLICATION:

EXAMPLE 1 $(2.3 \times 10^5)(5.0 \times 10^{-12}) =$

SOLUTION

Multiplying the first number in each, you get 11.5, and the addition of the exponents gives 10^{-7}. Now changing to a number with only one digit to the left of the decimal point gives you 1.15×10^{-6} for the answer.

EXAMPLE 2 $(5.1 \times 10^{-6})(2 \times 10^{-3}) =$

SOLUTION

$(5.1 \times 10^{-6})(2 \times 10^{-3}) = 10.2 \times 10^{-9} = 1.02 \times 10^{-8}$

EXAMPLE 3 $(3 \times 10^5)(6 \times 10^3) =$

SOLUTION

$(3 \times 10^5)(6 \times 10^3) = 18 \times 10^8 = 1.8 \times 10^9$

DIVISION:

(Notice that in division the exponents of 10 are subtracted.)

EXAMPLE 1 $(1.5 \times 10^3) \div (5.0 \times 10^{-2}) =$

SOLUTION

$(1.5 \times 10^3) \div (5.0 \times 10^{-2}) = 0.3 \times 10^5 = 3 \times 10^4$

EXAMPLE 2 $(2.1 \times 10^{-2}) \div (7.0 \times 10^{-3}) =$

SOLUTION

$(2.1 \times 10^{-2}) \div (7.0 \times 10^{-3}) = 0.3 \times 10^1 = 3$

ADDITION AND SUBTRACTION:

EXAMPLE 1 $4.2 \times 10^4 \text{ kg} + 7.9 \times 10^3 \text{ kg} =$

SOLUTION

4.2×10^4 kg + (0.79 $\times 10^4$ kg, so that the exponents of 10 are the same) = 4.2×10^4 kg + 0.79×10^4 kg = 4.99×10^4 kg = 5.0×10^4 kg

Apply the same process to subtraction problems.

DIMENSIONAL ANALYSIS

When you are working problems that involve numbers with units of measurement, it is convenient to use this method so that you do not become confused in the operation of multiplication or division. For example, if you are changing 0.001 kg to milligrams, you set up each conversion as a fraction so that all the units will factor out except the one you want in the answer.

EXAMPLE 1

$$1 \times 10^{-3} \text{ kg} \times \frac{1 \times 10^{3} \text{ g}}{1 \text{ kg}} \times \frac{1 \times 10^{3} \text{ mg}}{1 \text{ g}} = 1 \times 10^{3} \text{ mg}$$

SOLUTION

Notice that the kilogram in the denominator of the first fraction can be factored out with the original kilogram unit. The numerator is equal to the denominator except that it is expressed in smaller units. The second fraction has the gram unit in the denominator to be factored out with the gram unit in the preceding fraction. The answer is in milligrams since this is the only unit remaining and it assures you that the correct operations have been performed in the conversion.

EXAMPLE 2

1 foot (ft) = ? centimeters

SOLUTION

$$1 \text{ ft} \times \frac{12 \text{ in.}}{1 \text{ ft}} \times \frac{2.54 \text{ cm}}{1 \text{ in.}} = 30.48 \text{ cm}$$

This method is used in examples throughout this book.

PRECISION, ACCURACY, AND UNCERTAINTY

Two other factors to consider in measurement are **precision** and **accuracy**. *Precision* indicates the reliability or reproducibility of a measurement. *Accuracy* indicates how close a measurement is to its known or accepted value.

For example, suppose you are taking a reading of the boiling point of pure water at sea level. Using the same thermometer in three trials, you record 96.8°C, 96.9°C, and 97.0°C. Since these figures show a high reproducibility, you can say that they are precise. However, the values are considerably off from the accepted value of 100°C, and so we say they are not accurate. In this example we probably would suspect that the inaccuracy was the fault of the thermometer.

Regardless of precision and accuracy, all measurements have a degree of **uncertainty**. This is usually dependent on two factors—the limitation of the measuring instrument and the skill of the person making the measurement. This can best be shown by example.

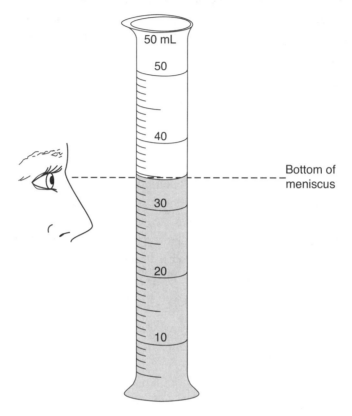

Graduated Cylinder Reading 34.3 mL

The graduated cylinder in the illustration contains a quantity of water to be measured. It is obvious that the quantity is between 30 and 40 mL because the **meniscus** lies between these two marked quantities. Now, checking to see where the bottom of the meniscus lies with reference to the ten intervening subdivisions, we see that it is between the fourth and fifth. This means that the volume actually lies between 34 and 35 mL. The next step introduces the uncertainty. We have to guess how far the reading is between these two markings. We can make an approximate guess, or estimate, that the level is more than 0.2 but less than 0.4 of the distance. We therefore report the volume as 34.3 mL. The last digit in any measurement is an estimate of this kind and is uncertain.

SIGNIFICANT FIGURES

Any time a measurement is recorded, it includes all the digits that are certain plus one uncertain digit. These certain digits plus the one uncertain digit are referred to as **significant figures**. The more digits you are able to record in a measurement, the less relative uncertainty there is in the measurement. The following table summarizes the rules of significant figures.

Rule	Example	Number of Significant Figures	
All digits other than zeros are significant.	25 g 5.471 g	2 4	
Zeros between nonzero digits are significant.	309 g 40.06 g	3 4	
Final zeros to the right of the decimal point are significant.	6.00 mL 2.350 mL	3 4	
In numbers smaller than 1, zeros to the left or directly to the right of the decimal point are not significant.	0.05 cm 0.060 cm	(1) (2)	The zeros merely mark the position of the decimal point. The first two zeros mark the position of the decimal point. The final zero is significant.

One last rule deals with final zeros in a whole number. These zeros may or may not be significant, depending on the measuring instrument. For instance, if an instrument that measures to the nearest mile (mi) is used, the number 3000 mi has four significant figures. If, however, the instrument in question records miles to the nearest thousands, there is only one significant figure. The number of significant figures in 3000 can be one, two, three, or four, depending on the limitation of the measuring device.

This problem can be avoided by using the system of scientific notation. For this example, the following notations indicate the numbers of significant figures:

$$3 \times 10^3 \qquad \text{one significant figure}$$

$$3.0 \times 10^3 \qquad \text{two significant figures}$$

$$3.00 \times 10^3 \qquad \text{three significant figures}$$

$$3.000 \times 10^3 \qquad \text{four significant figures}$$

CALCULATIONS WITH SIGNIFICANT FIGURES

When you do calculations involving numbers that do not have the same number of significant figures in each, it is important to keep these two rules in mind.

In multiplication and division, a simple rule that usually holds is that the number of significant figures in a product or a quotient obtained from manipulating figures of measured quantities is the same as the number of significant figures in the quantity having the smaller number of significant figures.

EXAMPLE
1
4.29 cm $\times$ 3.24 cm =

SOLUTION

Unrounded answer = 13.8996 cm^2.

Answer rounded to the correct number of significant figures = 13.9 cm^2.

Both measured quantities have three significant figures. Therefore, the answer should be rounded to three significant figures.

EXAMPLE
2
4.29 cm $\times$ 3.2 cm =

SOLUTION

Unrounded answer = 13.728 cm^2.

Answer rounded to the correct number of significant figures = 14 cm^2.

One of the measured quantities has only two significant figures. Therefore, the answer should be rounded to two significant figures.

EXAMPLE
3
8.47 cm^2/4.26 cm =

SOLUTION

Unrounded answer = 1.9882629 cm.

Answer rounded to the correct number of significant figures = 1.99 cm.

Both measured quantities have three significant figures. Therefore, the answer should be rounded to three significant figures.

In addition and subtraction, the simple rule is that when adding or subtracting measured quantities, the sum or difference should be rounded to the same number of decimal places as in the quantity having the least number of decimal places.

EXAMPLE
1

```
  3.56 cm
  2.6  cm
+ 6.12 cm
```
Total =

SOLUTION

Unrounded answer = 12.28 cm.

Answer rounded to the correct number of significant figures = 12.3 cm.

One of the quantities added has only one decimal place. Therefore, the answer should be rounded to only one decimal place.

EXAMPLE
2
3.514 cm

−2.13　cm

Difference =

SOLUTION

Unrounded answer = 1.384 cm.

Answer rounded to the correct number of significant figures = 1.38 cm.

One of the quantities has only two decimal places so the answer should be rounded to two decimal places.

Chapter Summary

The following terms summarize all the concepts and ideas that were introduced in this chapter. You should be able to explain their meaning and how you would use them in chemistry. They appear in boldface type in this chapter to draw your attention to them. The boldface type also makes the terms easier for you to look up, if you need to. You could also use a search engine on a computer to get a quick and expanded explanation of them.

TERMS YOU SHOULD KNOW

accuracy

activation energy

Celsius

chemical change

chemical properties

compound

density

derived quantity

element

endothermic

exothermic

gas

heterogeneous

homogeneous

inertia

Kelvin

joule

kinetic energy

Law of Conservation of Energy

Law of Conservation of Mass (Matter)

Law of Conservation of Mass and Energy

Law of Definite Composition (or Proportion)

liquid

mass

matter

meniscus

mixtures

physical change

physical property

potential energy

precision

scientific notation

SI system

significant figures

solid

uncertainty

weight

Chapter 1 Review Exercises

1. 1.2 mg   = _____ g

2. 6.3 cm   = _____ mm

3. 5.12 m   = _____ cm

4. 32°C   = _____ K

5. 6.111 mL   = _____ L

6. 1 km   = _____ mm

7. 1.03 kg   = _____ g

8. 0.003 g   = _____ kg

9. 22.4 L   = _____ mL

10. 10,013 cm   = _____ km

11. The density of CCl_4 (carbon tetrachloride) is 1.58 g/mL. What is the mass of 100. mL of CCl_4?

12. A piece of sulfur weighs 227 g. When it was submerged in a graduated cylinder containing 50 mL of H_2O, the level rose to 150 mL. What is the density (g/mL) of the sulfur?

13. (a) A box 20.0 cm × 20.0 cm × 5.08 in. has what volume in cubic centimeters?
    (b) What weight of H_2O at 4°C will the box hold?

14. Set up the following using dimensional analysis:

$$\frac{5 \text{ cm}}{\text{second}} = \frac{\text{kilometers}}{\text{hour}}$$

15. How many significant figures are in each of the following?

    (A) 1.01 g
    (B) 200.0 s
    (C) 0.0021 m
    (D) 0.0230 k

16. If the graphic representation of the energy levels of the reactants and products in a chemical reaction looks like this

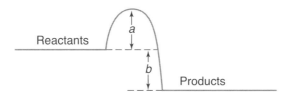

    (A) there is an exothermic reaction
    (B) there is an endothermic reaction
    (C) the *a* portion is the energy given off
    (D) the *b* portion is called the activation energy

17. The amount of mass per unit volume refers to the

    (A) density
    (B) specific weight
    (C) volume
    (D) weight

18. A baking powder can carry the statement, "Ingredients: corn starch, sodium bicarbonate, calcium hydrogen phosphate, and sodium aluminum sulfate." Therefore, this baking powder is

    (A) a compound
    (B) a mixture
    (C) a molecule
    (D) a mixture of elements

19. Which of the following is a physical property of sugar?

    (A) It decomposes readily.
    (B) Its composition is carbon, hydrogen, and oxygen.
    (C) It turns black with concentrated H_2SO_4.
    (D) It can be decomposed with heat.
    (E) It is a white, crystalline solid.

20. A substance that can be further simplified may be either

    (A) an element or a compound
    (B) an element or a mixture
    (C) a mixture or a compound
    (D) a mixture or an atom

21. A substance composed of two or more elements chemically united is called

(A) an isotope
(B) a compound
(C) an element
(D) a mixture

22. An example of a chemical change is the

(A) breaking of a glass bottle
(B) sawing of a piece of wood
(C) rusting of iron
(D) melting of an ice cube

23. A substance that cannot be further decomposed by ordinary chemical means is

(A) water
(B) air
(C) sugar
(D) silver

24. An example of a physical change is

(A) the fermenting of sugar to alcohol
(B) the rusting of iron
(C) the burning of paper
(D) a solution of sugar in water

25. A chemical action may involve all of the following EXCEPT

(A) combining of atoms of elements to form a molecule
(B) separation of the molecules in a mixture
(C) breaking down compounds into elements
(D) reacting a compound and an element to form a new compound and a new element

26. The energy of a system can be

(A) easily changed to mass
(B) transformed into a different form
(C) measured only as potential energy
(D) measured only as kinetic energy

27. If the ΔH of a reaction is a negative quantity, the reaction is definitely

    (A) endothermic
    (B) unstable
    (C) exothermic
    (D) reversible

In the following list, write E for the elements, C for the compounds, and M for the mixtures.

28. Water

29. Wine

30. Soil

31. Silver

32. Aluminum oxide

33. Hydrogen

34. Carbon dioxide

35. Air

36. Hydrochloric acid

37. Nitrogen

38. Tin

39. Potassium chloride

Answers and Explanations

1. 0.0012 g

$$1.2 \text{ mg} \times \frac{1 \text{ g}}{1000 \text{ mg}} = 0.0012 \text{ g}$$

2. 63 mm

$$6.3 \text{ cm} \times \frac{10 \text{ mm}}{1 \text{ cm}} = 63 \text{ mm}$$

3. 512 cm

$$5.12 \text{ m} \times \frac{100 \text{ cm}}{1 \text{ m}} = 512 \text{ cm}$$

4. 305 K
   $$32°C + 273 = 305 \text{ K}$$

5. 0.006111 L

$$6.111 \text{ mL} \times \frac{1 \text{ L}}{1000 \text{ mL}} = 0.006111 \text{ L}$$

6. 1,000,000 mm or 1×10^6 mm

$$1 \text{ km} \times \frac{1000 \text{ m}}{1 \text{ km}} \times \frac{1000 \text{ mm}}{1 \text{ m}} = 1,000,000 \text{ or } 1 \times 10^6 \text{ mm}$$

7. 1.03×10^3 g

$$1.03 \text{ kg} \times \frac{1000 \text{ g}}{1 \text{ kg}} = 1030 \text{ g or } 1.03 \times 10^3 \text{g}$$

8. 0.000003 kg or 3×10^{-6} kg

$$0.003 \text{ g} \times \frac{1 \text{ kg}}{1000 \text{ g}} = 0.000003 \text{ kg or } 3 \times 10^{-6} \text{ kg}$$

9. 22,400 mL or 2.24×10^4 mL

$$22.4 \text{ L} \times \frac{1000 \text{ mL}}{1 \text{ L}} = 22,400 \text{ or } 2.24 \times 10^4 \text{ mL}$$

10. 0.10013 km

$$10,013 \text{ cm} \times \frac{1 \text{ m}}{100 \text{ cm}} \times \frac{1 \text{ km}}{1000 \text{ m}} = 0.10013 \text{ km}$$

11. 158 g

Since the density is 1.58 g /mL and you want the mass of 100 mL, you use the formula density × volume = mass.

Inserting the values gives:

1.58 g /mL × 100 mL = 158 g

12. 2.27 g/mL

To find the volume of 227 g sulfur, subtract the volume of water before from the volume after.

150 mL − 50 mL = 100 mL

So, $\dfrac{227 \text{ g}}{100 \text{ mL}} = 2.27$ g/mL

13. (a) 5160 cm^3

    (b) 5160 g

    (a) Converting 5.08 in. to cubic centimeters,

$$5.08 \text{ in.} \times \frac{2.54 \text{ cm}}{1 \text{ in.}} = 12.9 \text{ cm}$$

    Then 20.0 cm × 20.0 cm × 12.9 cm = 5160 cm^3

    (b) Since 1 cm^3 of water at 4°C weighs
    1 g: 5160 cm^3 = 5160 g

14. $\dfrac{5 \text{ cm}}{\text{s}} \times \dfrac{1 \text{ m}}{100 \text{ cm}} \times \dfrac{1 \text{ km}}{1000 \text{ m}} \times \dfrac{60 \text{ s}}{1 \text{ min}} \times \dfrac{60 \text{ min}}{1 \text{ hr}} =$

15. **(A)** 3 significant figures—The zero nested between two nonzero (and hence significant) digits is significant.

    (B) 4 significant figures—The final zero is after the decimal point; therefore, it is significant. The other zeros are nested between significant digits and are therefore significant.

    (C) 2 significant figures—All zeros are just placeholders.

    (D) 3 significant figures—The first two zeros are placeholders. The last zero is significant as it is to the right of the decimal *and* ends the number.

16. **(A)** The reaction is exothermic because the reactants have a higher potential energy than the products.

17. **(A)** Density is defined as mass per unit volume.

18. **(B)** Baking powder is a physical blend of pure substances, making it a mixture.

19. **(E)** Physical properties deal with only the substance itself, not how it reacts.

20. **(C)** Elements cannot be further decomposed. An atom is best described as a particle, not a substance.

21. **(B)** The question defines a compound.

22. **(C)** The rusting of iron produces a new substance, iron (II) oxide. The production of new substances is what chemical change is about.

23. **(D)** Silver is an element; the other substances are not.

24. **(D)** Making a solution does not produce a new substance.

25. **(B)** A mixture is just a physical combination. No chemical action is required to separate the components.

26. **(B)** Energy can change form. Examples include light to heat and heat to electricity.

27. **(C)** When the heat of reaction (ΔH) is negative, the system is losing energy. When a system loses energy, the process is referred to as exothermic.

28. **(C)** Water is a chemical combination of hydrogen and oxygen in a particular ratio; it is a compound.

29. **(M)** Wine is a physical blend of many compounds and can be easily separated; it is a mixture.

30. **(M)** Soil is a physical blend of many compounds and can be easily separated; it is a mixture.

31. **(E)** Silver cannot be decomposed by ordinary chemical means; it is an element.

32. **(C)** Aluminum oxide is a chemical blend of aluminum and oxygen in a particular ratio; it is a compound.

33. **(E)** Hydrogen cannot be decomposed by ordinary chemical means; it is an element.

34. **(C)** Carbon dioxide is a chemical blend of carbon and oxygen in a particular ratio; it is a compound.

35. **(M)** Air is a physical blend of many gases (nitrogen and oxygen primarily) and can be easily separated; it is a mixture.

36. **(C)** Hydrochloric acid is the compound hydrogen chloride (generally in a water solution). The best answer is "compound" if you choose to focus on only the hydrogen chloride. The best answer is "mixture" if you choose to look at the hydrochloric acid as a solution.

37. **(E)** Nitrogen cannot be decomposed by ordinary chemical means; it is an element.

38. **(E)** Tin cannot be decomposed by ordinary chemical means; it is an element.

39. **(C)** Potassium chloride is a chemical blend of potassium and chlorine in a particular ratio; it is a compound.

PART II

THE NATURE OF MATTER

ATOMIC STRUCTURE AND THE PERIODIC TABLE

(2)

CHAPTER OBJECTIVES

Upon completing this chapter, you will be able to:

- Describe the history of the development of atomic theory

- Explain the structure of atoms

- Understand and predict the placement of electrons in principal energy levels and sublevels and the subsequent notation for such placement

- Place atoms in groups and periods based on their structure in the periodic table

- Explain how chemical and physical properties are related to position in the periodic table

History

The idea of small, invisible particles being the building blocks of matter can be traced back more than 2000 years to the Greek philosophers Democritus and Leucippus. These particles were supposed to be so small and indestructible that they could not be divided into smaller particles. The Greek word for "indivisible" is *atmos*. The English word *atom* comes from this Greek word. This early concept of atoms was not based upon experimental evidence but was simply a result of thinking and reasoning on the part of the philosophers. It was not until the eighteenth century that experimental evidence in favor of the atomic hypothesis began to accumulate. Finally, about 1805, John **Dalton** proposed some basic assumptions about atoms based on what was known through scientific experimentation and observation at that time. These assumptions were very closely related to what we presently know about atoms. Because of this, Dalton is often referred to as the father of modern atomic theory. Some of these basic ideas were:

1. All matter is made up of very small, discrete particles called atoms.

2. All atoms of an element are alike in weight and different from the weight of any other kind of atom.

3. Atoms cannot be subdivided, created, or destroyed.

4. Atoms of different elements combine in simple whole-number ratios to form chemical compounds.

5. In chemical reactions, atoms are combined, separated, or rearranged.

By the second half of the 1800s, many scientists believed that all the major discoveries related to the elements had been made. The only thing left for young scientists to do was to refine what was already known. This came to a suprising halt when J. J. Thomson discovered the electron beam in a cathode ray tube in 1897. Soon afterward, Henri Becquerel announced his work with radioactivity, and Marie Curie and her husband, Pierre, set about trying to isolate the source of radioactivity in their laboratory in France.

During the late nineteenth and early twentieth centuries, more and more physicists turned their attention to the structure of the atom. In 1913, the Danish physicist Niels Bohr published a theory explaining the line spectrum of hydrogen. He proposed a planetary model that quantized the energy of electrons to specific orbits. The work of Louis de Broglie and others in the 1920s and 1930s showed that quantum theory described a more probabilistic model of where the electrons could be found that resulted in the theory of orbitals. Enough was learned about nuclear structure to make practical use of atomic nuclei, as in nuclear power generators that use fission reactions and in fission and fusion bombs in the 1940s and 1950s. Today the search still goes on to study the particle physics of the atom and to attempt to control nuclear fusion reactions as a source of energy.

Electric Nature of Atoms

From the beginning of the twentieth century, scientists have been gathering evidence about the structure of atoms and fitting the information into a model of the atomic structure.

BASIC ELECTRIC CHARGES

The discovery of the electron as the first subatomic particle is credited to J. J. Thomson (England, 1897). He used an evacuated tube connected to a spark coil as shown in Figure 3. As the voltage across the tube was increased, a beam became visible. This was referred to as a cathode ray. Thomson found that the beam was deflected by both electrical and magnetic fields. From this, he concluded that the cathode rays were made up of very small negatively charged particles, which became known as **electrons**.

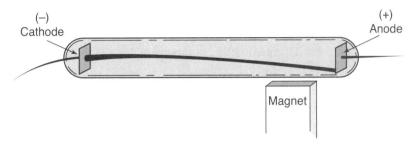

FIGURE 3. Cathode ray tube experiment.

Further experimentation led Thomson to find the ratio of the electrical charge of the electron to its mass. This was a major step toward understanding the nature of the particle. He was awarded a Nobel Prize in 1906 for his accomplishment.

It was an American scientist, Robert Millikan, who in 1909 was able to measure the charge on an electron using the apparatus pictured below.

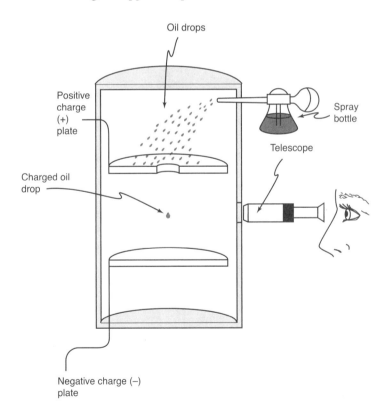

Oil droplets were sprayed into the chamber and in the process became negatively charged by passing a radioactive source. The electric field, produced by the plates shown, was adjusted so that a negatively charged drop would move slowly upward in front of the grid in the telescope. Knowing the rate at which the drop was rising, the strength of the field, and the weight of the drop, Millikan was able to calculate the charge on the drop. Combining the information with Thomson's results, he was able to calculate a value for the mass of a single electron. Eventually, this number was found to be 9.11×10^{-28} g.

Ernest Rutherford (England, 1911) performed a gold foil experiment (Figure 4) that had tremendous implications for atomic structure.

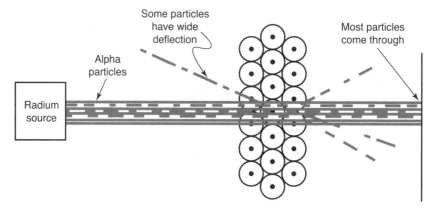

FIGURE 4. Rutherford experiment.

Alpha particles (helium nuclei) passed through the foil with few deflections. However, some deflections (1 per 8000) went almost directly back toward the source. This was unexpected and suggested an atomic model with mostly empty space between a **nucleus**, in which most of the mass of the atom was located and which was positively charged, and the electrons that defined the volume of the atom.

Further experiments showed that the nucleus was made up of still smaller particles called **protons**. Rutherford realized, however, that protons, by themselves, could not account for the entire mass of the nucleus. He predicted the existence of a new nuclear particle that would be neutral and would account for the missing mass. In 1932, James Chadwick (England) discovered this particle, the **neutron**.

BOHR MODEL

In 1913, Neils **Bohr** (Denmark) proposed his model of the atom. This pictured the atom as having a dense, positively charged nucleus and negatively charged electrons in specific spherical orbits, also called energy levels or shells, around this nucleus. These shells are arranged concentrically around the nucleus, and each shell is designated by a number: 1, 2, 3,. . . . The closer to the nucleus, the less energy an electron needs in one of these shells, but it has to gain energy to go from one shell to another that is farther away from the nucleus.

Because of its simplicity and general ability to explain chemical change, the Bohr model still has some usefulness today.

Principal Energy Level	Maximum Number of Electrons ($2n^2$)
1	2
2	8
3	18
4	32
5	32

COMPONENTS OF ATOMIC STRUCTURE

The following table lists the basic particles of the atom.

Particle	Charge	Symbol	Actual Mass	Relative Mass Compared to Proton	Discovery
Electron	$-(e^-)$	$_{-1}^{0}e$	9.109×10^{-28} g	1/1,837	J. J. Thomson–1897
Proton	$+ (p^+)$	$_{1}^{1}H$	1.673×10^{-24} g	1	_____ early 1900s
Neutron	$0 (n^0)$	$_{0}^{1}n$	1.675×10^{-24} g	1	J. C. Chadwick–1932

When these components are used in the Bohr model, we show the protons and neutrons in the nucleus. These particles are known as *nucleons*. The electrons are shown in the shells.

The number of protons in the nucleus of an atom determines the **atomic number**. All atoms of the same element have the same number of protons and therefore the same atomic number; atoms of different elements have different atomic numbers. Thus, the atomic number identifies the element. An English scientist, Henry Moseley, first determined the atomic numbers of the elements through the use of X-rays.

Since the actual masses of subatomic particles and atoms themselves are very small numbers when expressed in grams, scientists use atomic mass units instead. The unified **atomic mass unit** (as adopted by the International Union of Pure and Applied Chemists in 1961, symbolized by the letter u, and referred to as the *dalton*) is defined as 1/12 the mass of an atom of carbon-12. Thus, the mass of any atom can be expressed relative to the mass of one atom of carbon-12 in unified atomic mass units (u).

The sum of the number of protons and the number of neutrons in the nucleus is called the *mass number*.

Table 1 summarizes the relationships just discussed. Notice that the "outermost" energy level can contain no more than eight electrons. The explanation of this is given in the next section.

TABLE 1
TABLE OF THE FIRST 21 ELEMENTS*

Element	Atomic No.	Mass No.	Number of Protons	Number of Neutrons	Number of Electrons	Electrons in Energy Levels			
						1	2	3	4
Hydrogen	1	1	1	0	1	1			
Helium	2	4	2	2	2	2			
Lithium	3	7	3	4	3	2	1		
Beryllium	4	9	4	5	4	2	2		
Boron	5	11	5	6	5	2	3		
Carbon	6	12	6	6	6	2	4		
Nitrogen	7	14	7	7	7	2	5		
Oxygen	8	16	8	8	8	2	6		
Fluorine	9	19	9	10	9	2	7		
Neon	10	20	10	10	10	2	8		
Sodium	11	23	11	12	11	2	8	1	
Magnesium	12	24	12	12	12	2	8	2	
Aluminum	13	27	13	14	13	2	8	3	
Silicon	14	28	14	14	14	2	8	4	
Phosphorus	15	31	15	16	15	2	8	5	
Sulfur	16	32	16	16	16	2	8	6	
Chlorine	17	35	17	18	17	2	8	7	
Argon	18	40	18	22	18	2	8	8	
Potassium	19	39	19	20	19	2	8	8	1
Calcium	20	40	20	20	20	2	8	8	2
Scandium	21	45	21	24	21	2	8	9	2

*A complete list of the names and symbols of the known elements can be found in the Tables for Reference section.

There are cases in which different types of atoms of the same element have different masses. Three types of hydrogen atoms are known. The most common type is sometimes called protium and accounts for 99.985% of the hydrogen atoms found on earth. The nucleus of a protium atom consists of only one proton, and has one electron moving about it. There are two other known forms of hydrogen. One is called deuterium and accounts for 0.015% of the earth's hydrogen atoms. Each deuterium atom has a nucleus containing one proton and one neutron. The third form of hydrogen is known as tritium and is radioactive. It exists in very small amounts in nature, but it can be prepared artificially. Each tritium atom contains one proton, two neutrons, and one electron.

Protium, deuterium, and tritium are isotopes of hydrogen. **Isotopes** are atoms of the same element that have different masses. The isotopes of a particular element all have the same number of protons and electrons but different numbers of neutrons. In all three isotopes of hydrogen, the positive charge of the single proton is balanced by the negative charge of the electron. Most elements consist of mixtures of isotopes. Tin has ten stable isotopes, for example, the most of any element.

The percentage of each isotope in the naturally occurring element on earth is nearly always the same no matter where the element is found. The percentage at which each of an element's isotopes occurs in nature is taken into account when calculating the element's average atomic mass. *Average* **atomic mass** is the weighted average of the atomic masses of the naturally occurring isotopes of an element.

CALCULATING AVERAGE ATOMIC MASS

The average atomic mass of an element depends on both the mass and the relative abundance of each of the element's isotopes. For example, naturally occurring copper consists of 69.17% copper-63, which has an atomic mass of 62.919598 amu, and 30.83% copper-65, which has an atomic mass of 64.927793 amu. The average atomic mass of copper can be calculated by multiplying the atomic mass of each isotope by its relative abundance (expressed in decimal form) and adding the results:

$$0.6917 \times 62.929599 \text{ amu} + 0.3083 \times 64.927793 \text{ amu} = 63.55 \text{ amu}$$

Therefore, the calculated average atomic mass of naturally occurring copper is 63.55 amu. The average atomic mass is included for the elements listed in the periodic table rounded to one decimal place for use in calculations and to four decimal places in the table of elements in the reference section.

Valence

Each atom attempts to have its outer energy level complete. If the outer level is not complete, the atom accomplishes this by losing or gaining electrons. The electrons found in the outermost energy level are called **valence electrons**. The absolute number of electrons gained or lost is referred to as the **valence** of the atom. In a sense, the valence dictates the atom's combining capacity. When valence electrons are lost, the valence number is assigned a + sign. The particle has a positive charge. If valence electrons are gained, the valence number is assigned a – sign. The particle has a negative charge. These + and – signs go to the right of the number when describing the charge of the particle.

EXAMPLE 1

$$_{17}\text{Cl} = \overset{\text{nucleus}}{\bullet} \Bigg) 2 \Bigg| 8 \Bigg| 7 \leftarrow \text{valence electrons}$$

SOLUTION

This picture can be simplified to $\cdot \overset{\cdot\cdot}{\underset{\cdot\cdot}{\text{Cl}}} \colon$, showing only the valence electrons as dots. This is called the **Lewis dot structure** of the atom and was devised in

1916 by G. N. Lewis. To complete its outer energy level to eight electrons, Cl must gain one from another atom. Its valence number then is 1. When electrons are gained, we assign a – sign to this number so the charge of Cl with an extra electron is 1–.

EXAMPLE 2

$$_{11}Na = \overset{\text{nucleus}}{\bullet} \Big)2 \Big)8 \Big)1 \leftarrow \text{valence electrons}$$

Na • (Lewis dot structure)

SOLUTION

Since Na tends to lose its one valence electron, its charge will become 1+.

METALLIC, NONMETALLIC, AND NOBLE GAS STRUCTURES

On the basis of atomic structure, atoms are classified as **metals** if they tend to lose electrons, as **nonmetals** if they tend to gain electrons, and as *noble gases* if they tend neither to gain nor to lose electrons and have a complete outer energy level.

REACTIVITY

The fewer electrons an atom tends to gain, lose, or share to fill its outer energy level, the more reactive it tends to be in chemical reactions.

Atomic Spectra

The Bohr model was based on a simple postulate. Bohr applied to the hydrogen atom the concept that the electron can exist only in certain energy levels without an energy change, but that when the electron changes its state, it must absorb or emit the exact amount of energy that will bring it from the initial state to the final state. The *ground state* is the lowest energy state available to the electron. The *excited state* is any level higher than the ground state. The formula for changes in energy (ΔE) is

$$\Delta E_{\text{electron}} = E_{\text{final}} - E_{\text{initial}}$$

When the electron moves from the ground state to an excited state, it must absorb energy. When it moves from an excited state to the ground state, it emits energy. This release of energy is the basis for *atomic spectra*. (See Figure 5.)

The energy values were calculated from Bohr's equation

$$E_{\text{mole of electrons}} = \frac{-1312\,\text{kJ}}{n^2}$$

n (indicating each energy level) = $1, 2, 3, 4, \ldots$.

The equation for arriving at these values in joules per electron is given below. Although the derivation of the equation is not shown, it, too, comes from the Bohr model that describes the energy levels available to the electron in the hydrogen atom

$$E = -2.178 \times 10^{-18}\,\text{J}\left(\frac{Z^2}{n^2}\right)$$

where n again represents the energy level and Z is the nuclear charge (whose value for hydrogen is equal to 1). The values in joules per electron for the principal energy levels were calculated for Figure 5 by using this equation.

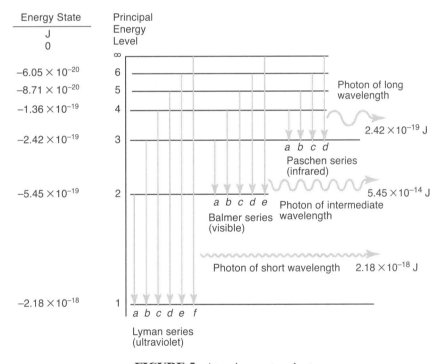

FIGURE 5. Atomic spectra chart.

When energy is released in the "allowed" values, it is released in the form of discrete radiant energy called *photons*. Each of the first three levels has a particular name associated with the emissions that occur when an electron drops to the energy state of that level. The emissions, consisting of ultraviolet radiation, that occur when an electron cascades from a level higher than the first level down to $n = 1$ are known as the Lyman series. Note in Figure 5 that the next two possible series of emissions results from electrons "dropping" to levels higher than $n = 1$ and have the names Balmer (for $n = 2$) and Paschen (for $n = 3$) series, respectively.

SPECTROSCOPY

When the light emitted by energized atoms is examined with an instrument called a *spectroscope*, the prism or diffraction grating in the spectroscope disperses the light to allow an examination of the *spectra* or distinct colored lines. Since only particular energy jumps are available in each type of atom, each element has its own unique emission spectra made up of only the lines of specific wavelength that correspond to its atomic structure.

EXAMPLE   Hydrogen can have an electron drop from the $n = 4$ to the $n = 2$ level. What visible spectral line in the Balmer series will result from this emission of energy?

SOLUTION

$$\Delta E_{\text{evolved}} = E_{n=2} - E_{n=4}$$

From Figure 5, $E_2 = -5.45 \times 10^{-19}$ J and $E_4 = -1.36 \times 10^{-19}$ J. Then,

$$E_{\text{evolved}} = -5.45 \times 10^{-19} \text{ J} - \left(-1.36 \times 10^{-19} \text{ J}\right) = -4.09 \times 10^{-19} \text{ J}$$

Since ΔE is negative, energy is released. The formula for the relationship of ΔE to the emission wavelength is

$$\Delta E = \frac{h \text{ (Planck's constant)} \, c \text{ (velocity of light)}}{\lambda \text{ (wavelength)}}$$

Substituting, we have

$$-4.09 \times 10^{-19} \text{ J} = \frac{\left(6.626 \times 10^{-34} \text{ J} \cdot \text{s}\right)\left(2.9979 \times 10^8 \text{ m/s}\right)}{\lambda}$$

$$\lambda = \frac{\left(6.626 \times 10^{-34} \text{ J} \cdot \text{s}\right)\left(2.9979 \times 10^8 \text{ m/s}\right)}{4.09 \times 10^{-19} \text{ J}}$$

$$= 4.87 \times 10^{-7} \text{ m}$$

This line is in the blue-green sector of the spectrum.

Visible Light Spectrum Wavelengths

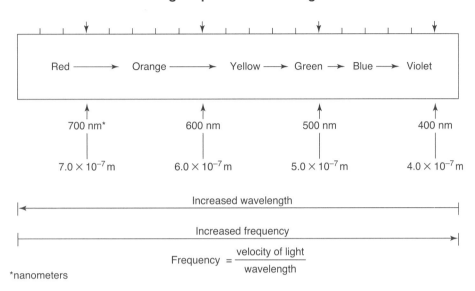

A partial atomic spectrum for hydrogen would look like this:

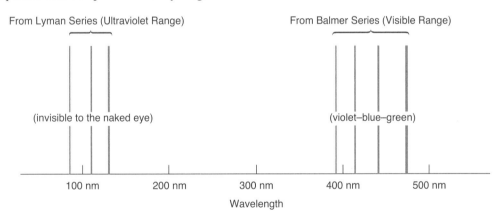

The right-hand group is in the visible range and is part of the Balmer series. The left-hand group is in the ultraviolet region and belongs to the Lyman series.

Investigating spectral lines like these can be used in the identification of unknown specimens.

MASS SPECTROSCOPY

Another tool used to identify specific atomic structures is mass spectroscopy, which is based on the concept that differences in mass cause differences in the degree of bending that occurs in a beam of ions passing through a magnetic field. This is shown in Figure 6.

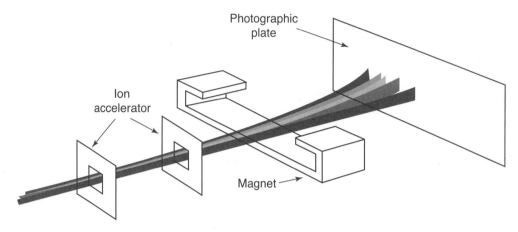

FIGURE 6. Mass spectroscope.

The intensity on the photographic plate indicates the amount of each particular *isotope*. Other collectors may be used in place of the photographic plate to collect and interpret these data.

The Wave-Mechanical Model

In the early 1920s, it was becoming apparent that there were some difficulties with the Bohr model of the atom. One difficulty was that although Bohr used classical mechanics (which is the branch of physics that deals with the motion of bodies under the influence of forces) to calculate the orbits of the hydrogen atom, it could not be used to explain the ability of electrons to stay in only certain energy levels without the loss of energy. Nor could it explain that the only change of energy occurred when an electron "jumped" from one energy level to another and could not exist in the atom at any energy level between these levels. According to Newton's laws, the kinetic energy of a body always changed smoothly and continuously, not in sudden jumps. The idea of only certain "quantized" energy levels being available in the Bohr atom was a very important one. The energy levels explained the existence of atomic spectra in the previous sections.

Another difficulty with the Bohr model was that it only worked well for the hydrogen atom with its single electron. It did not work with atoms that had more electrons. A new approach to the laws governing the behavior of electrons inside the atom was needed and such an approach was developed in the 1920s by the combined work of many scientists. Their work dealt with a more mathematical model usually referred to as *quantum mechanics* or *wave mechanics*. By this time Albert Einstein had already proposed a relativity mechanics to deal with the relative nature of mass as its speed approached the speed of light. In the same manner a quantum/wave mechanics was now needed to fit the data of the atomic model. Max Planck suggested in his quantum theory of light that it had particle-like properties as well as wavelike characteristics. In 1924, Louis de Broglie, a young French physicist, suggested that if light can have both wavelike and particle-like characteristics as Planck had suggested, then perhaps particles of matter can also have wavelike characteristics. In 1927, de Broglie's ideas were proven to be true experimentally when investigators showed that electrons could produce diffraction patterns, a property associated with waves. Diffraction patterns are patterns produced by waves as they pass through small holes or narrow slits.

In 1927, Werner Heisenberg stated what is now called the *uncertainty principle*. This principle states that it is impossible to know both the precise location and precise velocity of a subatomic particle at the same time. Heisenberg, in conjunction with the Austrian physicist Erwin Schrödinger, joined in the de Broglie concept that the electron was bound to the nucleus similarly to a standing wave, and they developed the complex equations that describe the **wave-mechanical model** of the atom. The solution of these equations gave specific wave functions called *orbitals*. These were not related at all to the Bohr orbits. The electron was not moving in a circular orbit in this model. Rather, the orbital is a three-dimensional region around the nucleus that indicates the probable location of an electron. They give no information on the pathway. Notice that the drawings in Figures 7 and 8 are only probability distribution representations of where the electrons in these orbitals might be found.

QUANTUM NUMBERS

Each electron orbital of the atom may be described by a set of four **quantum numbers** in this model. They give the position with respect to the nucleus, the shape of the orbital, its spatial orientation, and the spin of the electron in the orbital.

PRINCIPAL QUANTUM NUMBER (n)

1, 2, 3, 4, 5, etc.

refers to average distance of the orbital from the nucleus. 1 is closest to the nucleus and has the least energy. These numbers correspond to the orbits in the previous model. They are called energy levels.

ANGULAR MOMENTUM QUANTUM NUMBER (ℓ)

$0, 1, 2, 3, \ldots, n - 1$

(in order of increasing energy)

refers to the shape of the orbital. The number of possible shapes is limited by the principal quantum number. The first energy level, with an n value of 1, has only one possible shape, the s orbital. An s orbital is the only possibility when $n = 1$ because the only value ℓ could have is 0. The value of ℓ dictates the spherically shaped orbital labeled s. (Note that the designation s does not stand for spherical. The term s is used for another reason that coincidentally represents a spherical orbital.) In turn, the second energy level has two possible orbital shapes, s and p. The shape of a p orbital resembles a dumbbell or an hourglass. This shape becomes available when $n = 2$ because the possible ℓ values are 0 and 1. As described above, the s orbital is dictated when $\ell = 0$, but the p orbital is prescribed when $\ell = 1$. Other orbital shapes described by the letters d and f become available when the n value is large enough to have ℓ values of 2 and 3, respectively.

MAGNETIC QUANTUM NUMBER (m_ℓ)

$-\ell, \ldots, 0, \ldots, +\ell$

The drawings in Figure 7 show the s orbital shape, for the first three n values, and those of p orbitals, which have dumbbell shapes with three possible orientations on the axis shown. The number of spatial orientations possible for a given orbital is related back to the number of m_ℓ values. Accordingly, based on the scheme shown above for the magnetic quantum number (m_ℓ) as $-\ell, \ldots, 0, \ldots, +\ell$, the number of m_ℓ values is related to the ℓ value. That explains why there is only one spatial orientation when $\ell = 0$; m_ℓ can have only one value, 0. This complements the idea that when $\ell = 0$, a spherical orbital is realized that has only one "look" to it as dictated by the symmetry of the sphere. This system then indicates that when $\ell = 1$, there are 3 possible m_ℓ values (-1, 0, and $+1$). These dictate that 3 spatial orientations are possible for the p orbital. Figure 8 represents the d orbitals.

s orbital shape:

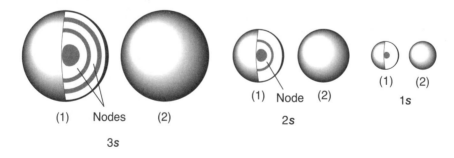

(1)   Nodes   (2)

3s

(1)   Node   (2)

2s

(1)   (2)

1s

The preceding drawings show representations of the hydrogen $1s$, $2s$, and $3s$ orbitals. Each drawing is called a *boundary surface diagram* whose outer surface and interior contain 90% of the total electron probability (the size of the orbital, by definition).

p orbital shapes:

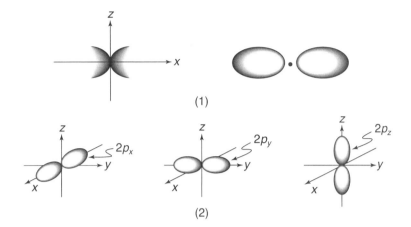

(1)

(2)

Representation of the 2*p* orbitals. (1) The electron probability distribution and (2) the boundary surface representations of all three orbitals.

FIGURE 7. Representations of *s* and *p* orbitals.

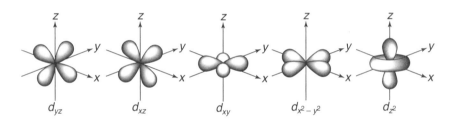

Representation of the 3*d* orbitals in terms of their boundary surfaces. The subscripts of the first four orbitals indicate the plane in which the four lobes are centered.

FIGURE 8. Representation of *d* orbitals.

SPIN QUANTUM NUMBER (m_s)

$$+\frac{1}{2} \text{ or } -\frac{1}{2}$$

Electrons are assigned one more quantum number, called the spin quantum number. This describes the spin in either of two possible directions. Each orbital can be filled by only two electrons, each with an opposite spin. The main significance of electron spin is explained by the postulate of Wolfgang Pauli. It states that in a given atom no two electrons can have the same set of four quantum numbers (n, ℓ, m_ℓ, and m_s). This is referred to as the **Pauli Exclusion Principle**. Therefore each orbital in Figures 7 and 8 can hold only two electrons.

Quantum numbers are summarized in the table below.

SUMMARY OF QUANTUM NUMBERS FOR THE FIRST FOUR LEVELS OF ORBITALS IN MULTIELECTRONIC ATOMS

Principal Quantum No., n	Angular Momentum Quantum No., ℓ	Orbital Shape Designation	Magnetic Quantum No., m_ℓ	Number of Orbitals	Total Electrons
1	0	1s	0	1	2
2	0	2s	0	1	2
	1	2p	−1, 0, +1	3	6
3	0	3s	0	1	2
	1	3p	−1, 0, +1	3	6
	2	3d	−2, −1, 0, +1, +2	5	10
4	0	4s	0	1	2
	1	4p	−1, 0, +1	3	6
	2	4d	−2, −1, 0, +1, +2	5	10
	3	4f	−3, −2, −1, 0, +1, +2, +3	7	14

Limits of Quantum Numbers

$n = 1, 2, 3, \ldots$ $\qquad$ $\ell = 0, 1, \ldots (n-1)$ $\qquad$ $m_\ell = -\ell, \ldots, 0, \ldots, +\ell$

AUFBAU PRINCIPLE

When "filling" the orbitals of multielectronic atoms, ground-state configurations are realized when the lowest energy orbital available is filled first. Therefore, the 1s orbital is filled first with up to two electrons spinning in opposite directions. If a third electron is called for, it would be placed into the 2s orbital. Each additional electron follows this filling principle.

HUND'S RULE OF MAXIMUM MULTIPLICITY

It is important to remember that, when there is more than one orbital at a particular energy level, such as three p orbitals or five d orbitals, only one electron will fill each orbital until each has one electron. After this, pairing will occur with the addition of one more electron to each orbital. This is called **Hund's Rule of Maximum Multiplicity** and is shown in Table 2, where each arrow indicates an electron (↑).

Sublevels and Electron Configuration

ORDER OF FILLING AND NOTATION

The sublevels do not fill up in numerical order, and the pattern of filling is shown on the right side of the approximate relative energy levels diagram (Figure 9). In the first instance of failure to follow numerical order, the $4s$ fills before the $3d$. (Study Figure 9 carefully before going on.)

TABLE 2
ORBITAL NOTATIONS

Chemical Symbol	Atomic No.	Orbital Notation			Electronic Configuration Notation
		1s	2s	2p	
H	1	⊡	☐	☐☐☐	$1s^1$
He	2	⊡	☐	☐☐☐	$1s^2$
Li	3	⊡	⊡	☐☐☐	$1s^2\,2s^1$
Be	4	⊡	⊡	☐☐☐	$1s^2\,2s^2$
B	5	⊡	⊡	⊡☐☐	$1s^2\,2s^2\,2p^1$
C	6	⊡	⊡	⊡⊡☐	$1s^2\,2s^2\,2p^2$
N	7	⊡	⊡	⊡⊡⊡	$1s^2\,2s^2\,2p^3$
O	8	⊡	⊡	⊡⊡⊡	$1s^2\,2s^2\,2p^4$
F	9	⊡	⊡	⊡⊡⊡	$1s^2\,2s^2\,2p^5$
Ne	10	⊡	⊡	⊡⊡⊡	$1s^2\,2s^2\,2p^6$

Maximum electrons in orbitals at a particular sublevel:

s = 2 (one orbital)

p = 6 (three orbitals)

d = 10 (five orbitals)

f = 14 (seven orbitals)

If each orbital is indicated in an energy diagram as a square (☐), we can show relative energies in a chart such as Figure 9. If this drawing represented a ravine with the energy levels as ledges onto which stones could come to rest only in numbers equal to the squares for orbitals, then pushing stones into the ravine would cause the stones to lose their potential energy as they dropped to the lowest level available to them. Much the same is true for electrons.

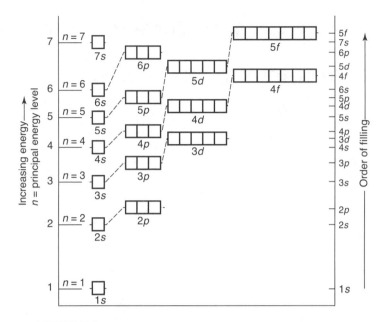

FIGURE 9. Approximate relative energy levels of subshells.

Examples of selected elements' electron configurations are shown below.

$_{19}K$ $\quad$ $1s^2\,2s^2\,2p^6\,3s^2\,3p^6\,4s^1$

$_{20}Ca$ $\quad$ $1s^2\,2s^2\,2p^6\,3s^2\,3p^6\,4s^2$

$_{21}Sc$ $\quad$ $1s^2\,2s^2\,2p^6\,3s^2\,3p^6\,4s^2\,3d^1$ (note that $4s$ filled before $3d$)

There is a more stable configuration at a half-filled or filled sublevel, and so at atomic number 24 the $3d$ sublevel becomes half-filled by taking a $4s$ electron

$_{24}Cr$ $\quad$ $1s^2\,2s^2\,2p^6\,3s^2\,3p^6\,3d^5\,4s^1$

and at atomic number 29 the $3d$ becomes filled by taking a $4s$ electron

$_{29}Cu$ $\quad$ $1s^2\,2s^2\,2p^6\,3s^2\,3p^6\,3d^{10}\,4s^1$

Table 3 on the following page shows the electron configurations of the elements. A triangular mark ▼ indicates the phenomenon of an outer-level electron dropping back to a lower unfilled orbital. These are exceptions to the Aufbau Principle. By following the atomic numbers through this chart, you can establish the same order of filling as shown in Figure 9.

TABLE 3
ELECTRON CONFIGURATIONS OF THE ELEMENTS

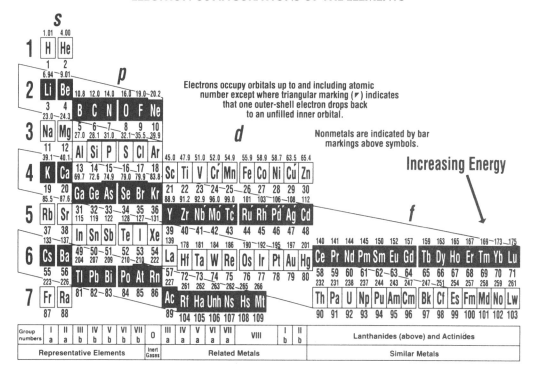

By following the atomic numbers in numerical order in Table 3, you can plot the order of filling of the orbitals for every element shown. A simplified method of showing the order of filling of the orbitals is to use the following diagram. It works for all naturally occurring elements through lanthanium, atomic number 88.

Start by drawing the diagonal arrows through the diagram as shown. The order of filling can be charted by following each arrow from tail to head and then to the tail of the next one. In this way you get the same order of filling as shown in Figure 9 and Table 3. It is

$$1s^2\ 2s^2\ 2p^6\ 3s^2\ 3p^6\ 4s^2\ 3d^{10}\ 4p^6\ 5s^2\ 4d^{10}\ 5p^6\ 6s^2\ 4f^{14}\ 5d^{10}\ 6p^6\ 7s^2$$

ELECTRON DOT NOTATION (LEWIS DOT STRUCTURES)

The Lewis dot structure can be used here to simplify the electron configuration notation. The electron dot notation shows only the chemical symbol surrounded by dots to represent the electrons in the incomplete outer level. Examples are:

$$\overset{\bullet}{K}, \ \cdot\overset{\bullet\bullet}{\underset{\bullet}{As}}\cdot, \ \overset{\bullet\bullet}{Sr}, \ \overset{\bullet\bullet}{:}\overset{}{I}\overset{\bullet\bullet}{:}, \ \overset{\bullet\bullet}{:}\overset{}{Rn}\overset{\bullet\bullet}{\underset{\bullet\bullet}{:}}$$

The symbol denotes the nucleus and all electrons except the valence electrons. The dots are arranged at the four sides of the symbol and are paired when appropriate. In the examples, the dots shown correspond to the following:

$4s^1$ is shown for potassium (K).

$4s^2, 4p^3$ are shown for arsenic (As).

$5s^2$ is shown for strontium (Sr).

$5s^2, 5p^5$ are shown for iodine (I).

$6s^2, 6p^6$ are shown for radon (Rn).

Transition Elements

The elements involved with the filling of a d sublevel with electrons after two electrons are in the s sublevel of the next principal energy level are often referred to as the **transition elements**. The first examples of these are the elements between calcium, atomic number 20, and gallium, atomic number 31. Their electron configurations are the same in the $1s$, $2s$, $2p$, $3s$, and $3p$ sublevels. It is the filling of the $3d$ and changes in the $4s$ sublevels that are of interest. This is shown in the following table.

Element	Atomic No.	Electron Configuration	
		$1s^2\ 2s^2\ 2p^6\ 3s^2\ 3p^6\ 3d$	$4s$
Scandium	21	1	2
Titanium	22	2	2
Vanadium	23	All 3	2
Chromium	24	the 5	1*
Manganese	25	same 5	2
Iron	26	6	2
Cobalt	27	7	2
Nickel	28	8	2
Copper	29	10	1*
Zinc	30	10	2

The asterisk (*) shows where a $4s$ electron is promoted into the $3d$ sublevel. This is due to the fact that the $3d$ and $4s$ sublevels are very close in energy and that there is a state of greater stability in half-filled and filled sublevels. There, chromium gains stability by the movement of an electron from the $4s$ sublevel into the $3d$ sublevel to give a half-filled $3d$ sublevel. It then has one electron in each of the five orbitals of the $3d$ sublevel. In copper, the movement of one $4s$ electron into the $3d$ sublevel gives the $3d$ sublevel a completely filled configuration.

The fact that the electrons in the $3d$ and $4s$ sublevels are so close in energy levels leads to the possibility of some or all of the $3d$ electrons being involved in chemical bonding. With the variable number of electrons available for bonding, it is not surprising that transition elements can exhibit variable charges.

The transition elements in the other periods of the table show this same type of anomaly, as they have d sublevels filling in the same manner.

Periodic Table of the Elements

HISTORY

The history of the development of a systematic pattern for the elements includes the work of a number of scientists such as John Newlands who, in 1863, proposed the idea of repeating octaves of properties.

Dimitry I. **Mendeleev**, in 1869, proposed a table containing seventeen columns and is usually given credit for the first periodic table, since he arranged elements in groups according to their atomic weights and properties. It is interesting to note that Lothar Meyer proposed a similar arrangement at about the same time. In 1871, Mendeleev rearranged some elements and proposed a table of eight columns, obtained by splitting each of the long periods across into a period of seven elements, an eighth group containing the three central elements (such as Fe, Co, Ni), and a second period of seven elements. The first and second periods of seven across were later distinguished by use of the letters a and b attached to the group symbols, which are Roman numerals. This nomenclature of periods (Ia, IIa, etc.) appears slightly revised in the present periodic table, even in the extended form of assigning Arabic numbers from 1 through 18 as shown in Table 4.

Mendeleev's table had the elements arranged by atomic weights with properties recurring in a periodic manner. Where atomic weight placement disagreed with the properties that should occur in a particular spot in the table, he gave preference to the element with the correct properties. He even predicted elements for places that were not yet occupied in the table. These predictions proved to be amazingly accurate.

TABLE 4
PERIODIC TABLE—PROPERTIES

PERIODIC TABLE—PROPERTIES showing periods 1–7, groups 1–18 (IA–O former designation), light metals, heavy metals (brittle, ductile), nonmetals, noble gases, s subshell, d subshell, p subshell, and trends: Acid properties increase, Base properties increase, Atomic radii decrease, Ionization energy increases, Nonmetallic properties increase. Amphoteric elements along the metalloid line are called metalloids.

PERIODIC LAW

Henry **Moseley** stated, after his work with X-ray spectra in the early 1900s, that the properties of elements are a periodic function of their atomic numbers, thus changing the basis of the periodic law from atomic weight to atomic number. This is the present statement of the **periodic law**.

THE TABLE

The horizontal rows of the periodic table are called **periods** or **rows**. There are seven periods, each of which begins with an atom having only one valence electron and ends with a complete outer-shell structure of an inert gas. The first three periods are short, consisting of 2, 8, and 8 elements, respectively. Periods 4 and 5 are long, with 18 each, while period 6 has 32 elements, and period 7 is incomplete with 22 elements, most of which are radioactive and do not occur in nature.

In Table 4, you should note the relationship of the length of periods to the orbital structure of the elements. In the first period, the $1s$ orbital is filled with the noble gas helium. The second period begins with the filling of the $2s$ orbital and then fills the $2p$ and again ends with a noble gas neon, Ne. The same pattern is repeated in period 3 going from $3s^1$ to $3p^6$. These eight elements from sodium, Na, to argon, Ar, complete the filling of the $n = 3$ energy level with $3s^2$ and $3p^6$. In the fourth period, the first two elements indicate the filling of the $4s$ orbital. Beyond calcium, Ca, the pattern becomes more complicated. As we discussed in the section "Order of Filling and Notation," the next orbitals to be filled are the five $3d$ orbitals whose elements represent transition elements. Then the three $4p$ orbitals are filled and end with the noble gas krypton, Kr. The fifth period is similar to the fourth period. The $5s$ orbital filling is represented by Rb and Sr, both of which resemble the elements directly above them on the table. Next come the transition elements to fill the five $4d$ orbitals before the next group of elements from In to Xe complete the three $5p$ orbitals. (Table 3 should be consulted for the irregularities that occur as the d orbitals fill.) The sixth period follows much the same pattern and has the order of filling as $6s^2$, $4f^{14}$, $5d^{10}$, $6p^6$. Here again irregularities occur, and these can best be followed by using Table 3.

The vertical columns of the periodic table are called **groups** or **families**. The elements in a group exhibit similar or related properties. The Roman numeral group number gives an indication of the number of electrons probably found in the outer shell of the atom and thus an indication of one of its possible valence numbers. In 1984, the International Union of Pure and Applied Chemistry (IUPAC) agreed that the Roman numerals of groups would be replaced by Arabic numbers 1 through 18. Most periodic tables today show this method of identifying groups, but some still show both.

Properties Related to the Periodic Table

Metals are found on the left of the chart (see Table 4) with the most active metal in the lower left corner. Nonmetals are found on the right side with the most active nonmetal in the upper right corner. The noble gases are on the far right. Since the most active metals react with water to form bases, the Group 1 metals are called alkali metals. As you proceed to the right, the base-forming property decreases and the acid-forming properties increase. The metals in the first two groups are the light metals, and those toward the center are heavy metals.

The elements found along the dark line in the periodic chart (Table 4) are called metalloids. These elements have certain characteristics of metals and other characteristics of nonmetals. Some examples of metalloids are boron, silicon, arsenic, and tellurium.

Here are some important general summary statements:

- Acid-forming properties increase moving to the right side of the table.
- Base-forming properties are high on the left side and decrease moving to the right.
- The atomic radii of elements decrease from left to right across a period.
- First ionization energies increase from left to right across a period.
- Metallic properties are greatest on the left side of the table and decrease to the right.
- Nonmetallic properties are greatest on the right side of the table and decrease to the left.

Study Table 4 carefully because it summarizes many of these properties. For a more detailed description of metals, alloys, and metalloids, see pages 294–299 in Chapter 13.

RADII OF ATOMS

The size of an atom is difficult to describe because atoms have no definite outer boundary. Unlike a volleyball, the atom does not have a definite circumference.

To overcome this problem, the size of an atom is estimated by describing its radius. In metals, this is done by measuring the distance between two nuclei in the solid state and dividing this distance by 2. Such measurements can be made with X-ray diffraction. For nonmetallic elements that exist in pure form as molecules, such as chlorine, measurements can be made of the distance between nuclei for two atoms covalently bonded together. Half of this distance is referred to as the **covalent radius**. The method for finding the covalent radius of the chlorine atom is illustrated below.

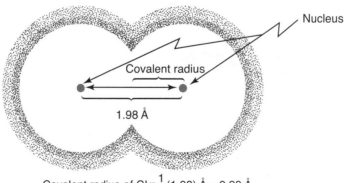

Covalent radius of $Cl = \frac{1}{2}(1.98)$ Å $= 0.99$ Å

Figure 10 shows the relative **atomic and ionic radii** for some elements. As you review this chart, you should note two trends:

1. Atomic radii decrease from left to right across a period in the periodic chart (until the noble gases).

2. Atomic radii increase from top to bottom in a group or family.

The reason for these trends will become clear in the following discussions.

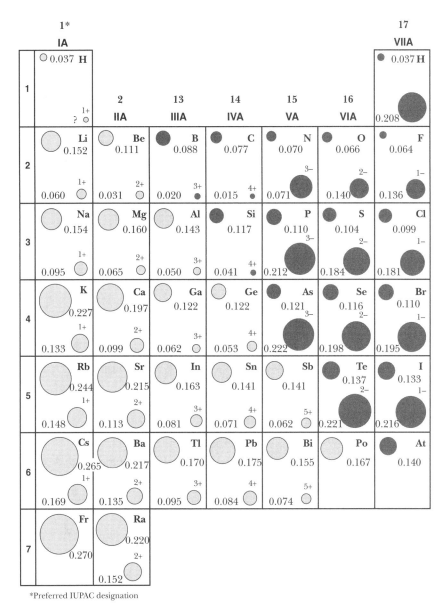

*Preferred IUPAC designation

FIGURE 10. Radii of some atoms and ions (in nanometers).

Note: The atomic radius is usually given for metal atoms, which are shown in gray, and the covalent radius is usually given for atoms of nonmetals, which are shown in black.

ATOMIC RADII IN PERIODS

Since the number of electrons in the outer principal energy level increases as you go from left to right in each period, the corresponding increase in the nuclear charge due to the additional protons pulls the electrons more tightly around the nucleus. This attraction more than balances the repulsion between the added electrons and the other electrons, and the radius is generally reduced. The inert gas has a slight increase because of the electron repulsion in the filled outer principal energy level.

ATOMIC RADII IN GROUPS

For a group of elements, the atoms of each successive member have another outer principal energy level in the electron configuration, and the electrons there are held less tightly by the nucleus. This is so because of their increased distance from the nuclear positive charge and the shielding of this positive charge by all the core electrons. Therefore the atomic radius increases down a group. See Figure 10.

IONIC RADIUS COMPARED TO ATOMIC RADIUS

Metals tend to lose electrons in forming positive ions. Enough electrons are generally lost in a way that makes the new electron arrangement similar to the noble gas (elements in column 18 of the periodic table) that precedes it. Therefore, the ion's outermost electron configuration is now lower by a principal energy level. The lower energy level is naturally closer to the nucleus, making the ion's radius smaller than the atom's.

Nonmetals tend to gain electrons in forming negative ions. With this added negative charge, which increases the inner electron repulsion, the ionic radius is increased beyond the atomic radius. See Figure 10 for relative atomic and ionic radii values.

ELECTRONEGATIVITY

The **electronegativity** of an element is a number that measures the relative strength with which the atoms of the element attract valence electrons in a chemical bond. This electronegativity number is based on an arbitrary scale going from 0 to 4. In general, a value of less than 2 indicates a metal.

Notice in Table 5 that the electronegativity decreases down a group and increases across a period. The lower the electronegativity number, the more electropositive an element is said to be. The most electronegative element is in the upper right corner—F. The most electropositive element is in the lower left corner—Fr.

IONIZATION ENERGY

Atoms hold on to their electrons to different degrees. At times, enough energy is supplied to an outer electron to remove it from its atom. The amount of energy needed to remove the first electron is called the **first ionization energy**. With the first electron gone, the removal of succeeding electrons generally becomes more difficult because of the decrease of repulsion between electrons, the loss of shielding from core electrons, and the lowering of the outermost electron's energy level. The lowest first ionization energy is found with the least electronegative atom.

TABLE 5

ELECTRONEGATIVITIES OF THE ELEMENTS

AND IONIZATION ENERGIES

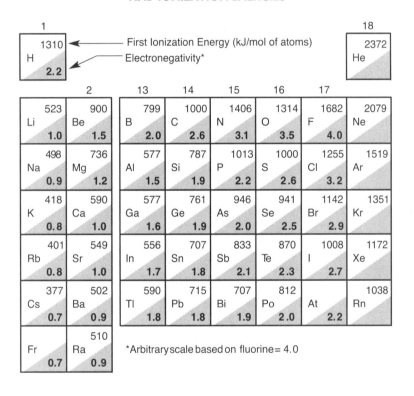

It is not surprising then that the highest peaks on the chart shown on the next page occur with the ionization energy needed to remove the first electron from the outer energy level of the noble gases, that is, He, Ne, Ar, Kr, Xe, and Rn, because of the stability of the filled p orbitals in the outer energy level. Notice that even among these elements there is a gradual decline in the energy needed. This can be explained by considering the distance the energy level involved is from the positively charged nucleus. With each succeeding noble gas, a more distant p orbital is involved, therefore making it easier to remove an electron from the positive attraction of the nucleus. Besides this consideration, as more energy levels are added to the atomic structure as the atomic number increases, the additional negative fields associated with the additional electrons screen out some of the positive attraction of the nucleus. Within a period such as from Li to Ne, the ionization energy generally increases. The lowest occurs when there is a lone electron in the outer *s* orbital as in Li. Once the 2*s* orbital fills with two electrons at atomic number 4, Be, added repulsions help overcome the nuclear pull and lower the ionization energy for the outermost electron. At atomic number 5, B, one lone electron is in a 2*p* orbital. This electron can be removed with less energy due to *shielding* of the nuclear pull from the 2*s* electrons. Therefore, a dip occurs in the ionization energy. With the 2*p* orbitals filling according to Hund's Rule (refer to Table 2) and with only one electron in each orbital before pairing occurs (and, consequently, no repulsions), an increase in the number of protons causes an increase in ionization energy up to element 7, N. After this peak, a dip due to repulsive effects occurs at element 8, O. Then a continual increase in ionization energy occurs until the 2*p* orbitals are completely filled with paired electrons at the noble gas Ne. As you continue to associate the atomic number with the line in the chart, you will find peaks occurring in the same general pattern and always related to the state of filling the orbitals involved and the number of protons in the nucleus.

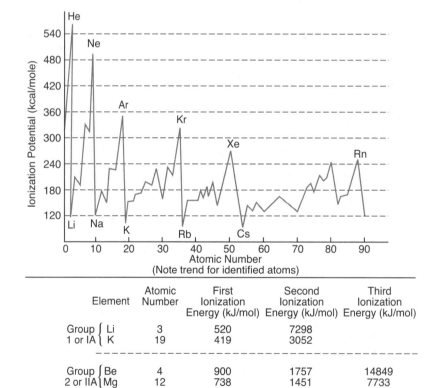

	Element	Atomic Number	First Ionization Energy (kJ/mol)	Second Ionization Energy (kJ/mol)	Third Ionization Energy (kJ/mol)
Group 1 or IA	Li	3	520	7298	
	K	19	419	3052	
Group 2 or IIA	Be	4	900	1757	14849
	Mg	12	738	1451	7733

Sample Ionization Energies for
Second and Third Electronic Removal

Chapter Summary

The following terms summarize all the concepts and ideas that were introduced in this chapter. You should be able to explain their meaning and how you would use them in chemistry. They appear in boldface type in this chapter to draw your attention to them. The boldface type also makes the terms easier for you to look up, if you need to. You could also use a search engine on a computer to get a quick and expanded explanation of them.

TERMS YOU SHOULD KNOW

atomic mass unit

atomic number

atomic radii

Aufbau Principle

Bohr model

covalent radius

Dalton's atomic theory

electron

electronegativity

first ionization energy

group or family

Hund's Rule

ionic radii

isotopes

Lewis dot structure

Mendeleev

metals

Moseley

neutron

nonmetals

nucleus

Pauli Exclusion Principle

periodic law

period or row

protron

quantum numbers

s, p, d, and f orbitals

transition elements

valence

valence electrons

wave-mechanical model

Chapter 2 Review Exercises

1. The two main aspects of an atom are the
   (A) principal energy levels and energy sublevels
   (B) nucleus and kernel
   (C) nucleus and energy levels
   (D) planetary electrons and energy levels

2. The lowest principal quantum number that an electron can have is
   (A) 0
   (B) 1
   (C) 2
   (D) 3

3. The sublevel that has only one orbital is identified by the letter

   (A) *s*
   (B) *p*
   (C) *d*
   (D) *f*

4. The sublevel that can be occupied by a maximum of ten electrons is identified by the letter

   (A) *d*
   (B) *f*
   (C) *p*
   (D) *s*

5. An orbital may never be occupied by

   (A) 1 electron
   (B) 2 electrons
   (C) 3 electrons
   (D) 0 electrons

6. An atom of beryllium consists of 4 protons, 5 neutrons, 4 electrons. The mass number of this atom is

   (A) 13
   (B) 9
   (C) 8
   (D) 5

7. The number of orbitals in the second principal energy level, $n = 2$, of an atom is

   (A) 1
   (B) 9
   (C) 16
   (D) 4

8. An electron-dot symbol consists of the symbol representing the element and an arrangement of dots that shows

   (A) the atomic number
   (B) the atomic mass
   (C) the number of neutrons
   (D) the electrons in the outermost energy level

9. Chlorine is represented by the electron-dot symbol $\cdot\overset{\displaystyle\cdot\cdot}{\underset{\displaystyle\cdot\cdot}{Cl}}\colon$. The atom that would be represented by an identical electron-dot arrangement has the atomic number

   (A) 7
   (B) 9
   (C) 15
   (D) 19

Using the periodic chart, answer the following questions:

10. Name the element of the first 20 whose atom gives up an electron the most readily.

   (A) Li
   (B) F
   (C) Cl
   (D) K

11. Name the element whose atom shows the greatest affinity for an additional electron.

   (A) Li
   (B) F
   (C) N
   (D) O

12. Name the most active nonmetal in period 3.

   (A) Na
   (B) Cl
   (C) Ar
   (D) S

13. Which of the following atoms would be smaller than that of oxygen?

   (A) N
   (B) S
   (C) Se
   (D) F

14. What is the most probable charge for aluminum in a compound?

   (A) 1+
   (B) 2+
   (C) 3+
   (D) 4+

15. Where in a periodic series do you find strong base formers?

  (A) left
  (B) right
  (C) middle
  (D) inert gases

Answers and Explanations

1. **(C)** The two main parts of the atom are the nucleus and its energy levels (where electrons reside).

2. **(B)** The principal quantum numbers start with the value of 1 to represent the first level.

3. **(A)** The letter s is used to represent the first orbital, which can hold 2 electrons.

4. **(A)** The sublevel d has five orbitals that each can hold 2 electrons, totaling 10 electrons.

5. **(C)** Each orbital can only hold 2 electrons.

6. **(B)** The mass number is the total of the number of protons and neutrons, which is 9 in this case.

7. **(D)** The second principal energy level has an s orbital and 3 p orbitals, making a total of 4.

8. **(D)** The Lewis electron-dot notation shows the symbol and the outermost energy level electrons, which are the valence electrons.

9. **(B)** Chlorine is a member of the halogen family found in Group 17. Another element in the same family would be fluorine, with the atomic number 9.

10. **(D)** Potassium has the lowest first ionization energy of the first 20 elements.

11. **(B)** Fluorine has seven electrons in its outer energy level and only needs one more to complete its outer energy level octet. The greater number of protons in fluorine (compared with the other period 2 elements listed) causes a greater amount of energy to be released when an electron is added to a fluorine atom. This occurs because of a larger nuclear attraction in fluorine than in the other elements listed.

12. **(B)** The most active nonmetal of the period is the element found farthest to the right in the row, chlorine.

13. **(D)** The size of atoms generally decreases to the right and up on the periodic chart.

14. **(C)** Aluminum loses 3 electrons relatively easy because it has fairly low first, second, and third ionization energies. The loss of a fourth electron is hard because of fairly large fourth ionization energy.

15. **(A)** Metals tend to form basic compounds; they are found on the left side of the periodic table.

BONDING

CHAPTER OBJECTIVES

Upon completing this chapter, you will be able to:

- Define ionic and covalent bonds and explain how they form
- Identify the differences in the continuum that exists between ionic and covalent bonding
- Describe the implications of the type of bond on the structure and properties of the compound
- Describe the three types of intermolecular bonds and how the bond type influences properties
- Explain how VSEPR theory and hybridization can be used to describe the known shapes of molecules

Some elements show no tendency to combine with either like atoms or other kinds of elements. These elements are said to be monatomic molecules; three examples are helium, neon, and argon. A *molecule* is defined as the smallest particle of an element or a compound that retains the characteristics of the original substance. Water is a triatomic molecule since two hydrogen atoms and one oxygen atom must combine to form the substance water with its characteristic properties. When atoms do combine to form molecules, there is a shifting of valence electrons, that is, the electrons in the outer energy level of each atom. Usually, this results in completion of the outer energy level of each atom. This more stable form may be achieved by the gain or loss of electrons or by the sharing of pairs of electrons. The resulting attraction of the atoms involved is called a *chemical bond*. When a chemical bond forms, energy is released; when this bond is broken, energy is absorbed.

63

This relationship of bonding to the valence electrons of atoms can be further explained by studying the electron structure of the atoms involved. As already mentioned, the noble gases are monatomic molecules. The reason for this can be seen in the electron distribution of these noble gases as shown below:

Noble Gas	Electron Distribution	Electrons in Valence Energy Level
Helium	$1s^2$	2
Neon	$1s^2\ 2s^2\ 2p^6$	8
Argon	$1s^2\ 2s^2\ 2p^6\ 3s^2\ 3p^6$	8
Krypton	$1s^2\ 2s^2\ 2p^6\ 3s^2\ 3p^6\ 3d^{10}\ 4s^2\ 4p^6$	8
Xenon	$1s^2\ 2s^2\ 2p^6\ 3s^2\ 3p^6\ 3d^{10}\ 4s^2\ 4p^6\ 4d^{10}\quad 5s^2\ 5p^6$	8
Radon	$1s^2\ 2s^2\ 2p^6\ 3s^2\ 3p^6\ 3d^{10}\ 4s^2\ 4p^6\ 4d^{10}\ 4f^{14}\ 5s^2\ 5p^6\ 5d^{10}\ 6s^2\ 6p^6$	8

The distinguishing factor in these very stable configurations is the arrangement of two *s* electrons and six *p* electrons in the valence energy level in five of the six atoms. (Note that helium, He, has only a single *s* valence energy level that is filled with two electrons, making He a very stable atom.) This arrangement is called a **stable octet**. All other elements, other than the noble gases, have one to seven electrons in their outer energy level. These elements are reactive to varying degrees. When they do react to form chemical bonds, usually the electrons shift in such a way that stable octets form. In other words, in bond formation, atoms usually attain the stable electron structure of one of the noble gases. The type of bond formed is directly related to whether this structure is achieved through gaining, losing, or sharing of electrons.

Types of Bonds

IONIC BONDS

When the electronegativity values of two kinds of atoms differ by 1.7 or more (especially greater differences than 1.7), the more electronegative atom will borrow the electrons it needs to fill its energy level, and the other kind of atom will lend electrons until it, too, has a complete energy level. Because of this exchange, the borrower becomes negatively charged and the lender becomes positively charged. They are now referred to as *ions*, and the bond or attraction between them is called an **ionic** or **electrovalence bond**. These ions do not retain the properties of the original atoms. An example can be seen in Figure 11.

The reaction:

Li (atom) + F (atom) $\longrightarrow$ Li$^+$ + F$^-$ (Ionic compound formed)

e$^-$

Electron notation representations:

Li $1s^2, 2s^1$ + F $1s^2, 2s^2, 2p^5$ $\longrightarrow$ Li$^+$ $1s^2$ + F$^-$ $1s^2, 2p^2, 2p^6$

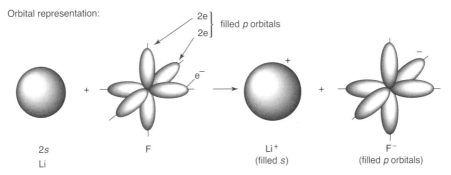

Orbital representation:

FIGURE 11. Three representations of the ionic bonding of LiF.

These ions do not form an individual molecule in the liquid or solid phase but are arranged into a crystal lattice or giant ion-molecule containing many of these ions. Ionic solids like this tend to have high melting points and do not conduct a current of electricity until they are in the molten state.

COVALENT BONDS

When the electronegativity difference between two or more atoms is zero or very small (not greater than about 0.4), they tend to share the valence electrons in the respective outer energy levels. This attraction is called a *nonpolar* **covalent bond**. Here is an example using electron-dot notation and orbital notation.

fluorine atoms $\longrightarrow$ fluorine molecule

These covalent bonded molecules do not have an electrostatic charge like the ionic bonded substances. In general, covalent compounds are gases, or liquids having fairly low boiling points, or solids that melt at relatively low temperatures. Unlike ionic compounds, they do not conduct electric currents.

When the electronegativity difference is between 0.4 and 1.7, there is no equal sharing of electrons between the atoms involved. The shared electrons will be more strongly attracted to the atom of greater electronegativity. As the difference in the electronegativities of the two elements increases above 0.4, the polarity or degree of ionic character increases. At a difference of more than 1.7, the bond has more than 50% ionic character. In fact, such a bond is then considered to be ionic. However, when the difference is between 0.5 and 1.7, the bond is called a *polar covalent bond*. Two examples are:

H :Cl: • hydrogen electron   H :O: • hydrogen electrons
              ○ chlorine electrons   ○ oxygen electrons
                                                     H

Hydrogen chloride   Water

Notice that the electron pair in the bond is shown closer to the more electronegative atom. Because of this unequal sharing, the bonds shown are said to be polar bonds. If these polar bonds are nonsymmetrically placed around a central atom, the overall molecule is polar as well. In the examples above, the chlorine (in HCl) and oxygen (in H_2O) are considered the central atoms. It is important to note that both **bonds** and **molecules** could be described as *polar*. Polar molecules are referred to as **dipoles** as the whole molecule itself has two distinct ends from a *charge* perspective. There are cases of polar covalent bonds existing in nonpolar molecules. Some examples are CO_2, CH_4, and CCl_4. (See Figure 12.)

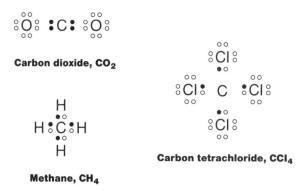

Carbon dioxide, CO_2

Methane, CH_4

Carbon tetrachloride, CCl_4

FIGURE 12. Polar covalent bonds in nonpolar molecules.

In all these examples, the bonds are polar covalent bonds, but the important thing is that they are symmetrically arranged in the molecule. This results in a nonpolar molecule.

In the *covalent* bonds described so far, the shared electrons in the pair are contributed one each from the atoms bonded. In some cases, however, both electrons for the shared pair are supplied by only one of the atoms. These bonds are called **coordinate covalent bonds**. Two examples are NH_4^+ and H_2SO_4. (See Figure 13.)

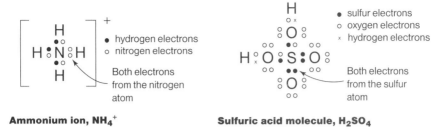

Ammonium ion, NH$_4^+$ Sulfuric acid molecule, H$_2$SO$_4$

FIGURE 13. Coordinate covalent bonds.

The formation of a covalent bond can be described in a graphic form and related to the potential energy of the atoms involved. Using the formation of the hydrogen molecule as an example, we can show how the potential energy changes as the two atoms approach and form a covalent bond. In the illustration that follows, (1), (2), and (3) show the effect on potential energy as the atoms move closer to each other. In (3), the atoms have reached the condition of lowest potential energy, but their inertia pulls them even closer, as shown in (4). The repulsion between them then forces the two nucleii into a stable position, as shown in (5).

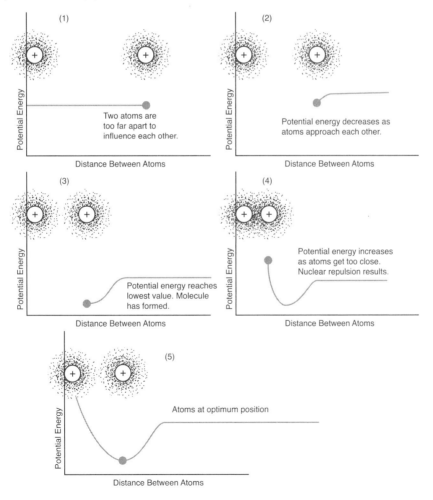

METALLIC BONDS

In the case of most metals, one or more of the valence electrons become detached from the atom and migrate into a "sea" of free electrons among the positive metal ions. The metallic bond strength varies with the nuclear charge of the metal atoms and the number of electrons in this electron sea. Both of these factors are reflected in the amount of heat required to vaporize a metal. The strong attraction between these differently charged particles forms a **metallic bond**. Because of this firm bonding, metals usually have high melting points, show great strength, and are good conductors of electricity.

Intermolecular Forces of Attraction

The term **intermolecular forces** refers to attractions *between* molecules. Although it is proper to refer to all intermolecular forces as **van der Waals forces**, this concept should be expanded for clarity.

DIPOLE-DIPOLE ATTRACTION

One type of van der Waals force is **dipole-dipole attraction**. It was shown in the discussion of polar covalent bonding that the unsymmetrical distribution of electronic charges leads to positive and negative charges in the molecules, which are referred to as *dipoles*. In polar molecular substances, the dipoles line up so that the positive pole of one molecule attracts the negative pole of another. This is much like the lineup of small bar magnets. The force of attraction between polar molecules is called *dipole-dipole attraction*. These attractive forces are less than the full charges carried by ions in ionic crystals.

LONDON DISPERSION FORCES

Another type of van der Waals force is called the **London dispersion force**. Found between both polar and nonpolar molecules, it can be attributed to the fact that an atom that usually is nonpolar sometimes becomes polar because the probability distribution of its electrons (the electron cloud) may become distorted and cause an uneven charge distribution at any one instant. When this occurs, the atom has a temporary dipole. This dipole can then cause a second, adjacent atom to be distorted and to have its nucleus attracted to the negative end of the first atom. London dispersion forces are about one-tenth the force of most dipole interactions and are the weakest of all the electrical forces that act between atoms or molecules. These forces help to explain why nonpolar substances such as noble gases and the halogens condense into liquids and then freeze into solids when the temperature is lowered sufficiently. In general, they also explain why liquids composed of discrete molecules with no permanent dipole attraction have low boiling points relative to their molecular masses. As molecular mass increases, however, these forces of attraction can become more significant. A greater molecular mass means a greater number of electrons and a greater possibility for temporary polarization. It is also true that compounds in the solid state that are bound mainly by this type of attraction have rather soft crystals, are easily deformed, and vaporize easily. Because of the low intermolecular forces, the melting points are low and evaporation takes place so easily that it may occur at room temperature. Examples of such solids are iodine crystals and mothballs (paradichlorobenzene and naphthalene).

HYDROGEN BONDS

A proton or hydrogen nucleus has a high concentration of positive charge. When a hydrogen atom is bonded to a highly electronegative atom, its positive charge will have an attraction for neighboring electron pairs. This special kind of dipole-dipole attraction is called a **hydrogen bond**. The more strongly polar the molecule is, the more effective the hydrogen bonding is in binding the molecules into a larger unit. As a result, the boiling points of such molecules are higher than those of similar polar molecules. Good examples are water and hydrogen fluoride. Studying Figure 14 shows that in the series of compounds consisting of H_2O, H_2S, H_2Se, and H_2Te there is an unusual rise in the boiling point of H_2O that is not in keeping with the slow increase in boiling point with the increase in molecular mass. Instead of an expected slope between H_2O and H_2S, shown here as a dashed line, the actual boiling point is quite a bit higher at 100°C. The explanation is that hydrogen bonding occurs in H_2O but not to any significant degree in the other compounds.

This same phenomenon occurs with the hydrogen halides (HF, HCl, HBr, and HI). Note in Figure 14 that hydrogen fluoride, which has strong hydrogen bonding, shows an unexpectedly high boiling point.

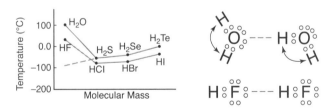

FIGURE 14. Boiling points of hydrogen compounds
with similar electron-dot structures.

Hydrogen bonding also explains why some substances have unexpectedly low vapor pressure, high heats of vaporization, and high melting points. In order for vaporization or melting to take place, molecules must be separated. Energy must be expended to break hydrogen bonds and thus break down the larger clusters of molecules into separate molecules. Like the boiling point, the melting point of H_2O is abnormally high when compared to the melting point of the hydrogen compounds of the other elements having six valence electrons, which are chemically similar but have no apparent hydrogen bonding. The hydrogen bonding effect in water is covered on pages 164–165.

Double and Triple Bonds

To achieve the *octet* structure, which is an outer energy level resembling the noble gas configuration of eight electrons, it is necessary for some atoms to share two or even three pairs of electrons. Sharing two pairs of electrons produces a *double bond*. An example is

$$\overset{x}{\underset{x\ x}{:}}\!O\!\overset{x}{\underset{x}{:}}\ \overset{o}{\underset{o}{:}}C\overset{o}{\underset{o}{:}}\ \overset{x}{\underset{x\ x}{:}}\!O\!\overset{x}{\underset{x}{:}}$$
carbon dioxide

, and by a line formula $O\!=\!C\!=\!O$

In the line formula, only the shared pair of electrons is indicated by a bond ($-$). The sharing of three electron pairs results in a *triple* bond. An example is:

$$H\overset{x}{\underset{o}{:}}C\overset{o}{\underset{o}{:}}\overset{o}{\underset{o}{:}}C\overset{x}{\underset{o}{:}}H$$
acetylene

, and by a line formula $H\!-\!C\!\equiv\!C\!-\!H$

It can be assumed from these structures that there is a greater electron density between the nuclei involved and hence a greater attractive force between the nuclei and the shared electrons. Experimental data verify that greater energy is required to break double bonds than single bonds, and triple bonds than double bonds. Also, since these stronger bonds tend to pull atoms closer together, the atoms joined by double and triple bonds have smaller interatomic distances and greater bond strengths, respectively.

Resonance Structures

It is not always possible to represent the bonding structure by one Lewis dot structure or the line drawing because data about the bonding distance and bond strength are between possible drawing configurations and really indicate a blended condition. To represent this, the possible alternatives are drawn with arrows in between. Classic examples are sulfur trioxide and benzene. These are shown in Chapters 13 and 14, respectively, but are repeated here as examples.

Sulfur trioxide resonance structures:

Benzene resonance structures:

Electrostatic Repulsion (VSEPR) and Hybridization

ELECTROSTATIC REPULSION (VSEPR)

Data have shown that bond angles for atoms in molecules with p orbitals in the outer energy level do not conform to the expected 90° separation of an x, y, z axis orientation. This variation can be explained by **electrostatic repulsion** *between valence electron charge clouds* and by the concept of **hybridization**. The valence energy level electron pair repulsion model is sometimes called the **VSEPR** model. It is an abbreviation of *valence shell electron pair repulsion*.

VSEPR uses as its basis the fact that like charges will orient themselves in such a way as to diminish the repulsion between them.

1. Mutual repulsion of two electron clouds forces them to the opposite sides of a sphere. This is called a *linear* arrangement.

• **EXAMPLE:** BeF$_2$, beryllium fluoride

2. Minimum repulsion between three electron pairs occurs when the pairs are at the vertices of an equilateral triangle inscribed in a sphere. This arrangement is called *trigonal planar*.

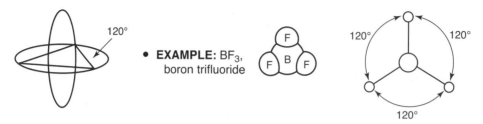

- **EXAMPLE:** BF_3, boron trifluoride

3. Four electron pairs are farthest apart at the vertices of a tetrahedron inscribed in a sphere. This is called a *tetrahedral* distribution of electron pairs.

- **EXAMPLE:** CH_4, methane

4. Mutual repulsion of six identical electron clouds directs them to the corners of an inscribed regular octahedron. This structure is said to have an arrangement called *octahedral*.

- **EXAMPLE:** SF_6, sulfur hexafluoride

VSEPR AND UNSHARED ELECTRON PAIRS

Ammonia, NH_3, and water, H_2O, are examples of molecules in which the central atom has both shared and unshared electron pairs. Here is how the VSEPR theory accounts for the geometries of these molecules.

The Lewis dot structure of ammonia shows that in addition to the three electron pairs it shares with three hydrogen atoms, the central nitrogen atom has one unshared pair of electrons.

VSEPR theory postulates that the lone pair (while occupying a bit more space than the bonding pairs) around the nitrogen atom repel just as the bonding pairs do. Thus, as in the methane molecule shown above, the electron pairs maximize their separation by assuming the four corners of a tetrahedron. Lone pairs occupy space, but our description of the observed shape of a molecule refers to the positions of only the atoms. Consequently, as shown in the drawing below, the molecular geometry of an ammonia molecule is that of a pyramid with a triangular base. The general VSEPR formula for molecules such as ammonia (NH_3) is AB_3E, where A replaces N, and B replaces H, and E represents the unshared electron pair.

A water molecule has two unshared electron pairs. It can be represented as an AB_2E_2 molecule. Here, the oxygen atom is at the center of a tetrahedron, with two corners occupied by hydrogen atoms and two by the unshared pairs as shown below. Again, VSEPR theory states that the lone pairs occupy space around the central atom but that the actual shape of the molecule is determined by the positions of only the atoms. In the case of water, this results in a "bent," or angular, molecule.

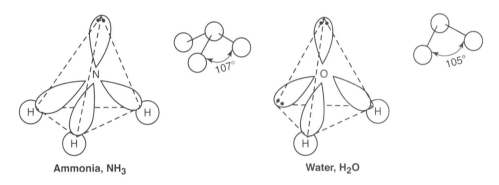

Ammonia, NH_3 Water, H_2O

VSEPR AND MOLECULAR GEOMETRY

The following table summarizes the molecular shapes associated with particular types of molecules. Notice that in VSEPR theory, double and triple bonds are treated in the same way as single bonds. It is helpful to use the Lewis dot structures and this table together to predict the shapes of molecules with double and triple bonds as well as the shapes of polyatomic ions.

Type of Molecule	Molecular Shape	Atoms Bonded to Central Atom	Lone Pairs of Electrons	Formula Example	Lewis Dot Structure
Linear		2	0	BeF_2	:F̈—Be—F̈:
Bent		2	1	$SnCl_2$	Sn / \ :C̈l C̈l:
Trigonal planar		3	0	BF_3	:F̈\ /F̈: B | :F̈:
Tetrahedral		4	0	CH_4	H | H—C—H | H
Trigonal pyramidal		3	1	NH_3	N̈ /|\ H H H
Bent		2	2	H_2O	Ö / \ H H
Trigonal bipyramidal		5	0	PCl_5	:C̈l: :C̈l: | /C̈l: :C̈l—P | \ :C̈l: C̈l:
Octahedral		6	0	SF_6	:F̈: :F̈\ | /F̈: S̈ :F̈/ | \F̈: :F̈:

HYBRIDIZATION

In order to arrive at these molecular shapes, chemists combine the envisioned shapes of individual atoms to form the known shapes of molecules. This concept is called **hybridization**. It describes the need to modify the atomic orbitals of isolated atoms to conform to the known shapes of molecules. Briefly stated, this means that two or more pure atomic orbitals (usually s, p, and d) can be mixed to form two or more hybrid atomic orbitals that are identical. This can be illustrated as follows:

sp hybrid orbitals

Beryllium fluoride spectroscopic measurements reveal a bond angle of 180° and equal bond lengths.

F—Be—F
↳ 180° ↲

The ground state of beryllium is

To accommodate the experimental data, we theorize that a $2s$ electron is excited to a $2p$ orbital; then the two orbitals hybridize to yield two identical orbitals called sp orbitals. Each contains one electron but is capable of holding two electrons.

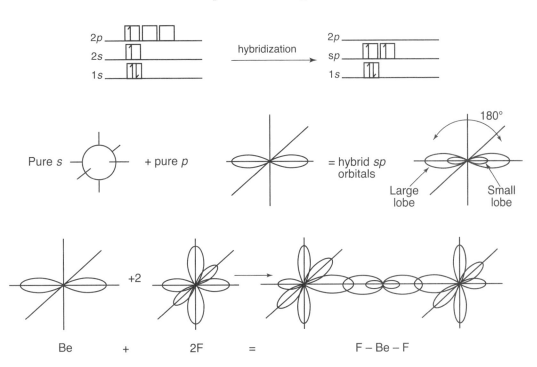

sp² hybrid orbitals

Boron trifluoride has bond angles of 120° of equal strength. To accommodate these data, the boron atom hybridizes from its ground state of $1s^2 2s^2 2p^1$ to

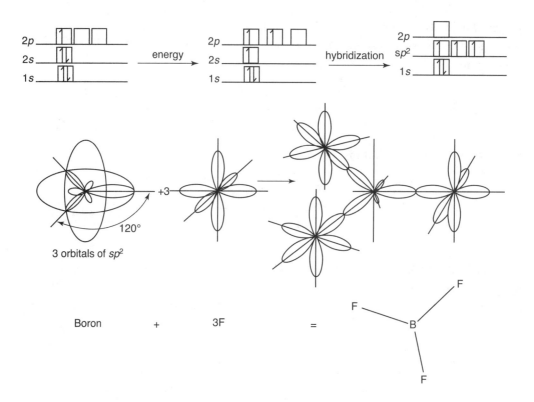

Boron   +   3F   =

sp³ hybrid orbitals

Methane, CH_4, can be used to illustrate this hybridization. Carbon has a ground state of $1s^2 2s^2 2p^2$. One $2s$ electron is excited to a $2p$ orbital, and the four involved orbitals then form four new identical sp^3 orbitals.

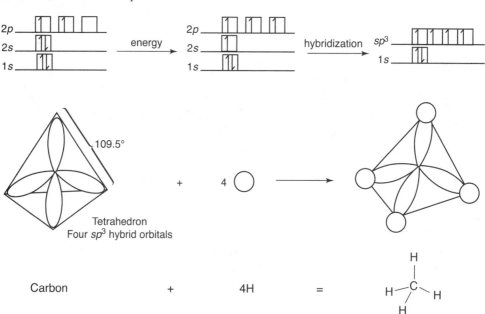

Tetrahedron
Four sp^3 hybrid orbitals

Carbon   +   4H   =

In some compounds where only certain sp^3 orbitals are involved in bonding, distortion in the bond angle occurs because the unbonded electron pairs spread out in space to a greater degree than the bonding pairs. Examples are:

a. Water, H_2O.

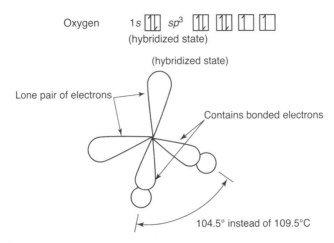

b. Ammonia, NH_3.

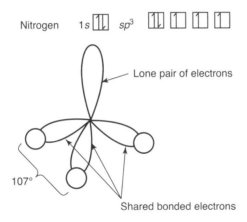

sp^3d^2 hybrid orbitals

These orbitals are formed from the hybridization of an s and a p electron promoted to d orbitals and transformed into six equal sp^3d^2 orbitals. The spatial form is

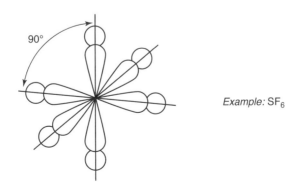

Number of Bonds	Number of Lone Pairs	Type of Hybrid Orbital	Angle Between Bonded Atoms	Geometry	Example
2	0	sp	180°	Linear	BeF_2
3	0	sp^2	120°	Trigonal planar	BF_3
4	0	sp^3	109.5°	Tetrahedral	CH_4
3	1	sp^3	107°	Pyramidal	NH_3
2	2	sp^3	109.5°	Angular	H_2O
6	0	sp^3d^2	90°	Octahedral	SF_6

Sigma and Pi Bonds

When bonding occurs between s and p orbitals, each bond is identified by a special term. A **sigma bond** is a bond between s orbitals or between an s orbital and another orbital such as a p orbital. It includes bonding between hybrids of s orbitals such as sp, sp^2, and sp^3.

In the methane molecule, the sp^3 orbitals are each bonded to hydrogen atoms. These are sigma bonds.

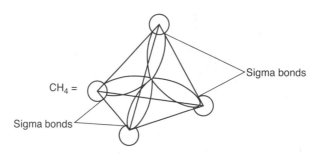

When two p orbitals share electrons in a covalent bond, this is called a **pi bond**. An example is found in ethene.

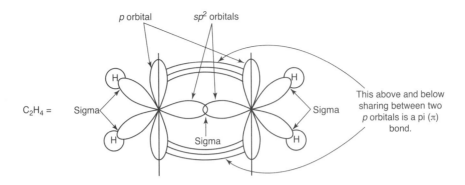

Chapter 14 gives more examples of sigma and pi bonding.

Properties of Ionic Substances

Laboratory experiments reveal that, in general, ionic substances are characterized by the following properties:

1. In the solid phase at room temperature, they do not conduct appreciable electric current.

2. In the liquid phase, they are relatively good conductors of electric current. The conductivity of ionic substances is much smaller than that of metallic substances.

3. They have relatively high melting and boiling points. There is a wide variation in the properties of different ionic compounds. For example, potassium iodide (KI) melts at 686°C and boils at 1330°C, while magnesium oxide (MgO) melts at 2800°C and boils at 3600°C. Both KI and MgO are ionic compounds.

4. They have relatively low volatilities and low vapor pressures. In other words, they do not vaporize readily at room temperature.

5. They are brittle and easily broken when stress is exerted on them.

6. Those that are soluble in water form electrolytic solutions which are good conductors of electricity. There is, however, a wide range in the solubilities of ionic compounds. For example, at 25°C, 92 g of sodium nitrate ($NaNO_3$) dissolves in 100 g of water, while only 0.0002 g of $BaSO_4$ dissolves in the same mass of water.

Properties of Molecular Crystals and Liquids

Experiments have shown that the following are general properties of these substances:

1. Neither the liquids nor the solids conduct electric current appreciably.

2. Many exist as gases at room temperature and atmospheric pressure, and many solids and liquids are relatively volatile.

3. The melting points of solid crystals are relatively low.

4. The boiling points of the liquids are relatively low.

5. The solids are often soft and have a waxy consistency.

6. A large amount of energy is often required to decompose the substance chemically into simpler substances.

Chapter Summary

The following terms summarize all the concepts and ideas that were introduced in this chapter. You should be able to explain their meaning and how you would use them in chemistry. They appear in boldface type in this chapter to draw your attention to them. The boldface type also makes the terms easier for you to look up, if you need to. You could also use a search engine on a computer to get a quick and expanded explanation of them.

TERMS YOU SHOULD KNOW

bonds
coordinate covalent bonds
covalent bond
dipoles
dipole-dipole attraction
electrostatic attraction
electrostatic repulsion
hybridization
hydrogen bond
intermolecular forces

ionic bond
London dispersion force
metallic bond
molecules
pi bond
resonance structure
stable octet
sigma bond
van der Waals forces
VSEPR theory

Chapter 3 Review Exercises

Use the following choices for questions 1 through 7:

    (A)   ionic

    (B)   covalent

    (C)   polar covalent

    (D)   coordinate covalent

    (E)   metallic

    (F)   van der Waals forces

    (G)   hydrogen bonding

1. When the electronegativity difference between two atoms is 2, what type of bond can be predicted?

2. If two atoms are bonded in such a way that one member of the pair is supplying both electrons that are shared, what is this type of bond called?

3. If the seven choices above were ordered from the strongest bond to the weakest, which would be the last one?

4. Which of the above bonds explains water's abnormally high boiling point?

5. When electron pairs are shared equally between two atoms, what is the bond called?

6. If the sharing of an electron pair is unequal, what is this sharing called?

7. Temporary dipole-dipole attraction created as a result of the momentary polarization of electron clouds that temporarily occurs in atoms falls into what category of bonding?

8. The VSEPR model of the BF_3 molecule results in the same trigonal planar structure as that of the hybridized form represented by

    (A) sp

    (B) sp^2

    (C) sp^3

    (D) sp^3d^2

Answers and Explanations

1. **(A)** When the electronegativity difference between two atoms is greater than 1.7, the bond between them is considered more than 50% ionic.

2. **(D)** When one atom contributes both of the electrons in a covalent bond, the bond is also referred to as coordinate or coordinate covalent.

3. **(F)** Van der Waals forces are generally the weakest of all the bonds.

4. **(G)** Because of the strong hydrogen bonding in water, a large amount of energy is needed to cause the molecules to break away from each other in the liquid state and change to vapor.

5. **(B)** Covalent is the name for the bond when electrons are shared.

6. **(C)** Unequal sharing of electrons in a bond means that one end of the bond has a higher electron density than the other. Therefore, the bond has two distinct ends (like a pole) from a charge perspective. This bond is called polar covalent.

7. **(F)** Temporary dipole-dipole attractions are also known as London dispersion forces. London dispersion forces fall in the van der Waals bonding category.

8. **(B)** The trigonal planar arrangement requires a hybridization of $1s$ and $2p$ orbitals to create three identical orbitals that are $120°$ apart. Each of these orbitals is referred to as sp^3.

PART III

USING ATOMS AND MOLECULES

CHEMICAL FORMULAS

CHAPTER OBJECTIVES

Upon completing this chapter, you will be able to:

- Articulate and apply the Laws of Definite Composition, Multiple Proportions, and Conservation of Mass
- Write chemical formulas for common chemical substances from their names
- Write the names of common chemical substances from their formulas
- Write the names of common acids from their formulas and vice versa
- Write a simple balanced reaction equation using chemical formulas including phase information

Laws of Definite Composition and Multiple Proportions

Near the end of the eighteenth century, the French chemist Joseph Proust argued that the chemical elements combined only in particular ways by mass to produce chemical compounds. As chemical knowledge grew, this concept became generally accepted and is known today as the **Law of Definite Composition** (sometimes referred to as the Law of Definite Proportions). This idea was one of the foundations of the atomic theory. In the early nineteenth century, chemist John Dalton theorized that all matter is comprised of atoms (small, discrete bits of matter). This is the atomic theory.

The Law of Definite Composition, with its specificity in describing the particular masses of the elements that combine to form compounds, naturally hints at the particulate nature of matter. John Dalton himself developed the **Law of Multiple Proportions**, embellishing the idea that matter is specific, discrete, and particular. Dalton noted that when *two* elements combined to form *two different* compounds, they did so in such a way that the ratio of the masses of the second element, when combined with equal amounts of the first, would always turn out to be small whole numbers. Both of these laws then pointed to the *fixedness* or *discreteness* (or the *atomic nature*) of matter.

Examples of both laws can be seen in the familiar compounds carbon monoxide and carbon dioxide. The masses of carbon and oxygen required to make 100.0 grams of both compounds are shown:

carbon monoxide: **42.9 g C and 57.1 g O**
carbon dioxide: **27.3 g C and 72.7 g O**

No matter where in the universe we find the substance carbon monoxide, every sample contains the *exact same ratio* by mass of carbon and oxygen as shown. In other words, anytime a sample of carbon monoxide is properly analyzed, the results will always show 42.9% of the compound by mass as carbon and 57.1% by mass as oxygen, a reflection of the Law of Definite Composition!

As the figure below describes, if a sample of carbon monoxide contains only 27.3 g of carbon, 36.3 grams of oxygen would have to be in the sample to keep the 42.9% carbon to 57.1% oxygen ratio intact. A comparison of these values with the amount of each element in the original carbon dioxide sample above shows that when a sample of carbon monoxide contains the same amount of carbon as the sample of carbon dioxide (27.3 g C), the ratio of the amounts of oxygen present in each sample reduces to a small whole number. That is, the 72.7 grams of oxygen in the carbon dioxide sample is essentially 2 times the amount of oxygen in the carbon monoxide sample, 36.3 g, when each sample contains the same amount of carbon, confirming the Law of Multiple Proportions.

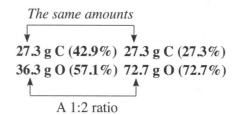

A sample of   A sample of
carbon monoxide   carbon dioxide
could contain: could contain:

The same amounts

27.3 g C (42.9%)   27.3 g C (27.3%)
36.3 g O (57.1%)   72.7 g O (72.7%)

A 1:2 ratio

Writing Chemical Formulas

The Laws of Definite Composition and Multiple Proportions indicate that compounds have *recipes* associated with them. Chemists refer to these recipes as **chemical formulas**. The rules by which elements combine have been discovered and are easily followed by placing compounds into particular categories and applying the rules for that category to the compound in question. Although many compounds do not fit into the categories shown below, the vast majority of the compounds encountered by first-year chemistry students do. The categorization is first based on the number of elements in the compound and then on the type of bond between them. (Bond types were discussed in Chapter 3.)

As the table that follows shows, compounds can be differentiated as *binary* or *other than binary*. As the name implies, **binary compounds** contain two elements. **Other than binary compounds** contain at least three elements. In the table:

- Three of the categories are *binary* and only one category is *other than binary*.
- The first two types of *binary* compounds are *ionically* bonded substances while the third kind is *covalently* bonded.
- The *other than binary* compounds, for which first-year chemistry students need to be able to write chemical formulas, generally contain ionic bonds and **polyatomic ions**.

Examples of writing formulas for compounds in each category follow the table.

FORMULA WRITING—CATEGORIES AND RULES

Categories		Rules
Binary Compounds	Type I: Ionic (containing a metal element with only one charge possibility and a nonmetal element)	1. Write down the elemental symbols.* 2. Use subscripts to balance the charges of the ions.**
	Type II: Ionic (containing a metal element that has more than one charge possibility and a nonmetal element)	
	Type III: Covalent (containing two nonmetal elements)	1. Write down the elemental symbols. 2. Use the prefixes in the compound name to write the subscripts for each element.
Other Than Binary Compounds	Ionic compounds containing at least three elements	1. Write down the elemental symbols or polyatomic ion formulas.*** 2. Use subscripts to balance the charges of the ions.****

*The most metallic element's symbol is written first.
**Charges for many ionic substances are shown In Table 6.
***Selected polyatomic ion formulas are shown In Table 6.
****Parentheses are needed if more than one polyatomic ion is required.

EXAMPLE 1

Aluminum Chloride (A Type I Binary Ionic Compound)

Aluminum chloride is a type I binary ionic compound because aluminum is a metal that has only one charge possibility, 3+. The name chloride implies that the nonmetal element chlorine is also present in the compound (the reason for the adjusted *-ide* ending will be described shortly). The chloride ion has a charge of 1–. Therefore, to write the formula correctly, write the symbols for aluminum and chlorine first (rule 1):

$$Al\ Cl$$

In order to balance the charge from one aluminum ion (rule 2), 3 chloride ions are needed so that the total negative and positive charges are equal. That is what is meant by "use subscripts to balance the charges" in rule 2. Therefore, a subscript of 3 is placed after the symbol for chlorine:

$$AlCl_3$$

One way always to balance the charges appropriately is to take the value of the charge (not the sign) on the first element and make it the subscript on the second element. The subscript on the first element will then be the value of the charge on the second element. This is called the **crisscross and reduce method** for formula writing. When the subscript is a 1, the subscript is not written as it is understood to be that value in the absence of any other number. Since ionic substances' formulas always represent the simplest ratio of the ions in the compound, the subscripts must also be reduced to the smallest whole numbers possible.

EXAMPLE 2

Iron (III) Oxide (A Type II Binary Ionic Compound)

Iron (III) oxide is a type II binary ionic compound because iron is a metal that has more than one charge possibility, 2+ and 3+. The charge on the iron ion in this case is 3+ as indicated by the Roman numeral after the metal's name. Sometimes Latin or Greek names for the elements are used with special suffixes to indicate the charge on the metal ion. Table 6 lists the common metallic elements that exhibit more than one charge possibility and their alternative names. The use of Roman numerals in parentheses after the metal's name is referred to as the stock system. The name oxide implies that the nonmetal element oxygen is also present in the compound. The oxide ion has a charge of 2–. Therefore, to write the formula correctly, write the symbols for iron (III) and oxygen first (rule 1):

$$Fe\ O$$

By using the crisscross and reduce method outlined previously, the subscript on the oxygen will be the value of the charge on iron, 3. The subscript on the iron will be the value of the charge on the oxygen, 2. The formula for iron (III) oxide is

$$Fe_2O_3$$

TABLE 6
SOME COMMON IONS AND THEIR CHARGES

Monatomic Ions (fixed charge)				Monatomic Ions (multicharged)			Polyatomic Ions	
Positive Ions					Stock System	Latin/Greek Name	NO_3^-	nitrate
H^+	hydrogen	Li^+	lithium				NO_2^-	nitrite
Na^+	sodium	K^+	potassium				ClO_4^-	perchlorate
Rb^+	rubidium	Cs^+	cesium	Cu^+	copper (I)	cuprous	ClO_3^-	chlorate
Be^{2+}	beryllium	Mg^{2+}	magnesium	Cu^{2+}	copper (II)	cupric	ClO_2^-	chlorite
Ca^{2+}	calcium	Sr^{2+}	strontium	Fe^{2+}	iron (II)	ferrous	ClO^-	hypochlorite
Ba^{2+}	barium	Ra^{2+}	radium	Fe^{3+}	iron (III)	ferric	CrO_4^{2-}	chromate
Ag^+	silver	Zn^{2+}	zinc	Pb^{2+}	lead (II)	plumbous	$Cr_2O_7^{2-}$	dichromate
				Pb^{4+}	lead (IV)	plumbic	CN^-	cyanide
				Sn^{2+}	tin (II)	stannous	MnO_4^-	permanganate
Negative Ions				Sn^{4+}	tin (IV)	stannic	OH^-	hydroxide
H^-	hydride	F^-	fluoride	Au^+	gold (I)	aurous	IO^-	hypoiodite
Cl^-	chloride	Br^-	bromide	Au^{3+}	gold (III)	auric	O_2^{2-}	peroxide
I^-	iodide	O^{2-}	oxide				NH_2^-	amide
S^{2-}	sulfide	Se^{2-}	selenide				CO_3^{2-}	carbonate
Te^{2-}	telluride	N^{3-}	nitride				HCO_3^-	hydrogen carbonate (bicarbonate)
P^{3-}	phosphide							
							SO_4^{2-}	sulfate
							SO_3^{2-}	sulfite
							$C_2O_4^{2-}$	oxalate
							PO_4^{3-}	phosphate
							HPO_4^{2-}	hydrogen phosphate
							PO_3^{3-}	phosphate
							$H_2PO_4^-$	dihydrogen phosphate
							$S_2O_3^{2-}$	thiosulfate
							HS^-	hydrogen sulfide
							$C_4H_4O_6^{2-}$	tartrate
							$C_2H_3O_2^-$	acetate

Know These Prefixes

mono—1	*tri*—3	*penta*—5
di—2	*tetra*—4	*hexa*—6

EXAMPLE 3

Sulfur Hexafluoride (A Type III Binary Covalent Compound)

Sulfur hexafluoride is a type III binary covalent compound because both sulfur and fluorine are nonmetals. Therefore, to write the formula correctly, write the symbols for sulfur and fluorine first (rule 1):

$$S\ F$$

The prefix *hexa*, in front of the term fluoride, indicates 6 fluorine atoms will be needed for every 1 sulfur atom. (No prefix on the sulfur implies only one atom.) The formula for sulfur hexafluoride is

$$SF_6$$

EXAMPLE 4

Potassium Nitrate (An Other Than Binary Ionic Compound)

Potassium nitrate is an other than binary ionic compound because three elements are present in the compound. Potassium is a metal that has only one charge possibility, 1+. The other elements come from the polyatomic ion nitrate. Table 6 lists the common polyatomic ions. Polyatomic ions are like molecules with a charge. They can be treated as an individual unit. The name nitrate indicates the polyatomic ion NO_3 with a 1– charge. Therefore, to write the formula correctly, write the symbols for potassium and nitrate (rule 1):

$$K\ NO_3$$

As with the other ionic compounds, using the crisscross and reduce method mandates the subscript on the potassium will be the value of the charge on nitrate, 1, and the subscript on the nitrate will be the value of the charge on the potassium, 1. The formula for potassium nitrate is

$$KNO_3$$

Naming Compounds from Their Formulas

Naming a compound from its formula (the opposite of what was described in the previous section) also requires adherence to a set of rules. The categorizations used in the previous section are now used to differentiate the types of compounds to which first-year chemistry students will find themselves exposed. The table that follows describes the rules for naming compounds from their chemical formulas. The examples afterward describe how the rules are used.

EXAMPLE 1

Na_2S (A Type I Binary Ionic Compound)

Na_2S is a type I binary ionic compound because sodium, whose symbol is Na, is a metal that has only one charge possibility, 1+. That is why this compound is type I. The symbol S implies that the nonmetal element sulfur is also present in the compound. By using the rules for this category, the name of the compound is

sodium sulfide

EXAMPLE 2

Pb_3N_2 (A Type II Binary Ionic Compound)

Pb_3N_2 is a type II binary ionic compound because lead, whose symbol is Pb, is a metal element that has more than one charge possibility, 2+ and 4+. That is what makes the compound type II. The charge on the lead ion in this case is 2+ because three lead ions are needed to balance the charge from two nitrogen atoms, which each have a 3– charge. The nitrogen's charge is determined by looking up the information in Table 6. By using the rules for this category, the name of the compound is

lead (II) nitride

NAMING COMPOUNDS—CATEGORIES AND RULES

Categories		Rules
Binary Compounds	Type I: Ionic (containing a metal element with only one charge possibility and a nonmetal element)	1. Write the name of the metal first.* 2. Write the name of the nonmetal with an adjusted *-ide* ending.**
	Type II: Ionic (containing a metal element that has more than one charge possibility and a nonmetal element)	
	Type III: Covalent (containing two nonmetal elements)	1. Write the name of the first nonmetal with a prefix to describe the number of atoms if more than one atom is indicated. 2. Write the name of the second nonmetal with a prefix and an adjusted *-ide* ending even if only one atom is indicated.
Other Than Binary Compounds	Ionic compound containing at least three elements	Name the parts of the compound in the order presented in the formula.

*Use the stock system to indicate the charge on the metal if the metal has more than one charge possibility.

**Adjusted *-ide* endings generally replace the last syllable of the nonmetal's elemental name.

 EXAMPLE 3 P_2Cl_5 **(A Type III Binary Covalent Compound)**

P_2Cl_5 is a type III binary covalent compound because both phosphorous and chlorine, whose symbols are P and Cl, respectively, are nonmetals. By using the rules for this category, the name of this compound is

diphosphorous pentachloride

EXAMPLE 4 $MgClO_3$ **(An Other Than Binary Ionic Compound)**

$MgClO_3$ is an other than binary ionic compound because three elements are present in the compound: magnesium, chlorine, and oxygen. When this is the case, a grouping of atoms that represent a polyatomic ion must be recognized. Familiarity with the polyatomic ions listed in Table 6 allows for the identification of the chlorate ion, ClO_3^-, in this compound bound to the metal magnesium. By using the rules for this category, the name for this compound is

magnesium chlorate

Acids

Although the definition of an acid will deepen as your chemical knowledge increases, an appropriate way to define the most common acids which first-year chemistry students should be familiar with, is as a *hydrogen compound dissolved in water*. The term *hydrogen compound* implies that hydrogen is combined with either a nonmetal element or a polyatomic ion. In the name or the formula for these compounds, the hydrogen will always be listed first. These should not be confused with compounds that have hydrogen in other sections of the name or formula.

The naming of acids and the writing of their formulas, once again, requires knowledge of special rules that allows all chemists to speak the same language. The fundamental categorization depends on the presence of oxygen (or lack thereof) in the acid. The table that follows summarizes these rules and examples of their use follow the table.

NAMING ACIDS—CATEGORIES AND RULES

Categories	Rules
Aqueous hydrogen compounds without oxygen. (Generally binary hydrogen compounds in which hydrogen is combined with a nonmetal element)	1. The prefix *hydro-* begins the name of the acid . . . 2. followed by the *root name of the nonmetal element* that is combined with hydrogen . . . 3. followed by the suffix *-ic* . . . 4. followed by the word *acid*.
Aqueous hydrogen compounds containing oxygen. (Generally other than binary compounds in which hydrogen is combined with a polyatomic ion)	1. The *root name of the polyatomic ion* begins the name of the acid . . . 2. followed by the suffix *-ic* (if the polyatomic ion's name ends in *-ate*) or *-ous* (if the polyatomic ion's name ends in *-ite*) . . . 3. followed by the word *acid*.

 EXAMPLE 1

$HCl_{(aq)}$ (An Aqueous Hydrogen Compound Without Oxygen)

The lack of oxygen in this hydrogen compound is evident. The fact that it is dissolved in water is indicated by the subscript (aq). The (aq) subscript stands for the term *aqueous solution*. By definition, aqueous solutions are those in which a solute, in this case a hydrogen compound, is dissolved in water. (More information on aqueous solutions of all types is presented in Chapter 7.) Using the rules for this category necessitates the use of the prefix *hydro-*, followed by the root name of the element *chlor*ine, followed by the suffix *-ic*, to write the name of the acid:

hydrochloric acid

EXAMPLE 2 **HNO$_{3(aq)}$ (An Aqueous Hydrogen Compound With Oxygen)**
The presence of oxygen in this hydrogen compound is evident. Therefore a grouping of atoms that represent a polyatomic ion containing oxygen must be recognized. Familiarity with Table 6 allows the identification of the polyatomic ion nitrate, NO$_3^-$. Using the rules for this category necessitates the use of the root name of the polyatomic ion *nitr*ate followed by the suffix *-ic* to write the name of the acid:

nitric acid

The names and the formulas for some common acids are listed in Table 7.

TABLE 7
SOME COMMON ACIDS

Without Oxygen		With Oxygen	
Names	**Formulas**	**Names**	**Formulas**
hydrofluoric acid	HF$_{(aq)}$	nitrous acid	HNO$_{2(aq)}$
hydrochloric acid	HCl$_{(aq)}$	nitric acid	HNO$_{3(aq)}$
hydrobromic acid	HBr$_{(aq)}$	hypochlorous acid	HClO$_{(aq)}$
hydroiodic acid	HI$_{(aq)}$	chlorous acid	HClO$_{2(aq)}$
hydrosulfuric acid	H$_2$S$_{(aq)}$	chloric acid	HClO$_{3(aq)}$
hydrocyanic acid	HCN$_{(aq)}$	perchloric acid	HClO$_{4(aq)}$
		sulfurous acid	H$_2$SO$_{3(aq)}$
		sulfuric acid	H$_2$SO$_{4(aq)}$
		phosphoric acid	H$_3$PO$_{4(aq)}$
		carbonic acid	H$_2$CO$_{3(aq)}$
		acetic acid	HC$_2$H$_3$O$_{2(aq)}$
		oxalic acid	HC$_2$O$_{4(aq)}$

Putting It All Together

Identifying the type of substance categorically, along with *strict* adherence to the rules for each category, allows even novice chemistry students to be successful. First-year chemistry students would do well to ask themselves routinely a series of questions that would help place the compound into the proper category. The following flow chart identifies the questions that should be asked when writing the formula for a compound.

To Be Successful in Writing Formulas, You Should Ask Yourself:

For writing formula purposes, hydrogen compounds are considered ionic even though they are really covalently bonded molecules.

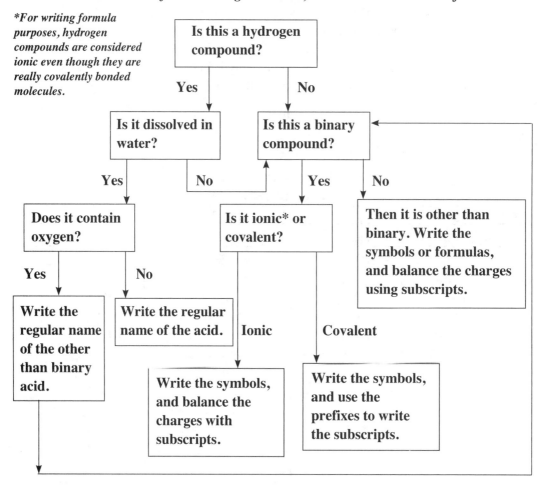

Now use the flow chart to answer three practice problems.

PROBLEM 1

Write the formula for the compound **tin (IV) sulfate**.

First of all, this is not a hydrogen compound. The name hydrogen does not begin the name and the compound is not designated as an acid. Second, it is not binary. This is evident because the polyatomic ion sulfate is present. Therefore, this is an other than binary ionic compound. Its formula is written by following the rules in the appropriate box in the flow chart:

$$Sn(SO_4)_2$$

PROBLEM 2

Write the formula for the compound **hydrogen fluoride**.

First, this is a hydrogen compound. However, it is not dissolved in water because no information is given to describe it as such. Second, the compound is binary. Only two elements, hydrogen and fluorine, are present. (An -*ide* ending is generally indicative of a single nonmetal element, except in the case of a few polyatomic ions. The suffix -*ate* and -*ite* generally indicate the presence of a polyatomic ion.) As indicated in the flow chart, hydrogen compounds are actually covalently bonded substances whose formulas are written as if they were ionic. Therefore, we use the rules for binary ionic substances given in the appropriate box in the flow chart:

$$HF$$

PROBLEM 3 Write the formula for the compound **acetic acid**.

First, this is a hydrogen compound dissolved in water because it is an acid. Second, the compound contains oxygen because the prefix *hydro-* does not begin the name. Therefore, the regular name of this compound is *hydrogen acetate* as the hydrogen must be combined with a polyatomic ion whose root name *acet-* must end in the suffix -*ate* since the name of the acid ends in -*ic*. As indicated by the arrows in the flow chart, the next question is whether or not the compound (as described by its *regular* name as opposed to its *acid* name) is binary or not. Since acetate is a polyatomic ion, the answer is no. Therefore, we use the rules for other than binary ionic substances in the appropriate box in the flow chart:

$$HC_2H_3O_{2(aq)}$$

The (aq) is added since, as described above, the term *acid* indicates that this is a hydrogen compound dissolved in water.

The same flow chart questions used to write formulas, given the names of compounds, can be asked in the reverse order. This is done to determine the names of compounds, if the formulas are given, to help place a compound into its appropriate category.

Chemical Formulas

As you have seen, the chemical formula is an indication of the makeup of a compound in terms of the kinds of atoms and their relative numbers. It also has some quantitative applications. By using the atomic masses assigned to the elements, we can find the **formula mass** of a compound. If we are sure that the formula represents the actual makeup of one molecule of the substance, the term **molecular mass** may be used as well. In some cases the formula represents an ionic lattice and no discrete molecule exists, as in the case of table salt, $NaCl$, or the formula merely represents the simplest ratio of the combined substances and not a molecule of the substance. For example, CH_2 is the simplest ratio of carbon and hydrogen united to form the actual compound ethylene, C_2H_4. This simplest ratio formula is called the **empirical formula**, and the actual formula is the **true formula**. The formula mass is determined by multiplying the atomic mass (in whole numbers) by the subscript for that element in the formula.

EXAMPLE $Ca(OH)_2$ (one calcium atomic mass + two hydrogen and two oxygen atomic masses = formula mass).

SOLUTION

$$1 \text{ Ca (at. mass} = 40) = 40$$
$$2 \text{ O (at. mass} = 16) = 32$$
$$\underline{2 \text{ H (at. mass} = 1) = 2}$$
$$\text{Formula mass } Ca(OH)_2 = 74$$

Fe_2O_3

$$2 \text{ Fe (at. mass} = 56) = 112$$
$$\underline{3 \text{ O (at. mass} = 16) = 48}$$
$$\text{Formula mass } Fe_2O_3 = 160$$

It is sometimes useful to know what percentage of the total mass of a compound is made up of a particular element. This is called finding the **percentage composition**. The simple formula for this is

$$\frac{\text{Total mass of the element in the compound}}{\text{Total formula mass}} \times 100\% = \begin{array}{c} \text{Percentage composition} \\ \text{of that element} \end{array}$$

EXAMPLE To find the percentage composition of calcium in calcium hydroxide in the example above, we set the formula up as follows:

$$\frac{Ca = 40}{\text{Formula mass} = 74} \times 100\% = 54\% \text{ calcium}$$

To find the percentage composition of oxygen in calcium hydroxide:

$$\frac{O = 32}{\text{Formula mass} = 74} = 100\% = 43\% \text{ oxygen}$$

To find the percentage composition of hydrogen in calcium hydroxide:

$$\frac{H = 2}{\text{Formula mass} = 74} \times 100\% = 2.7\% \text{ hydrogen}$$

Find the percentage compositions of Cu and H_2O in the compound $CuSO_4 \cdot 5H_2O$ (the dot is read "with").

SOLUTION

First, we calculate the formula mass

$$1 \text{ Cu} = 64$$
$$1 \text{ S} = 32$$
$$4 \text{ O} = 64 \ (4 \times 16)$$
$$\underline{5 \text{ H}_2\text{O} = 90 \ (5 \times 18)}$$
$$250$$

and then find the percentages
Percentage Cu:

$$\frac{\text{Cu} = 64}{\text{Formula mass} = 250} \times 100\% = 26\%$$

Percentage H_2O:

$$\frac{5 \text{ H}_2\text{O} = 90}{\text{Formula mass} = 250} \times 100\% = 36\%$$

When you are given the percentage of each element in a compound, you can find the empirical formula as shown with the following example:

EXAMPLE 1 Given that a compound is composed of 60.0% Mg and 40.0% O, find the empirical formula of the compound.

SOLUTION

It is easiest to think of 100 mass units of this compound. In this case, the 100 mass units are composed of 60 mass units of Mg and 40 mass units of O. Since you know that 1 unit of Mg is 24 mass units (from its atomic mass) and, likewise, 1 unit of O is 16, you can divide 60 by 24 to find the number of units of Mg in the compound and divide 40 by 16 to find the number of units of O in the compound.

Mg	O
$24\overline{)60}$	$16\overline{)40}$
2.5 units Mg	2.5 units O

Now, since we know formulas are made up of whole-number units of the elements that are expressed as subscripts, we must manipulate these numbers to get whole numbers. This is usually accomplished by dividing these numbers by the smallest quotient. In this case they are equal, and so we divide by 2.5.

$$
\begin{array}{cc}
\text{Mg} & \text{O} \\
2.5\overline{)2.5} & 2.5\overline{)2.5} \\
1 & 1
\end{array}
$$

Our empirical formula, then, is one Mg and one O. Therefore, MgO is the formula.

EXAMPLE 2

Given: Ba = 58.81%, S = 13.73%, and O = 27.46%.
Find the empirical formula.

SOLUTION

Divide each percentage by the atomic weight of the element.

$$
\begin{array}{ccc}
\text{Ba} & \text{S} & \text{O} \\
137\overline{)58.8} & 32\overline{)13.7} & 16\overline{)27.5} \\
0.43 & 0.43 & 1.72
\end{array}
$$

Manipulate the numbers to get small whole numbers. Try dividing them all by the smallest first.

$$
\begin{array}{ccc}
\text{Ba} & \text{S} & \text{O} \\
0.43\overline{)0.43} & 0.43\overline{)0.43} & 0.43\overline{)1.72} \\
1 & 1 & 4
\end{array}
$$

The formula is $BaSO_4$.

In some cases you may be given the true formula mass of the compound. To check if your empirical formula is correct, add up the formula mass of the empirical formula and compare it to the given formula mass. If it is *not* the same, multiply the empirical formula by the small whole number that gives you the correct formula mass. For example, if your empirical formula is CH_2 (which has a formula mass of 14) and the true formula mass is given as 28, you can see that you must double the empirical formula by doubling all the subscripts. The true formula is C_2H_4.

Balancing Simple Reaction Equations

In the middle of the eighteenth century, another French chemist, Antoine Lavoisier, recognized that the masses of reacting substances in a chemical reaction equal the total mass of the substances produced. This observation is known as the **Law of Conservation of Mass**. It is another piece of information that John Dalton was aware of that pointed toward the particulate nature of matter. As with the Laws of Definite Composition and Multiple Proportions, the Law of Conservation of Mass hinted at the specificity and discreteness of matter. The tenet in Dalton's atomic theory (discussed in Chapter 2) based on the conservation law states that *during a chemical change, atoms are simply rearranged*. Since this law must always be satisfied, chemical reactions must always be balanced in terms of the number of atoms found in the reacting substances and the number of atoms in the substances produced.

A chemical reaction equation is a simplified manner of expressing a chemical change. Reaction equations use chemical symbols and formulas to represent the substances reacting (the **reactants**) and those produced (the **products**). Once words describing the reactants and products are replaced with their proper chemical formulas, reactions can be balanced by inspecting the number of atoms found on each side of the equation and changing the number of each if needed. The number of the atoms can be changed by placing a coefficient in front of the formulas in the equation such that the number of atoms on each side of the arrow (the **yields** sign) is the same. First-year chemistry students increase their success in balancing reaction equations by systematically attacking each problem in a consistently similar manner. The following suggestions tend to breed success:

1. Write the formulas for the compounds correctly. Once written, do not mess with the subscripts!

2. To change the number of atoms for any element, change the coefficient in front of the formula containing that element from the number 1, which is understood in the absence of a number shown, to the number of atoms desired. Be aware that the coefficient multiplies every atom in the formula and, in doing so, multiplies the subscript already shown in the formula.

   a. Always start on the left-hand side of the equation with the first element you encounter. See how many atoms of that element are on both sides of the equation and make adjustments to the coefficients so that each side has the same number.

   b. The only exception to the previous rule is that oxygen and hydrogen should be saved until the end.

   c. Balance groups of atoms (polyatomic ions), if they remain intact on each side of the equation, as individual units.

   d. View water as HOH if the polyatomic ion hydroxide is found on the other side of the equation. This way the polyatomic ion can be balanced on both sides as it remains intact.

   e. Use fractions if doing so makes it easier to balance the numbers of atoms needed. Then multiply all coefficients by the denominator of the fraction used so as to clear the fraction.

EXAMPLE 1 **Balance the reaction equation:**

hydrogen plus oxygen yields water

By replacing the words with chemical formulas, we have

$$H_2 + O_2 \rightarrow H_2O$$

We replace hydrogen and oxygen with the formulas for their diatomic molecular states (as they exist under normal conditions) and write the appropriate formula for water. Seven elements commonly exist diatomically in their elemental states. The word "yields" is replaced with an arrow.

The diatomic elements

hydrogen	**nitrogen**	**fluorine**	**iodine**
oxygen	**chlorine**	**bromine**	

Once the formulas are written correctly, we begin inspecting the reaction from the left. Since hydrogen and oxygen are the only elements in this reaction, suggestion "2b" in the previous list is moot. The number of hydrogen atoms on each side is 2, so the hydrogen is balanced. To balance the oxygen, a 2 must be placed in front of the water in the products so that each side has 2 oxygen atoms. We now have

$$H_2 + O_2 \rightarrow 2H_2O$$

Placing the coefficient 2 in front of the water also multiplies the number of hydrogen atoms that were present by 2 as well. Therefore, 4 hydrogen atoms are now present in the products. A coefficient of 2 must be placed in front of the hydrogen in the reactants to compensate. Now we have the balanced equation

$$2H_2 + O_2 \rightarrow 2H_2O$$

EXAMPLE 2 **Balance the reaction equation:**

aqueous silver nitrate reacts with solid copper to form
aqueous copper (II) nitrate plus solid silver

By replacing the words with chemical formulas, we have:

$$AgNO_{3(aq)} + Cu_{(s)} \rightarrow Cu(NO_3)_{2(aq)} + Ag_{(s)}$$

We replace the compounds' chemical names with their chemical formulas and the elemental names with their chemical symbols. Note that neither copper nor silver are diatomic elements. Therefore, no subscript is shown for either. Additionally, no charges are placed on their symbols as they are simply representing atoms of their respective elements and are not ions. The ionic nature of copper and silver arises only when they enter into compounds. The additional information of *phase* has also been presented as it was supplied in the word description of the reaction. Phases are indicated as subscripts in parentheses to the right of the symbol or formula. It is noteworthy that many ionic compounds do not react appreciably in the solid state. Therefore, they are often dissolved in water, an aqueous solution, to enhance their rate of reaction. Phase information is summarized in the following chart.

Phase information

| (s) solid | (l) liquid | (g) gas | (aq) aqueous |

Once the formulas are written correctly, we begin inspecting the reaction from the left. Each side of the equation has 1 silver atom. The nitrate group remains intact on both sides of the equation, so that should be balanced next by placing a 2 in front of the $AgNO_3$ in the reactants. That gives us

$$2AgNO_{3(aq)} + Cu_{(s)} \rightarrow Cu(NO_3)_{2(aq)} + Ag_{(s)}$$

Placing the coefficient 2 in front of the $AgNO_3$ also multiplies the number of silver atoms that were present by 2 as well. Therefore, 2 Ag atoms are now present in the reactants. A coefficient of 2 must be placed in front of the Ag in the products to compensate. Since the number of Cu atoms on both sides is 1, we now have the balanced equation:

$$2AgNO_{3(aq)} + Cu_{(s)} \rightarrow Cu(NO_3)_{2(aq)} + 2Ag_{(s)}$$

Writing Ionic Equations

An example of phase notation in an equation is

$$2HCl(aq) + Zn(s) \rightarrow ZnCl_2(aq) + H_2(g)$$

In words, this says that a water solution of hydrogen chloride (called hydrochloric acid) reacts with solid zinc to produce zinc chloride dissolved in water plus hydrogen gas.

There are times that chemists choose to show only the substances that react in the chemical action. These are called net *ionic* equations because they stress the reaction and production of ions. If we look at the preceding equation, we see that the complete cast of "actors" in this equation is

$$2HCl(aq) \rightarrow 2H^+(aq) + 2Cl^-(aq)$$
$$Zn(s) \rightarrow Zn(s) \text{ particles}$$
$$ZnCl_2(aq) \rightarrow Zn^{2+}(aq) + 2Cl^-(aq)$$
$$H_2(g) \rightarrow H_2(g)$$

Writing the complete reaction using these results, we have

$$2H^+(aq) + 2Cl^-(aq) + Zn(s) \rightarrow Zn^{2+}(aq) + 2Cl^-(aq) + H_2(g)$$

Notice that nothing happened to the chloride ion. It appears the same on both sides of the equation. It is referred to as a spectator ion. In writing the net ionic equation, spectator ions are omitted. So, the net ionic equation is

$$2H^+(aq) + Zn(s) \rightarrow Zn^{2+}(aq) + H_2(g)$$

Chapter Summary

The following terms summarize all the concepts and ideas that were introduced in this chapter. You should be able to explain their meaning and how you would use them in chemistry. They appear in boldface type in this chapter to draw your attention to them. The boldface type also makes the terms easier for you to look up, if you need to. You could also use a search engine on a computer to get a quick and expanded explanation of them.

TERMS YOU SHOULD KNOW

binary compound
chemical formulas
coefficient
crisscross and reduce method
empirical formula
formula mass
Law of Conservation of Mass
Law of Definite Composition

Law of Multiple Proportions
molecular mass
percentage composition
polyatomic ion
products
reactants
true formula
yields

Chapter 4 Review Exercises

Write the formula *or the* name *of each in items 1–16:*

1. silver chloride _____

2. calcium sulfate _____

3. oxalic acid _____

4. gold (III) oxide _____

5. sulfur trioxide _____

6. ammonium nitrate _____

7. hydrogen peroxide _____

8. hydroselenic acid _____

9. Na_2F _____

10. $MgClO_3$ _____

11. $H_2CO_{3(aq)}$ _____

12. FeN _____

13. CCl$_4$ _____

14. Zn(ClO$_2$)$_2$ _____

15. HBr$_{(aq)}$ _____

16. HCN$_{(aq)}$ _____

17. Find the percentage of sulfur in H$_2$SO$_4$.

18. What is the empirical formula and the true formula of a compound composed of 85.7% C and 14.3% H with a true formula mass of 42 u?

19. Write a balanced equation for the reaction in which aqueous sodium hydroxide reacts with aqueous iron (III) nitrate to form solid iron (III) hydroxide and aqueous sodium nitrate. Afterward, write the complete ionic equation.

Answers and Explanations

1. **AgCl** Binary ionic compound; wrote symbols and balanced the charges.

2. **CaSO$_4$** Other than binary ionic compound; wrote symbols and balanced the charges.

3. **H$_2$C$_2$O$_{4(aq)}$** Regular name is hydrogen oxalate. Even though it is a molecular compound, the formula is written as if it is an other than binary ionic compound dissolved in water; wrote symbols and balanced the charges; included (aq) due to acid.

4. **Au$_2$O$_3$** Binary ionic compound; wrote symbols and balanced the charges.

5. **SO$_3$** Covalent binary compound; wrote symbols and used prefixes to write subscripts.

6. **NH$_4$NO$_3$** Other than binary ionic compound; wrote symbols and balanced the charges.

7. **H$_2$O$_2$** Even though it is a molecular compound, the formula is written as if it is a binary ionic compound because it is a hydrogen compound; wrote symbols and balanced the charges.

8. **H$_2$Se$_{(aq)}$** Even though it is a molecular compound, the formula is written as if it is a binary ionic compound because it is a hydrogen compound; wrote symbols and balanced the charges.

9. **sodium fluoride** Binary ionic compound; named the metal and then the nonmetal with the adjusted -*ide* ending.

10. **magnesium chlorate** Other than binary ionic compound; named both parts.

11. **carbonic acid** Hydrogen compound dissolved in water; also known as hydrogen carbonate; named the root name of polyatomic ion (carbonate) with an -*ic* suffix because of the -*ate* ending of the polyatomic ion followed by the word *acid*.

12. **iron (III) nitride** Binary ionic compound; named the metal and then the nonmetal with the adjusted -*ide* ending.

13. **carbon tetrachloride** Binary covalent compound; named the first nonmetal then the second nonmetal with a prefix and the adjusted -*ide* ending.

14. **zinc chlorite** Other than binary ionic compound; named both parts.

15. **hydrobromic acid** Hydrogen compound dissolved in water; also known as hydrogen bromide; named the root name of the anion (bromine) with the prefix *hydro-* (no oxygen) and the suffix -*ic* followed by the word *acid*.

16. **hydrocyanic acid** Hydrogen compound dissolved in water; also known as hydrogen cyanide; named the root name of the polyatomic ion (cyanide) with the prefix *hydro-* (no oxygen) and the suffix -*ic* followed by the word *acid*.

17. 32.65%

    H_2SO_4 is composed of

    $$
    \begin{array}{r}
    2\,H = 4 \\
    1\,S = 32 \\
    \underline{4\,O - 64} \\
    \text{Total} = 98
    \end{array}
    $$

    Percentage of S: $\dfrac{S = 32}{\text{Total} = 98} \times 100\% = 32.65 \text{ or } 33\%$

18. CH_2 and C_3H_6

    C = 12)85.7% C   H = 1)14.3% H
    $\quad\quad\quad$ 7.14 $\quad\quad\quad\quad\quad$ 14.3

    To find the lowest ratio of the whole numbers:

    7.14)7.14   7.14)14.3
    $\quad\quad$ 1.0 $\quad\quad\quad\quad$ 2.0

    So the empirical formula is CH_2.

    Since the formula mass is given as 42, the empirical formula CH_2, which totals 14 amu, divides into 42 three times. Therefore, the true formula is C_3H_6.

19. Formula equation:

    $$3\,NaOH_{(aq)} + Fe(NO_3)_{3(aq)} \rightarrow 3\,NaNO_{3(aq)} + Fe(OH)_{3(s)}$$

    Complete ionic equation:

    $$3\,Na^+{}_{(aq)} + 3\,OH^-{}_{(aq)} + Fe^{3+}{}_{(aq)} + 3\,NO_3{}^-{}_{(aq)} \rightarrow 3\,Na^+{}_{(aq)} + 3\,NO_3{}^-{}_{(aq)} + Fe(OH)_{3(s)}$$

    Net ionic equation:

    $$3\,OH^-{}_{(aq)} + Fe^{3+}{}_{(aq)} \rightarrow Fe(OH)_{3(s)}$$

PART IV

THE STATES AND PHASES OF MATTER

GASES AND THE GAS LAWS

CHAPTER OBJECTIVES

Upon completing this chapter, you will be able to:

- Describe the physical and chemical properties of oxygen and hydrogen

- Explain how atmospheric pressure is measured, read the pressure in a manometer, and use the units associated with pressure measurement

- Read and explain a graphic distribution of the number of molecules versus kinetic energy at different temperatures

- Know and use the following laws to solve gas problems: Graham's, Charles's, Boyle's, Dalton's, Combined Gas, and Ideal Gas

Introduction—Gases in the Environment

When we discuss gases today, the most pressing concern is the gases in our atmosphere. These are the gases that are held against the earth by the gravitational field. The mixture of gases in the air today has had 4.5 billion years in which to evolve. The earliest atmosphere must have consisted of volcanic discharges alone. Gases erupted by volcanoes today are mostly a mixture of water vapor, carbon dioxide, sulfur dioxide, and nitrogen, with almost no oxygen. If this is the same mixture that existed in the early atmosphere, various processes would have had to operate to produce the mixture we have today. One of these processes was condensation. As it cooled, much of the volcanic water vapor condensed to fill the earliest oceans. Chemical reactions would also have occurred. Some carbon dioxide would have reacted with the rocks of the earth's crust to form carbonate minerals, and some would have dissolved in the new oceans. Later, as primitive life capable of photosynthesis evolved in the seas, new marine organisms began producing oxygen. Almost all the free oxygen in the air today is believed to have formed this way. About 570 million years ago, the oxygen content of the atmosphere and oceans became high enough to permit marine life capable of respiration. Later, about 400 million years ago, the atmosphere contained enough oxygen for air-breathing animals to emerge from the seas.

The principal constituents of the atmosphere of the earth today are nitrogen (78%) and oxygen (21%). The gases in the remaining 1% are argon (0.9%), carbon dioxide (0.03%), varying amounts of water vapor, and trace amounts of hydrogen, ozone, methane, carbon monoxide, helium, neon, krypton, and xenon. Oxides and other pollutants added to the atmosphere by factories and automobiles have become a major concern because of their damaging effects in the form of acid rain (discussed further in Chapter 11). In addition, a strong possibility exists that the steady increase in atmospheric carbon dioxide, mainly attributed to fossil fuel combustion over the past century, may affect Earth's climate by causing a greenhouse effect, resulting in a steady rise in temperatures worldwide.

Studies of air samples show that up to 55 miles above sea level the composition of the atmosphere is substantially the same as at ground level; continuous stirring produced by atmospheric currents counteracts the tendency of the heavier gases to settle below the lighter ones. In the lower atmosphere, ozone is normally present in extremely low concentrations. The atmospheric layer 12 to 30 miles up contains more ozone that is produced by the action of ultraviolet radiation from the sun. In this layer, however, the percentage of ozone is only 0.001 by volume. Human activity adds to the ozone concentration in the lower atmosphere where it can be a harmful pollutant.

The ozone layer became a subject of concern in the early 1970s when it was found that chemicals known as fluorocarbons, or chlorofluoromethanes, were rising into the atmosphere in large quantities because of their use as refrigerants and as propellants in aerosol dispensers. The concern centered on the possibility that these compounds, through the action of sunlight, could chemically attack and destroy stratospheric ozone, which protects Earth's surface from excessive ultraviolet radiation. As a result, U.S. industries and the Environmental Protection Agency phased out the use of certain chlorocarbons and fluorocarbons. Despite these and other international efforts, there is still ongoing concern about both these environmental problems: the greenhouse effect and the deterioration of the ozone layer.

Some Representative Gases

OXYGEN

The composition of the air varies slightly from place to place because air is a mixture of gases. However, of the gases that occur in the atmosphere, the most important one to us is oxygen. Although it makes up only approximately 21% of the atmosphere by volume, the oxygen found on the earth is equal in weight to all the other elements combined. About 50% of Earth's crust (including the waters on earth and the air surrounding it) is oxygen. (See Figure 15.)

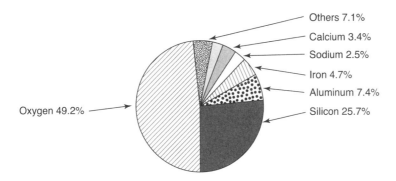

FIGURE 15. Composition of Earth's crust.

PREPARATION OF OXYGEN

In 1774, an English scientist named Joseph Priestley discovered oxygen by heating mercuric oxide in an enclosed container with a magnifying glass. That mercuric oxide decomposes into oxygen and mercury and the reaction can be expressed in an equation: $2HgO \rightarrow 2Hg + O_2$. After his discovery, Priestley visited one of the greatest of all scientists, Antoine Lavoisier, in Paris. As early as 1773, Lavoisier had carried out experiments involving burning, and they had caused him to doubt the phlogiston theory (that a substance called phlogiston was released when a substance burned; the theory went through several modifications before it was finally abandoned). By 1775, Lavoisier had demonstrated the true nature of burning and called the vital part of the air needed for it to occur "oxygen."

Today oxygen is usually prepared in the lab by heating an easily decomposed oxygen compound such as potassium chlorate ($KClO_3$). The equation for this reaction is

$$2KClO_3 + MnO_2 \rightarrow 2KCl + 3O_2(g) + MnO_2$$

A possible laboratory setup is shown in Figure 16.

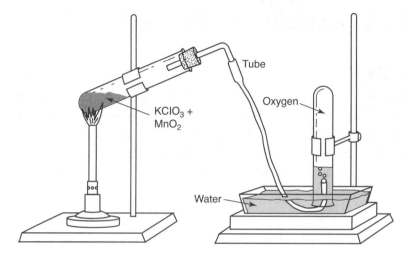

FIGURE 16. A possible laboratory preparation of oxygen.

In this preparation, manganese dioxide (MnO_2) is often used. This compound is not used up in the reaction and can be shown to have the same composition as it had before the reaction occurred. The only effect it seems to have is that it lowers the temperature needed to decompose the $KClO_3$ and thus speeds up the reaction. Substances that behave in this manner are referred to as *catalysts*. The mechanism by which a catalyst acts is not completely understood in all cases, but it is known that in some reactions the catalyst does change its structure temporarily. Its effect is shown graphically in the reaction graphs in Figure 17.

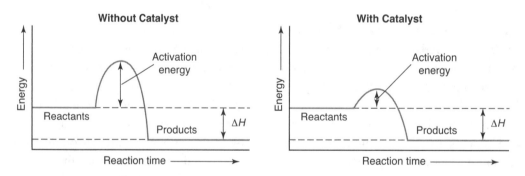

FIGURE 17. Catalyzed versus uncatalyzed reaction.

Notice that in the reaction with the catalyst the activation energy necessary to make the reaction take place is lower so that the reaction can take place at a more rapid rate.

Commercially, oxygen is prepared by either the fractional distillation of liquid air or the electrolysis of water. The fractional distillation of liquid air is typical of all fractional distillation processes in that it attempts to separate different components of a mixture in solution by taking advantage of the fact that each component has a distinct boiling point. By raising the temperature slowly and holding it constant when any boiling is taking place, the various components can be separated. With liquid air at –200°C, the temperature can be controlled so that nitrogen will boil off at −196°C. Then the oxygen is boiled off at −183°C.

Electrolysis of water will be explained more completely in Chapter 12, but essentially it consists of passing an electric current through water to cause it to decompose into hydrogen gas (H_2) and oxygen gas (O_2).

OXIDATION

When oxygen combines with an element slowly so that *no* noticeable heat and light are given off, the process is referred to as slow **oxidation**. A common example of this is rusting of iron. When the combination of oxygen with an element occurs so rapidly that the amount of energy released is visible as light and felt as heat, the process is referred to as rapid oxidation or normal burning. It is not always necessary to have oxygen present for burning or combustion to take place. A jet of hydrogen gas will burn in an atmosphere of chlorine.

PROPERTIES OF OXYGEN

That oxygen is a gas under ordinary conditions of temperature and pressure, and that it is a gas that is colorless, odorless, tasteless, and slightly heavier than air, are all physical properties of this element. Oxygen is only slightly soluble in water, thus making it possible to collect the gas over water, as shown in Figure 16. Oxygen at −183°C changes to a pale blue liquid with magnetic properties. This attraction by a magnet (paramagnetism) is believed to be due to unpaired electrons found in the molecule. Although the standard Lewis dot structure model does not include unpaired electrons as a feature in the diatomic oxygen model, a more advanced model (called molecular orbital theory) does. This theory can explain the paramagnetic property seen in oxygen. Use of molecular orbital theory is generally beyond the scope of this text and is not needed to explain the properties of most of the molecules we will discuss.

Although oxygen will support combustion, as noted above, it will not burn. This is one of its chemical properties. The usual test for oxygen is to lower a glowing splint into the gas and see if the oxidation increases in its rate to reignite the splint. (*Note*: This is not the only gas that does this. N_2O reacts the same.)

OZONE

Ozone is another form of oxygen that contains three atoms in its molecular structure (O_3). Since ordinary oxygen and ozone differ in energy content and form, they have slightly different properties. They are called allotropic forms of oxygen. Ozone occurs, as explained in the introduction, in small quantities in the upper layers of Earth's atmosphere and can be formed in the lower atmosphere where high-voltage electricity in lightning passes through the air. This formation of ozone also occurs around machinery using high voltages. The reaction can be shown by the equation

$$3O_2 + elec. \rightarrow 2O_3$$

Because of its higher energy content, ozone is more reactive chemically than oxygen.

HYDROGEN

Of all the 109 elements in the periodic chart, hydrogen is number 1 because it is the lightest element known. Its lightness makes it useful in balloons, but care must be taken because of its extreme combustibility. Its ability to burn well makes it an important part of most fuels, and its powerful attraction to oxygen makes it a good reducing agent in some important industrial reactions. Hydrogen can be added to liquid fats, under the proper conditions, to make solid shortenings, or it can be added to certain fuels to improve their burning properties. Its isotopes were used in making the hydrogen bomb.

PREPARATION OF HYDROGEN

Although there is evidence of the preparation of hydrogen before 1766, Henry Cavandish was the first person to recognize this gas as a separate substance. He observed that whenever it burned it produced water. Lavoisier named it *hydrogen*, which means "water former."

Electrolysis of water, which is the process of passing an electric current through water to cause it to decompose, is one method of obtaining hydrogen. This is a widely used commercial method as well as a laboratory method.

Another method of producing hydrogen is to displace it from the water molecule by using a metal. To choose the metal, you must be familiar with its activity with respect to hydrogen. The activities of common metals are shown in Table 8.

TABLE 8
ACTIVITY CHART OF METALS
COMPARED TO HYDROGEN

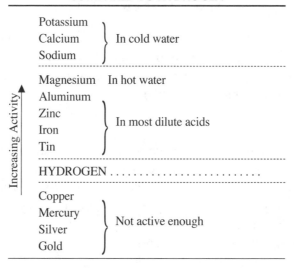

As noted in Table 8, the first three metals will react with water in the following manner:

Very active metal + water = Hydrogen + metal hydroxide

Using sodium as an example,

$$2Na + 2HOH \rightarrow H_2(g) + 2NaOH$$

With metals that react slower, a dilute acid reaction is needed to produce hydrogen in sufficient quantities to collect in the laboratory. This general equation is

Active metal + dilute acid $\rightarrow$ Hydrogen + salt of the acid

An example:

$$Zn + \text{dil. } H_2SO_4 \rightarrow H_2(g) + ZnSO_4$$

This equation shows the usual laboratory method of preparing hydrogen. Mossy zinc is used in a setup as shown in Figure 18. The acid is introduced down the thistle tube after the zinc is placed in the reacting bottle. In this sort of setup, you do not begin collecting the gas that bubbles out of the delivery tube for a few minutes so that the air in the system has a chance to be expelled and you can collect a rather pure volume of the gas generated.

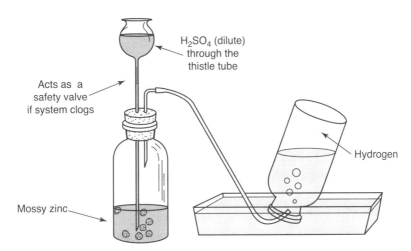

FIGURE 18. Preparation of an insoluble gas by the addition of a liquid to the other reactant.

In industry, hydrogen is produced by (1) the electrolysis of water, (2) passing steam over red-hot iron or through hot coke, or (3) decomposing natural gas (mostly methane, CH_4) with heat ($CH_4 + H_2O \rightarrow CO + 3H_2$).

PROPERTIES OF HYDROGEN

Hydrogen has the following important physical properties:

1. It is ordinarily a gas; colorless, odorless, and tasteless when pure.

2. It weighs 0.9 g/liter at 0°C and 1 atmosphere (atm) pressure. This is 1/14 as heavy as air.

3. It is slightly soluble in water.

4. It becomes a liquid at a temperature of –240°C and a pressure of 13 atm.

5. It diffuses (moves from place to place in gases) more rapidly than any other gas. This property can be demonstrated as shown in Figure 19.

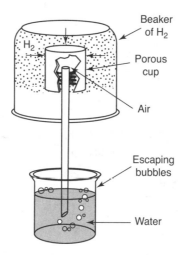

FIGURE 19. Diffusion of hydrogen.

Here the H_2 in the beaker that is placed over the porous cup diffuses faster through the cup than the air can diffuse out. Consequently, there is a pressure buildup in the cup, which pushes the gas out through the water in the lower beaker.

The chemical properties of hydrogen are:

1. It burns in air or in oxygen, giving off large amounts of heat. Its high heat of combustion makes it a good fuel.

2. It does not support ordinary combustion.

3. It is a good reducing agent in that it withdraws oxygen from many hot metal oxides.

MEASURING THE PRESSURE OF A GAS

Pressure is defined as force per unit area. With respect to the atmosphere, pressure is the result of the weight of a mixture of gases. This pressure, which is called *atmospheric pressure*, *air pressure*, or *barometric pressure*, is approximately equal to the weight of a kilogram mass on every square centimeter of surface exposed to it. This weight is about 10 newtons (N).

The pressure of the atmosphere varies with altitude. At higher altitudes, the weight of the overlying atmosphere is less, and so the pressure is less. Air pressure also varies somewhat with weather conditions as low- and high-pressure areas move with weather fronts. On the average, however, the air pressure at sea level can support a column of mercury 760 mm in height. This average sea-level air pressure is known as **normal atmospheric pressure**, also called **standard pressure**.

The instrument most commonly used for measuring air pressure is the **mercury barometer**. The following diagram shows how it operates. Atmospheric pressure is exerted on the mercury in the dish, and this in turn holds the column of mercury up in the tube. This column at standard pressure will measure 760 mm above the level of the mercury in the dish below.

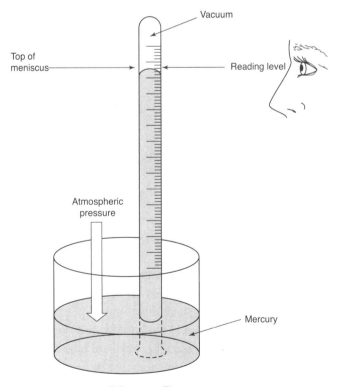

Mercury Barometer

In gas-law problems, pressure may be expressed in various units. One standard atmosphere (1 atm) is equal to 760 mm Hg or 760 **torr**, a unit named for Evangelista Torricelli. In the SI system, the unit of pressure is the **pascal** (Pa), named in honor of the scientist of the same name. Standard pressure in pascals is 101,325 Pa or 101.325 kPa.

SUMMARY OF UNITS OF PRESSURE

Unit	Abbreviation	Unit Equivalent to 1 atm
Atmosphere	atm	1 atm
Millimeters of Hg	mm Hg	760 mm Hg
Torr	torr	760 torr
Inches of Hg	in. Hg	29.9 in. Hg
Pounds per square inch	lb/in.^{2} (psi)	14.7 lb/in.^{2}
Pascal	Pa	101,325 Pa

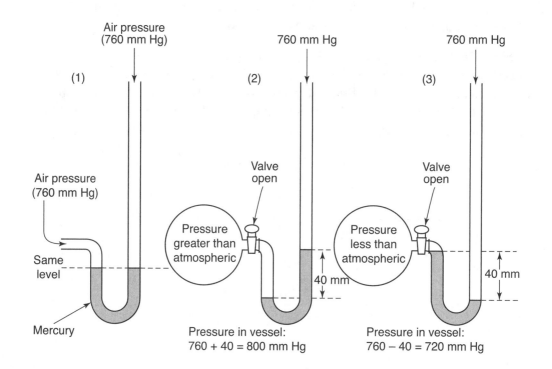

Manometer

A device similar to the barometer can be used to measure the pressure of a gas in a confined container. This apparatus, called a **manometer**, is illustrated above. A manometer is basically a U-tube containing mercury or some other liquid. When both ends are open to the air, as in (1) in the diagram, the level of the liquid will be the same on both sides since the same pressure is being exerted on both ends of the tube. In (2) and (3), a vessel is connected to one end of the U-tube. Now the height of the mercury column serves as a means of reading the pressure inside the vessel if the atmospheric pressure is known. When the pressure inside the vessel is the same as the atmospheric pressure outside, the levels of liquid are the same. When the pressure inside is greater than outside, the column of liquid will be higher on the side that is exposed to the air, as in (2). Conversely, when the pressure inside the vessel is less than the outside atmospheric pressure, the additional pressure will force the liquid to a higher level on the side near the vessel, as in (3).

KINETIC MOLECULAR THEORY

By indirect observations, the **Kinetic Molecular Theory** has been arrived at to explain the forces between molecules and the energy the molecules possess. There are three basic assumptions to the Kinetic Molecular Theory:

1. Matter in all its forms (solid, liquid, and gas) is composed of extremely small particles. In many cases these are called molecules. The space occupied by the gas particles themselves is ignored in comparison with the volume of the space they occupy.

2. The particles of matter are in constant motion. In solids, this motion is restricted to a small space. In liquids, the particles have a more random pattern but still are restricted to a kind of rolling over one another. In a gas, the particles are in continuous, random, straight-line motion.

3. When these particles collide with each other or with the walls of the container, there is no loss of energy.

SOME PARTICULAR PROPERTIES OF GASES

As the temperature of a gas is increased, its kinetic energy is increased, thereby increasing the random motion. At a particular temperature not all the particles have the same kinetic energy, but the temperature is a measure of the average kinetic energy of the particles. A graph of the various kinetic energies resembles a normal bell-shaped curve with the average found at the peak of the curve (see Figure 20).

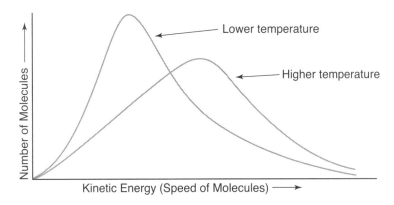

FIGURE 20. Molecular speed distribution in a gas at different temperatures.

When the temperature is lowered, the gas reaches a point at which the kinetic energy can no longer overcome the attractive forces between the particles (or molecules) and the gas condenses to a liquid. The temperature at which this condensation occurs is related to the type of substance the gas is composed of and the type of bonding in the molecules themselves. This relationship of bond type to condensation point (or boiling point) is pointed out in Chapter 3.

That gases are moving in a random motion so that they may move from one position to another is referred to as *diffusion*. You know that if a bottle of perfume is opened in one corner of a room, the perfume, that is, its molecules, will move or diffuse to all parts of the room in time. The rate of diffusion is the rate of the mixing of gases.

Effusion is the term used to describe the passage of a gas through a tiny orifice into an evacuated chamber. The rate of effusion measures the speed at which the gas is transferred into the chamber.

Gas Laws and Related Problems

GRAHAM'S LAW

This law relates the rate at which a gas diffuses (or effuses) to the type of molecule in the gas. It can be expressed as follows:

> **The rate of diffusion of a gas is inversely proportional to the square root of its molecular mass.**

Hydrogen, with the lowest molecular mass, can diffuse more rapidly than other gases under similar conditions. Here is a sample problem:

Compare the rate of diffusion of hydrogen to that of oxygen under similar conditions.

The formula is

$$\frac{\text{Rate A}}{\text{Rate B}} = \frac{\sqrt{\text{Molecular mass of B}}}{\sqrt{\text{Molecular mass of A}}}$$

Let A be H_2 and B be O_2

$$\frac{\text{Rate } H_2}{\text{Rate } O_2} = \frac{\sqrt{32}}{\sqrt{2}} = \frac{\sqrt{16}}{\sqrt{1}} = \frac{4}{1}$$

Therefore, hydrogen diffuses four times as fast as oxygen.

In dealing with the gas laws, a student must know what is meant by standard conditions of temperature and pressure (STP). Standard pressure is defined as the height of mercury that can be held in an evacuated tube by 1 atm of pressure (14.7 psi). This is usually expressed as 760 mm Hg or 101.3 Pa. **Standard temperature** is defined as 273 K or absolute (which corresponds to 0°C).

CHARLES'S LAW

Jacques Charles, a French chemist of the early nineteenth century, discovered that when a gas under constant pressure is heated from 0°C to 1°C, it expands 1/273 of its volume. It contracts this amount when the temperature is dropped 1°C to –1°C. Charles reasoned that if a gas at 0°C were cooled to –273°C (actually found to be –273.15°C), its volume would be zero. Actually, all gases are converted into liquids before this temperature is reached. By using the Kelvin scale to rid the problem of negative numbers, we can state **Charles's Law** as follows:

> **If the pressure remains constant, the volume of a gas varies directly as the absolute temperature.**

Then, initial $\dfrac{V_1}{T_1}$ = final $\dfrac{V_2}{T_2}$ at constant pressure.

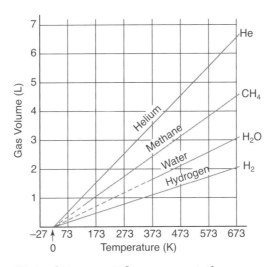

Plots of *V* versus *T* for representative gases

GRAPHIC RELATIONSHIP—CHARLES'S LAW

The dashed lines represent extrapolation of the data into regions where the gas would become liquid or solid. Extrapolation shows that each gas, if it remained gaseous, would reach zero volume at 0 K or –273°C.

EXAMPLE The volume of a gas at 20°C is 500. mL. Find its volume at standard temperature if the pressure is held constant.

SOLUTION

Convert temperatures:

$$20°C = 20° + 273° = 293\ K$$
$$0°C = 0° + 273° = 273\ K$$

If you know that cooling a gas decreases its volume, then you know that 500. mL will have to be multiplied by a fraction (made up of the Kelvin temperatures), which has a smaller numerator than the denominator. So

$$500.\ mL \times \frac{273}{293} = 465\ mL$$

Or you can use the formula

$$\frac{V_1}{T_1} = \frac{V_2}{T_2} \quad \text{so} \quad \frac{500.\ mL}{293} = \frac{X\ mL}{273}$$
$$X\ milliliters = 465\ mL$$

BOYLE'S LAW

Robert Boyle, a seventeenth-century English scientist, found that the volume of a gas decreases when the pressure on it is increased, and vice versa, when the temperature is held constant. **Boyle's Law** can be stated as follows:

> **If the temperature remains constant, the volume of a gas varies inversely as the pressure changes.**

Then $P_1V_1 = P_2V_2$ at a constant temperature.

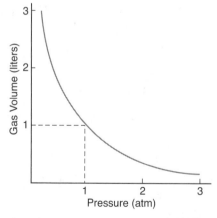

Graphic relationship of Boyle's Law

EXAMPLE Given the volume of a gas as 200. mL at 1.05 atm pressure, calculate the volume of the same gas at 1.01 atm. The temperature is held constant.

SOLUTION

If you know that this *decrease* in pressure will cause an increase in the volume, then you know 200. mL must be multiplied by a fraction (made up of the two pressures) that has a larger numerator than the denominator. So

$$200. \text{ mL} \times \frac{1.05 \text{ atm}}{1.01 \text{ atm}} = 208 \text{ mL}$$

Or you can use the formula

$$P_1 V_1 = P_2 V_2$$
$$\text{So, } V_2 = V_1 \times \frac{P_1}{P_2}$$
$$V_2 = 200. \text{ mL} \times \frac{1.05 \text{ atm}}{1.01 \text{ atm}} = 208 \text{ mL}$$

COMBINED GAS LAW

This is a combination of the two preceding gas laws and can be expressed as follows:

$$\frac{P_1 V_1}{T_1} = \frac{P_2 V_2}{T_2}$$

EXAMPLE The volume of a gas at 780 mm pressure and 30°C is 500. mL. What volume would the gas occupy at STP?

SOLUTION

Again you can use reasoning to determine the kind of fractions the temperatures and pressures must be to arrive at your answer. Since the pressure is going from 780 mm to 760 mm, the volume should increase. The fraction must then be $\frac{780}{760}$. Also, since the temperature is going from 30°C (303 K) to 0°C (273 K), the volume should decrease; this fraction must be $\frac{273}{303}$. So

$$500. \text{ mL} \times \frac{780}{760} \times \frac{273}{303} = 462 \text{ mL}$$

Or you can use the formula

$$\frac{P_1 V_1}{T_1} = \frac{P_2 V_2}{T_2}$$

$$\text{Solve for } V_2 = V_1 \times \frac{P_1}{P_2} \times \frac{T_2}{T_1}$$

$$V_2 = 500. \text{ mL} \times \frac{780 \text{ mm Hg}}{760 \text{ mm Hg}} \times \frac{273 \text{ K}}{303 \text{ K}} = 462 \text{ mL}$$

PRESSURE VERSUS TEMPERATURE

> At constant volume, the pressure of a given mass of gas varies directly with the absolute temperature.

Then $\dfrac{P_1}{T_1} = \dfrac{P_2}{T_2}$ at constant volume.

EXAMPLE A steel tank contains a gas at 27°C and a pressure of 12 atm. Determine the gas pressure when the tank is heated to 100°C.

SOLUTION

Reasoning that an increase in temperature will cause an increase in pressure at constant volume, you know the pressure must be multiplied by a fraction that has a larger numerator than denominator. The fraction must be $\dfrac{373 \text{ K}}{300 \text{ K}}$. So

$$12 \text{ atm} \times \frac{373 \text{ K}}{300 \text{ K}} = 14.9 \text{ atm}$$

Or you can use the formula

$$\frac{P_1}{T_1} = \frac{P_2}{T_2}$$

$$P_2 = P_1 \times \frac{T_2}{T_1}$$

$$\text{So } P_2 = 12 \text{ atm} \times \frac{373 \text{ K}}{300 \text{ K}} = 14.9 \text{ atm}$$

DALTON'S LAW OF PARTIAL PRESSURES

When a gas is made up of a mixture of different gases, the total pressure of the mixture is equal to the sum of the partial pressures of the components; that is, the partial pressure of the gas is the pressure of the individual gas if it alone occupied the volume.

$$P_{\text{total}} = P_{\text{gas 1}} + P_{\text{gas 2}} + P_{\text{gas 3}} + \cdots$$

EXAMPLE A mixture of gases at 760 mm Hg pressure contains 60% nitrogen, 15% oxygen, and 20% carbon dioxide by volume. What is the partial pressure of each gas?

SOLUTION

$$0.65 \times 760 = 494 \text{ mm pressure } (N_2)$$
$$0.15 \times 760 = 114 \text{ mm pressure } (O_2)$$
$$0.20 \times 760 = 152 \text{ mm pressure } (CO_2)$$

If the pressure was given as 1 atmosphere (atm), you would substitute 1 atm for 760 mm Hg. The answers would be:

$$0.65 \times 1 \text{ atm} = 0.65 \text{ atm } (N_2)$$
$$0.15 \times 1 \text{ atm} = 0.15 \text{ atm } (O_2)$$
$$0.20 \times 1 \text{ atm} = 0.20 \text{ atm } (CO_2)$$

CORRECTIONS OF PRESSURE

CORRECTION OF PRESSURE WHEN A GAS IS COLLECTED OVER WATER

When a gas is collected over a volatile liquid, such as water, some of the water vapor is present in the gas and contributes to the total pressure. Assuming that the gas is saturated with water vapor for that temperature, you can find the partial pressure due to the water vapor in a table of such water vapor values. (See Table Ⓜ in the Tables for Reference section at the back of the book.) This vapor pressure, which depends on only the temperature, must be subtracted from the total pressure to find the partial pressure of the gas being measured.

CORRECTION OF DIFFERENCE IN THE HEIGHT OF THE FLUID

When gases are collected in eudiometers (glass tube closed at one end), it is not always possible to get the level of the liquid inside the tube to equal the level on the outside. This deviation of levels must be taken into account when determining the pressure of the enclosed gas. There are then two possibilities: (1) When the level inside is higher than the level outside the tube, the pressure on the inside is less, by the height of fluid in excess, than the outside pressure. If the fluid is mercury, you simply subtract the difference from the outside pressure reading (also in height of mercury and in the same units) to get the corrected pressure of the gas. If the fluid is water, you must first convert the difference to an equivalent height of mercury by dividing the difference by 13.6 (since mercury is 13.6 times as heavy as water, the height expressed in Hg will be 1/13.6 the height of water). Again, care must be taken that this equivalent height of Hg is in the same units as the expression for the outside pressure before it is subtracted. This gives the corrected pressure for the gas in the eudiometer. (2) When the level inside is lower than the level outside the tube, a correction must be added to the outside pressure. If the difference in height between the inside and the outside is expressed in height of water, you must take 1/13.6 of this quantity to correct it to millimeters of mercury units. This relationship is shown in Figure 21. This quantity is then added to the expression of the outside pressure, which must also be in milliliters of mercury. If the tube contains mercury, then the difference between the inside and outside levels is merely added to the outside pressure to get the corrected pressure for the enclosed gas.

EXAMPLE   Hydrogen gas is collected in a eudiometer tube over water. It is impossible to level the outside water with that in the tube, and so the water level inside the tube is 40.8 mm higher than that outside. The barometric pressure is 730 mm Hg. The water vapor pressure at the room temperature of 29°C is found in a handbook to be 30.0 mm. What is the pressure of the dry hydrogen?

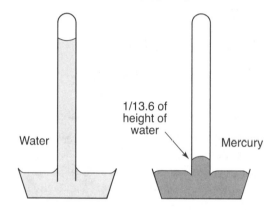

FIGURE 21. Same pressure exerted on both liquids.

SOLUTION

STEP 1: To find the true pressure of the gas, we must first subtract the water level difference expressed in mm Hg: $\frac{40.8}{13.6} = 3$ mm Hg, and so 730 mm – 3 mm = 727 mm total pressure of gases in the eudiometer.

STEP 2: Correcting for the partial pressure due to water vapor in the hydrogen, we subtract the vapor pressure (30.0 mm) from 727 mm and get our answer: 697 mm.

IDEAL GAS LAW

The laws discussed previously do not include the relationship of number of moles of a gas to the pressure, volume, and temperature of the gas. A law derived from the Kinetic Molecular Theory relates these variables. It is called the **Ideal Gas Law** and is expressed as

$$PV = nRT$$

P, V, and T retain their usual meanings, n stands for the number of moles of gas, and R represents the ideal gas constant.

Boyle's Law and Charles's Law are actually derived from the Ideal Gas Law. Boyle's Law applies when the number of moles and the temperature of the gas are constant. Then in $PV = nRT$, the number of moles, n, is constant; the gas constant, R, remains the same; and by definition T is constant. Therefore, $PV = k$. At the initial set of conditions of a problem, $P_1 V_1 = $ a constant (k). At the second set of conditions, the terms on the right side of the equation are equal to the same constant, and so $P_1 V_1 = P_2 V_2$. This matches the Boyle's Law equation introduced earlier.

The same can also be done with Charles's Law, since $PV = nRT$ can be expressed with the variables on the left and the constants on the right.

$$\frac{V}{T} = \frac{nR}{P}$$

In Charles's Law the number of moles and the pressure are constant. Substituting k for the constant term, $\frac{nR}{P}$, we have

$$\frac{V}{T} = k$$

The expression relating two sets of conditions can be written as

$$\frac{V_1}{T_1} = \frac{V_2}{T_2}$$

To use the Ideal Gas Law in the form $PV = nRT$, the gas constant, R, must be determined. This can be done mathematically as shown in the following example.

One mole of oxygen gas is collected in the laboratory at a temperature of 24.0°C and a pressure of exactly 1 atm. The volume is 24.35 L. Find the value of R.

$$PV = nRT$$

Rearranging the equation to solve for R gives

$$R = \frac{PV}{nT}$$

Substituting the known values on the right, we have

$$R = \frac{1\,\text{atm} \times 24.35\,\text{L}}{1\,\text{mol} \times 297\,\text{K}}$$

Calculating R, we get

$$R = 0.0821 \frac{\text{L} \cdot \text{atm}}{\text{mol} \cdot \text{K}}$$

Once R is known, the Ideal Gas Law can be used to find any of the variables given the other three.

For example, calculate the pressure, at 16.0°C, of 1.00 g of hydrogen gas occupying 2.54 L.

Rearranging the equation to solve for P, we get

$$P = \frac{nRT}{V}$$

The molar mass of hydrogen is 2.00 g/mol, and so the number of moles in this problem is

$$\frac{1.00 \text{ g}}{2.00 \text{ g/mol}} = 0.500 \text{ mol}$$

Substituting the known values, we have

$$P = \frac{(0.500 \text{ mol}) \left(0.0821 \frac{\text{L} \cdot \text{atm}}{\text{mol} \cdot \text{K}}\right) 289 \text{ K}}{2.54 \text{ L}}$$

Calculating the value, we get

$$P = 4.66 \text{ atm}$$

Another use of the Ideal Gas Law is to find the number of moles of a gas when P, T, and V are known.

For example, how many moles of nitrogen gas are there in 0.38 L of gas at 0°C and 380 mm Hg pressure?

Rearranging the equation to solve for n gives

$$n = \frac{PV}{RT}$$

Changing temperature to kelvins and pressure to atmospheres gives

$$T = 0°\text{C} + 273 = 273 \text{ K}$$

$$P = \frac{380 \text{ mm Hg}}{760 \text{ mm Hg/atm}} = 0.5 \text{ atm}$$

Substituting in the equation, we have

$$n = \frac{(0.5 \ \text{atm})(0.38 \ L)}{\left(0.0821 \ \dfrac{L \cdot \text{atm}}{\text{mol} \cdot K}\right)(273 \ K)} = 0.0085 \ \text{mol}$$

$$n = 0.0085 \ \text{mol of nitrogen gas}$$

IDEAL GAS DEVIATIONS

In the use of the gas laws, we have assumed that the gases involved were "ideal" gases. This means that the molecules of the gas were not taking up space in the gas volume and that no intermolecular forces of attraction were serving to pull the molecules closer together. You will find that a gas behaves like an ideal gas at low pressures and high temperatures that move the molecules as far as possible from conditions that would cause condensation. In general, pressures below a few atmospheres cause most gases to exhibit sufficiently ideal properties for application of the gas laws with a reliability of a few percent or better. If, however, high pressures are used, the molecules will be forced into closer proximity to each other as the volume decreases until the attractive force between molecules becomes a factor. This factor decreases the volume and, therefore, the PV values at high-pressure conditions will be less than that predicted by the Ideal Gas Law where the product of PV remains a constant.

Examining what occurs at very low temperatures creates a similar situation. Again, the molecules, because they have slowed down at low temperatures, come into closer proximity with each other and begin to feel the attractive forces between them. This tends to make the gas volume smaller, and therefore causes the product PV to be lower than that expected in the ideal gas situation. So, under conditions of very high pressures and low temperatures, there will be deviations from the expected results of the Ideal Gas Law.

Chapter Summary

The following terms summarize all the concepts and ideas that were introduced in this chapter. You should be able to explain their meaning and how you would use them in chemistry. They appear in boldface type in this chapter to draw your attention to them. The boldface type also makes the terms easier for you to look up, if you need to. You could also use a search engine on a computer to get a quick and expanded explanation of them.

TERMS YOU SHOULD KNOW

Boyle's Law

Charles's Law

Combined Gas Law

Dalton's Law of Partial Pressure

Graham's Law

Ideal Gas Law

Kinetic Molecular Theory

manometer

mercury barometer

normal atmospheric pressure

oxidation

ozone

pascal

standard pressure

standard temperature

torr

Chapter 5 Review Exercises

1. The most abundant element in Earth's crust is

   (A) sodium
   (B) oxygen
   (C) silicon
   (D) aluminum

2. A compound that can be decomposed to produce oxygen gas in the laboratory is

   (A) MnO_2
   (B) NaOH
   (C) CO_2
   (D) $KClO_3$

3. In the usual laboratory preparation equation for the reaction in question 2, what is the coefficient of the O_2?

   (A) 1
   (B) 2
   (C) 3
   (D) 4

4. In a graphic representation of the energy contents of reactants and resulting products, which would have a higher energy content in an exothermic reaction?

   (A) the reactants
   (B) the products
   (C) both the same
   (D) neither one

5. The process of separating components of a mixture by making use of the difference in their boiling points is called

(A) destructive distillation
(B) displacement
(C) fractional distillation
(D) filtration

6. When oxygen combines with an element to form a compound, the resulting compound is called a(n)

(A) salt
(B) oxide
(C) oxidation
(D) oxalate

7. According to the activity chart of metals, which metal would react most vigorously in a dilute acid solution?

(A) zinc
(B) iron
(C) aluminum
(D) magnesium

8. Graham's Law refers to

(A) boiling points of gases
(B) gaseous diffusion
(C) gas compression problems
(D) volume changes of gases when the temperature changes

9. When 200 mL of a gas at constant pressure is heated, its volume

(A) increases
(B) decreases
(C) remains unchanged

10. When 200 mL of a gas at constant pressure is heated from 0°C to 100°C, the volume must be multiplied by

(A) 0/100
(B) 100/0
(C) 273/373
(D) 373/273

11. If you wish to find the corrected volume of a gas which was at 20°C and 1 atm pressure and conditions are changed to 0°C and 0.92 atm pressure, by what fractions should you multiply the original volume?

(A) $\dfrac{293}{273} \times \dfrac{1}{0.92}$

(B) $\dfrac{273}{293} \times \dfrac{0.92}{1}$

(C) $\dfrac{273}{293} \times \dfrac{1}{0.92}$

(D) $\dfrac{293}{273} \times \dfrac{0.92}{1}$

12. When the level of mercury inside a gas tube is higher than the level in the reservoir, you can find the correct pressure inside the tube by taking the outside pressure reading and _?_ the difference in the height of mercury.

(A) subtracting
(B) adding
(C) dividing by 13.6
(D) doing both (C) and (A)

13. If water is the liquid in question 12 instead of mercury, you can change the height difference to an equivalent mercury expression by

(A) dividing by 13.6
(B) multiplying by 13.6
(C) adding 13.6
(D) subtracting 13.6

14. Standard conditions are

(A) 0°C and 14.7 mm
(B) 0°C and 76 cm
(C) 273°C and 760 mm
(D) 4°C and 7.6 m

15. When a gas is collected over water, the pressure is corrected by

(A) adding the vapor pressure of water
(B) multiplying by the vapor pressure of water
(C) subtracting the vapor pressure of water at that temperature
(D) subtracting the temperature of the water from the vapor pressure

16. At 5 atm pressure and 70°C, how many moles are present in 1.5 L of O_2 gas?

    (A) 0.036
    (B) 0.266
    (C) 0.536
    (D) 1.60

Answers and Explanations

1. **(B)** Oxygen is the most abundant element in Earth's crust with 21% composition.

2. **(D)** $KClO_3$ can be decomposed with heat to form KCl and O_2.

3. **(C)** The equation is $2\ KClO_3 \rightarrow 2\ KCl + 3\ O_2\uparrow$.

4. **(A)** Since energy is given off in an exothermic reaction, the heat content of the reactants is higher than that of the products.

5. **(C)** Fractional distillation can separate liquids of different boiling points.

6. **(B)** Oxygen compounds formed in combination reactions are called oxides.

7. **(D)** Magnesium, which is the most active of the given metals, will react to release hydrogen.

8. **(B)** Graham's Law refers to gaseous diffusion or effusion.

9. **(A)** When confined gases at a constant pressure are heated, they expand.

10. **(D)** The temperature must be changed to absolute temperature by adding 273 to the Celsius readings.

11. **(C)** Since the temperature is decreasing, the volume must decrease with the temperature fraction. Because the pressure is decreasing, the volume must increase. The correct answer shows this.

12. **(A)** The level is higher because the pressure inside the tube is less than outside. Therefore, you must subtract the height inside the tube from the outside pressure.

13. **(A)** If water is the liquid, you must divide by 13.6 to change the height to the equivalent height of mercury due to density differences.

14. **(B)** The only correct indication of standard conditions is B. Standard conditions are usually listed as 0°C and 760 mm of Hg.

15. **(C)** Since the vapor pressure of water is a part of the pressure reading, it must be subtracted to get the atmospheric pressure.

16. **(B)** To use the General Gas Law, $PV = nRT$, first convert 70°C to K by adding 273, 70°C + 273 = 343 K. Solve for n:

$$n = \frac{PV}{RT} = \frac{(5\text{ atm})(1.5\text{ L})}{\left(0.0821\dfrac{\text{L} \bullet \text{atm}}{\text{mol} \bullet \text{K}}\right)(343\text{ K})} = 0.266\text{ mol}$$

CHEMICAL CALCULATIONS (STOICHIOMETRY) AND THE MOLE CONCEPT

(6)

CHAPTER OBJECTIVES

Upon completing this chapter, you will be able to:

- Use the mole concept to find the molar mass of various chemical substances, recognizing how gas volumes are related to molar mass

- Solve stoichiometric problems involving Gay-Lussac's Law and density as well as classic mass-mass, volume-volume, mass-volume, limiting reactant, and percent yield problems

Solving Problems

This chapter deals with the solving of a variety of chemistry problems, which is often referred to as *stoichiometry*. Solving problems should be done in an organized manner, and it would be to your advantage to go back to the Introduction to this book and review the section called, "Problem Solving: A Thinking Skill." It describes a well-planned method for attacking the process of solving problems that you will find helpful in this chapter.

Although there are many specific methods for solving the types of problems in this chapter, we will focus primarily on the technique called **dimensional analysis**. Dimensional analysis provides a very clear problem-solving pathway for stoichiometry problems. It emphasizes not only the numerical values involved in the calculations but also the units describing the quantities in question. Dimensional analysis was introduced in Chapter 1 to make unit conversions and will be used here in the same manner but specifically to relate quantities of reactants and products in a chemical reaction.

The Mole Concept

Providing a name for a quantity of things taken as a whole is common in everyday life. Some examples are a dozen, a gross, and a ream. Each of these represents a specific number of items and is not dependent on the commodity. A dozen eggs, oranges, or bananas will always represent 12 items.

In chemistry we have a unit that describes a quantity of particles. It is called the **mole** (abbreviated *mol*). A mole is 6.02×10^{23} particles. Technically, that is the number of carbon atoms found in exactly 12 grams of carbon-12. Since the atomic masses of all the elements' atoms are related to the mass of carbon-12, a mole is also the number of atoms found in the atomic mass of *any* other element if it is *expressed in grams*. Keep in mind that the masses found on the periodic table for any element are actually weighted averages of all the isotopes that exist for that element (based on their relative natural abundances) and that if expressed in atomic mass units (amu) represent just one average atom for that element. If, however, the value for mass was expressed in grams, that sample of the element would contain 6.02×10^{23} atoms of that element. This value is also known as **Avogadro's Number** in honor of the Italian scientist whose hypothesis concerning the volumes of gases led to its determination. (Avogadro's Hypothesis will be discussed in an upcoming section on gas volumes and molar mass.) You should recognize that the value of Avogadro's Number needs to be very large because the items being counted (atoms) are very small. That means 6.02×10^{23} atoms of most elements represent samples of atoms that are conveniently sized for use in the laboratory.

Molar Mass and Moles

The mass of a mole of particles is referred to as its **molar mass**. For moles of atoms, the atomic mass found on the periodic table for that element expressed in grams is the molar mass for that element. However, some elements naturally exist as molecules. The molar mass of the element considered in this way takes into account the number of atoms of that element in the molecule *in an additive manner*. Most elements are considered in a monatomic way (one atom), while a few (hydrogen, nitrogen, oxygen, fluorine, chlorine, bromine, and iodine) are typically considered in a diatomic manner (two atoms) based on the way they are generally found to exist. This is not to say that you couldn't count moles of hydrogen atoms (H) as opposed to hydrogen molecules (H_2), but you should always be cognizant of the type of particle involved in any mole calculation.

EXAMPLE Determine the molar mass of calcium, oxygen, and carbon using the periodic table.

SOLUTION

The atomic mass of calcium is 40.1 amu as found on the periodic table. Therefore, the molar mass of calcium is 40.1 g and represents 6.02×10^{23} *atoms* of calcium or one mole of calcium atoms.

The atomic mass of oxygen is 16.0 amu as found on the periodic table. Oxygen (O_2) is a diatomic element, however. It has a molar mass of 32.0 grams which represents 6.02×10^{23} *molecules* of oxygen or one mole of oxygen molecules. A mole of oxygen (O) atoms would have a mass of 16.0 grams if they were the appropriate particle to be considered in a given circumstance.

The atomic mass of carbon is 12.0 amu. Therefore, carbon has a molar mass of 12.0 grams. This represents a sample of 6.02×10^{23} atoms of carbon or one mole of carbon atoms. Carbon is not typically found in nature as a diatomic molecule.

Elements are just one type of substance for which the molar mass can be found. The molar masses of compounds can be found in a way similar to that of diatomic elements by adding up the molar masses of the individual elements found in the compound, based on the compound's formula.

EXAMPLE 1 Find the molar mass of CO_2. (This is a molecular compound named as carbon dioxide.)

SOLUTION

As shown above, the molar mass of carbon is 12.0 g and the molar mass of oxygen is 16.0 g. Since there are one carbon and two oxygen atoms per molecule of carbon dioxide, the molar mass of carbon dioxide is 44.0 g and represents 6.02×10^{23} molecules of carbon dioxide or one mole of carbon dioxide molecules.

EXAMPLE 2 Find the molar mass of $CaCO_3$. (This is an ionic compound known as calcium carbonate.)

SOLUTION

Again, as shown above, the molar masses of calcium, carbon, and oxygen are 40.1 g, 12.0 g, and 16.0 g, respectively. Since there are three oxygen particles per formula unit of calcium carbonate in addition to the single particles of calcium and carbon, the molar mass of calcium carbonate is 40.1 g + 12.0 g + (3)16.0 g or 100.1 g. This represents 6.02×10^{23} formula units (the particle for an ionic compound) or one mole of calcium carbonate.

EXAMPLE 3   Find the molar mass of $MgSO_4 \cdot 7\ H_2O$. (This is a hydrated ionic compound known as magnesium sulfate heptahydrate and is commonly called Epsom salt.)

SOLUTION

The molar masses of magnesium, sulfur, oxygen, and water are 24.3 g, 32.1 g, 16.0 g, and 18.0 g, respectively. Since there are four oxygen particles and seven water molecules per formula unit in addition to the single particles of magnesium and sulfur, the molar mass of magnesium sulfate heptahydrate is 24.3 g + 32.1 g + (4)16.0 g + (7)18.0 g or 246.4 g. This represents 6.02×10^{23} formula units or one mole of magnesium sulfate heptahydrate.

Gas Volumes and Molar Mass

In 1811, Amedeo Avogadro made a far-reaching scientific assumption that also bears his name. **Avogadro's Hypothesis** states that equal volumes of different gases contain equal numbers of particles at the same temperature and pressure. This means that under the same conditions, the number of molecules of hydrogen in a 1-liter container is exactly the same as carbon dioxide (or any other gas) in a 1-liter container, even though the individual molecules of the different gases have different masses and sizes. Because of the substantiation of this hypothesis by much data since its inception, it is often referred to as **Avogadro's Law** and can be added to the list of gas laws discussed in Chapter 5. Avogadro's Law shows the relationship between the volume and the number of particles of a gas sample when the temperature and pressure are constant:

$$\frac{V}{n} = k$$

In other words, volume and the number of gas particles are directly related.

Because the volume of a gas may vary depending on the conditions of temperature and pressure, a standard is set for comparing gases. As stated in Chapter 5, the standard conditions of temperature and pressure (abbreviated **STP**) are 273 K and 1 atmosphere. Because the relationship between volume and number of particles of a gas is direct when the temperature and pressure are constant, the molar mass of a gas (which represents a *set* number of particles, namely 1 mole) occupies a *set* volume. The volume of 22.4 L is recognized as the molar volume of any gas at STP.

Molar mass and molar volume are typically used as conversion factors to change quantities of reactants expressed as masses or volumes to moles for use in stoichiometry problems via dimensional analysis.

EXAMPLE
1
Find the number of moles of calcium present in 4.01 g of calcium.

SOLUTION

Recall that calcium is an element that is not diatomic. It has a molar mass of 40.1 g. That mass of calcium contains 1.00 mol of calcium atoms if significant figures are kept in mind.

Use dimensional analysis:

$$\frac{4.01 \text{ g Ca}}{1} \times \frac{1.00 \text{ mol Ca}}{40.1 \text{ g Ca}} = 0.100 \text{ mol Ca}$$

Note that when using dimensional analysis, the given quantity is simply multiplied by a factor that is equal to the value 1 since the numerator and denominator in that factor are equal to each other. Consequently, the magnitude of the given *quantity* is not changed. What does change is the unit in which it is expressed as the unit of the given cancels out, leaving only the unit in the numerator of the factor to describe the quantity.

EXAMPLE
2
Find the number of moles of calcium carbonate in 0.750 g of calcium carbonate.

SOLUTION

Recall that calcium carbonate is an ionic compound. It has a molar mass of 100.1 g. That mass of calcium carbonate contains 1.00 mol of calcium carbonate formula units.

Use dimensional analysis:

$$\frac{0.750 \text{ g CaCO}_3}{1} \times \frac{1.00 \text{ mol CaCO}_3}{100.1 \text{ g CaCO}_3} = 0.00749 \text{ mol CaCO}_3$$

EXAMPLE
3
Find the number of moles of carbon dioxide in 4.48 L of carbon dioxide at STP.

SOLUTION

Recall that carbon dioxide is a gas at STP. Also, because the given quantity is supplied as a volume and not a mass, molar volume (not molar mass) will be used in the dimensional analysis equation:

$$\frac{4.48 \text{ L CO}_2}{1} \times \frac{1.00 \text{ mol CO}_2}{22.4 \text{ L CO}_2} = 0.200 \text{ mol CO}_2$$

Density and Molar Mass

Since the density of a gas is usually given in g/L of gas at STP, we can use the molar volume to molar mass relationship to solve the following types of problems.

EXAMPLE 1 Find the molar mass of a gas when the density is given as 1.25 g/L.

SOLUTION

Since it is known that 1 mol of a gas occupies 22.4 L at STP, we can solve this problem by multiplying the mass of 1 L by 22.4 L/mol.

$$\frac{1.25 \text{ g}}{\cancel{L}} \times \frac{22.4 \text{ } \cancel{L}}{1 \text{ mol}} = 28 \text{ g/mol}$$

Even if the mass given is not for 1 L, the same setup can be used.

EXAMPLE 2 If 3 L of a gas weighs 2 g, find the molar mass at STP.

SOLUTION

$$\frac{2 \text{ g}}{3 \text{ } \cancel{L}} \times \frac{22.4 \text{ } \cancel{L}}{1 \text{ mol}} = 14.9 \text{ g/mol}$$

You can also find the density of a gas if you know the molar mass. Since the molar mass occupies 22.4 L at STP, dividing the molar mass by 22.4 L will give you the mass per liter, or the density.

EXAMPLE 3 Find the density of oxygen at STP.

SOLUTION

Oxygen exists naturally as a diatomic molecule. The molar mass of O_2 is 32.0 g/mol.

Therefore,

$$\frac{32.0 \text{ g}}{1 \text{ } \cancel{mol}} \times \frac{1 \text{ } \cancel{mol}}{22.4 \text{ L}} = 1.43 \text{ g/L}$$

Example 3 shows that you can find the density of any gas at STP by dividing its molar mass by 22.4 L.

Mole-Mole Relationships

The types of mole problems investigated so far have been ones involving only one substance. Chemical calculations often involve more than one substance and take into consideration information found in **balanced reaction equations**, as discussed in Chapter 4. Recall that coefficients from a balanced equation can be used to describe the numbers of atoms, ions, or molecules involved in the chemical process. Also, recall that the quantities of each must balance on each side of the equation to satisfy the Law of Conservation of Matter. Those coefficients can also represent larger numbers of those particles, namely moles of those substances that are reacting or being produced. Therefore, the balanced chemical reaction

$$2NaClO_3(s) \rightarrow 2NaCl(s) + 3O_2(g)$$

can be interpreted in two ways:

1. Decomposing two **formula units** of sodium chlorate produces two **formula units** of sodium chloride and three **molecules** of oxygen.

2. Decomposing two **moles** of formula units of sodium chlorate produces two **moles** of formula units of sodium chloride and three **moles** of molecules of oxygen.

Coefficients in balanced chemical reaction equations can then be understood to provide **mole ratios** for reacting substances and substances produced.

EXAMPLE 1 How many moles of sodium chloride can be produced from 0.0450 moles of sodium chlorate via the reaction above?

SOLUTION
Use dimensional analysis:

$$\frac{0.0450 \text{ mol NaClO}_3}{1} \times \frac{2 \text{ mol NaCl}}{2 \text{ mol NaClO}_3} = 0.0450 \text{ mol NaCl}$$

EXAMPLE 2 How many moles of sodium chlorate are needed to produce 0.900 moles of oxygen?

SOLUTION
Use dimensional analysis:

$$\frac{0.900 \text{ mol O}_2}{1} \times \frac{2 \text{ mol NaClO}_3}{3 \text{ mol O}_2} = 0.600 \text{ mol NaClO}_3$$

Mass-Mass Problems

In order to work with substances in the laboratory, chemists must work with quantities they can easily measure. A mole is an amount impractical to actually count out in the lab. That's why the concept of molar mass was discussed in a previous section of this chapter. Molar mass, which relates moles to mass, allows for *counting by massing*. Using molar mass allows problems revolving around moles to be based on the *practically* measurable quantity of mass. **Mass-mass problems** often involve determining the masses of other substances needed to react with a given mass of a substance or the mass of other substances that can be produced from that given mass. Knowledge of mole ratios found in balanced reactions is central to solving each of these types of problems, with molar mass supplying the practical pathway to take advantage of that knowledge.

EXAMPLE Consider the following balanced chemical reaction equation:

$$CaCO_3(s) + 2HCl(aq) \rightarrow CaCl_2(aq) + H_2O(\ell) + CO_2(g)$$

If 5.18 g of $CaCO_3$ was reacted with enough HCl to use it up, what mass of HCl would be needed and what mass of CO_2 would be produced?

SOLUTION

Use dimensional analysis:

$$\frac{5.18 \text{ g CaCO}_3}{1} \times \underbrace{\frac{1 \text{ mol CaCO}_3}{100.1 \text{ g CaCO}_3}}_{\uparrow \textbf{ Factor \#1}} \times \underbrace{\frac{2 \text{ mol HCl}}{1 \text{ mol CaCO}_3}}_{\uparrow \textbf{ Factor \#2}} \times \underbrace{\frac{36.5 \text{ g HCl}}{1 \text{ mol HCl}}}_{\uparrow \textbf{ Factor \#3}} = 3.78 \text{ g HCl}$$

The first factor in the dimensional analysis equation converts the mass of the $CaCO_3$ to moles of $CaCO_3$ using information concerning molar mass as found on the periodic table. The second factor converts moles of $CaCO_3$ to moles of HCl needed to react with that quantity of $CaCO_3$. The mole ratio was found from the coefficients in front of each substance in the balanced reaction equation. The third factor converts moles of HCl to grams of that substance, once again using molar mass as found on the periodic table.

To find the mass of CO_2 produced, set up a similar dimensional analysis equation. First, convert the mass of $CaCO_3$ to moles as in the previous problem. This time, however, use the mole ratio between CO_2 and $CaCO_3$ as the second factor. Finally, use the molar mass of carbon dioxide to convert moles of the CO_2 to grams.

$$\frac{5.18 \text{ g CaCO}_3}{1} \times \frac{1 \text{ mol CaCO}_3}{100.1 \text{ g CaCO}_3} \times \frac{1 \text{ mol CO}_2}{1 \text{ mol CaCO}_3} \times \frac{44.0 \text{ g CO}_2}{1 \text{ mol CO}_2} = 2.28 \text{ g CO}_2$$

Volume-Volume Problems

In addition to being able to work with masses of substances in the laboratory, you can easily measure and will often use volumes of gases. Recall that the molar volume of any gas at STP is equal to 22.4 L. Therefore, the volume of a gas at standard temperature and pressure can be converted to moles via dimensional analysis. Combining those conversions with the mole ratios found in balanced equations, for reactions involving gases, allows for volume-volume stoichiometry problems to be solved. In volume-volume problems, you are given the volume of one gas (often at STP) and asked to determine the volume(s) of other gases involved in the reaction.

Consider the following balanced chemical reaction equation:

$$N_2(g) + 3H_2(g) \rightarrow 2NH_3(g)$$

EXAMPLE If 2.800 L of NH_3 was to be produced at STP, what volumes of nitrogen and hydrogen, also at STP, would be required?

SOLUTION
Use dimensional analysis:

$$\frac{2.800 \text{ L NH}_3}{1} \times \underbrace{\frac{1 \text{ mol NH}_3}{22.4 \text{ L NH}_3}}_{\uparrow \textbf{ Factor \#1}} \times \underbrace{\frac{1 \text{ mol N}_2}{2 \text{ mol NH}_3}}_{\uparrow \textbf{ Factor \#2}} \times \underbrace{\frac{22.4 \text{ L N}_2}{1 \text{ mol N}_2}}_{\uparrow \textbf{ Factor \#3}} = 1.400 \text{ L N}_2$$

The first factor in the dimensional analysis equation converts the volume of NH_3 to moles. The second factor uses the mole ratio from the balanced equation to convert moles of NH_3 to moles of N_2. Finally, the third factor converts moles of N_2 to the volume of N_2 at STP. Mathematically, Factors #1 and #3 simply undo each other. This displays a relationship that was suggested in Chapter 5 through the Ideal Gas Law as well as Avogadro's Hypothesis (discussed earlier in this chapter). Namely, there is a direct relationship between the volumes of gases, at the same temperature and pressure, and their numbers of particles (measured in moles). In other words, mole ratios can be construed as volume ratios between gases existing at the same temperature and pressure. The only factor needed to solve the previous problem mathematically was Factor #2. To express the dimensional analysis equation properly, though, requires using the mole ratios as volume ratios, as is shown below.

$$\frac{2.800 \text{ L NH}_3}{1} \times \frac{1 \text{ L N}_2}{2 \text{ L NH}_3} = 1.400 \text{ L N}_2$$

To find the amount of H_2 needed to produce the 2.800 L of NH_3 requires a similar equation but with a different ratio between the gases:

$$\frac{2.800 \text{ L NH}_3}{1} \times \frac{3 \text{ L H}_2}{2 \text{ L NH}_3} = 4.200 \text{ L H}_2$$

Using mole ratios, that is, the coefficients from the balanced equation, as volume ratios, saves time and makes volume-volume problems less cumbersome to solve. The relationship between the volumes of reacting gases was first noted by the French scientist Joseph Louis Gay-Lussac and is sometimes called **Gay-Lussac's Law of Combining Gases**. This law states that when only gases are investigated in a chemical reaction, the volumes of the reacting gases and the volumes of the gaseous products are in small whole-number ratio with each other. Those small whole numbers are found as the coefficients in the balanced reaction equation.

Often, reactions between gases do not occur at STP, but Gay-Lussac's Law still applies. The reason is fundamentally due to Avogadro's Law, which states that the only requirement for the volumes of gases to be related to the number of particles of that gas is that the temperature and pressure of the gases be the same. Whether or not the temperature and pressure of the gases are at STP is inconsequential; you are encouraged to use the coefficients from the balanced reaction equation as volume ratios between reacting gases in dimensional analysis/stoichiometry problems.

Mass-Volume or Volume-Mass Problems

Reactions often involve gases and other phases of matter. In those reactions, it is common to know the mass of one substance involved in the chemical process and yet need to determine the volume of a different substance, such as a gas. Likewise, it's not unusual to know the volume of a gas taking part in a reaction and yet need to determine the mass of another substance, often a solid or liquid, taking part in the reaction. Even if the reaction is taking place at STP, Gay-Lussac's Law cannot be used here as both of the substances under consideration are not gases and the information desired is not restricted to just volumes. In other words, Guy-Lussac's Law applies only when all the substances being considered are gases.

In the reaction below, what mass of zinc would be required to produce 3.68 L of H_2 at STP?

$$Zn(s) + 2\ HCl(aq) \rightarrow ZnCl_2(aq) + H_2(g)$$

SOLUTION

Use dimensional analysis considering both molar volume and molar mass:

$$\frac{3.68\ L\ H_2}{1} \times \frac{1\ mol\ H_2}{22.4\ L\ H_2} \times \frac{1\ mol\ Zn}{1\ mol\ H_2} \times \frac{65.3\ g\ Zn}{1\ mol\ Zn} = 10.7\ g\ Zn$$

Reactions involving gases and other phases of matter *not* at STP are very common. Obviously, Gay-Lussac's Law cannot be used to solve such problems because both of the substances in question are not gases. Additionally, the relationship 22.4 L = 1 mole of the gas cannot be used since the reaction is not taking place at STP. To solve problems such as these, the Ideal Gas Law, $PV = nRT$, must be considered. Recall from Chapter 5, that the Ideal Gas Law can be manipulated to solve for the moles of gas at any temperature and pressure as long as the volume is supplied ($n = PV/RT$). Likewise, the volume of a gas could be determined if the number of moles of the gas is known along with its temperature and pressure ($V = nRT/P$).

EXAMPLE 2

Based on the reaction below, what volume of oxygen is produced from the decomposition of 3.50 g of hydrogen peroxide if the oxygen produced was collected at 50.0°C and 1.75 atm pressure?

$$2H_2O_2(aq) \rightarrow 2H_2O(\ell) + O_2(g)$$

SOLUTION

First, use dimensional analysis to determine the number of moles of O_2 that would be produced.

$$\frac{3.50 \text{ g } H_2O_2}{1} \times \frac{1 \text{ mol } H_2O_2}{34.0 \text{ g } H_2O_2} \times \frac{1 \text{ mol } O_2}{2 \text{ mol } H_2O_2} = 0.0515 \text{ mol } O_2$$

The number of moles of O_2 produced can now be plugged into the Ideal Gas Law to find the volume of that amount at the temperature and pressure outlined in the problem.

$$V = \frac{(0.0515 \text{ mol})\left(0.0821\frac{\text{L atm}}{\text{mol K}}\right)(323 \text{ K})}{1.75 \text{ atm}}$$

$$V = 0.780 \text{ L}$$

Problems with an Excess of One Reactant

It will not always be true that the amounts given in a particular problem are exactly in the proportion required for the reaction to use up all of the reactants. In other words, at times some of one reactant will be left over after the other has been used up. This is similar to the situation in which two eggs are required to mix with one cup of flour in a particular recipe, and you have four eggs and four cups of flour. Since two eggs require only one cup of flour, four eggs can use only two cups of flour and two cups of flour will be left over.

A chemical equation is very much like a recipe.

Consider the following reaction:

$$2\ C_4H_{10}(g) + 13\ O_2(g) \rightarrow 8\ CO_2(g) + 10\ H_2O(g)$$

If you are given 25.0 grams of butane (C_4H_{10}) and 25.0 grams of oxygen (O_2), how many grams of carbon dioxide (CO_2) can be produced? Which reactant will be left over? How much of this reactant will not be used?

SOLUTION

Once again, dimensional analysis will be used to solve this problem; however, two equations will be required as it will be necessary to determine how much carbon dioxide can be produced from each reactant:

$$\frac{25.0\ g\ C_4H_{10}}{1} \times \frac{1\ mol\ C_4H_{10}}{58.0\ g\ C_4H_{10}} \times \frac{8\ mol\ CO_2}{2\ mol\ C_4H_{10}} \times \frac{44.0\ g\ CO_2}{1\ mol\ CO_2} = 75.9\ g\ CO_2$$

$$\frac{25.0\ g\ O_2}{1} \times \frac{1\ mol\ O_2}{32.0\ g\ O_2} \times \frac{8\ mol\ CO_2}{13\ mol\ O_2} \times \frac{44.0\ g\ CO_2}{1\ mol\ CO_2} = 21.2\ g\ CO_2$$

Since the oxygen can only produce 21.2 g CO_2 (the lesser of the quantities of CO_2 shown above), it is referred to as the **limiting reactant**. There is simply not enough of the oxygen to make what the butane has the potential to produce. That smaller amount of CO_2 is referred to as the **theoretical yield**. A portion of the C_4H_{10} (the **reactant in excess**) will be used, however, to produce the 10.3 g of CO_2. To determine that amount, another dimensional analysis will be needed.

$$\frac{25.0\ g\ O_2}{1} \times \frac{1\ mol\ O_2}{32.0\ g\ O_2} \times \frac{2\ mol\ C_4H_4}{13\ mol\ O_2} \times \frac{58.0\ g\ C_4H_{10}}{1\ mol\ C_4H_{10}} = 6.97\ g\ C_4H_{10}\ (needed)$$

Since only 6.97 g of the C_4H_{10} is needed, the amount of the C_4H_{10} in excess can be determined by subtraction.

(have)   **(need)**   **(left over)**

$$25.0\ g\ C_4H_{10} - 6.97\ g\ C_4H_{10} = 18.03\ g\ C_4H_{10}$$

(or 18.0 g when considering significant figures)

Percent Yield of a Product

In most stoichiometric problems, we assume that the results are exactly what we would theoretically expect. In reality, the theoretical yield is rarely the actual yield. This difference occurs for many reasons. For instance, some of the product is often lost during the purification or collection process.

Chemists are usually interested in the efficiency of a reaction. The efficiency is expressed by comparing the actual and the theoretical yields. The **percent yield** is the ratio of the actual yield to the theoretical yield, multiplied by 100%.

$$\text{Percent Yield} = \left(\frac{\text{Actual Yield}}{\text{Theoretical Yield}}\right) \times 100\%$$

EXAMPLE   Aluminum is commonly produced by the smelting of aluminum oxide into aluminum metal by the reaction below:

$$2Al_2O_3 \text{ (dissolved)} + 3C \text{ (s)} \rightarrow 4\,Al\,(\ell) + 3\,CO_2\,(g)$$

If 2.00 kg of aluminum oxide is smelted and the actual yield of Al is 0.87 kg, what is the percent yield associated with the process?

SOLUTION

The *theoretical yield* of aluminum from that amount of aluminum oxide can be found using dimensional analysis:

$$\frac{2000 \text{ g Al}_2O_3}{1} \times \frac{1 \text{ mol Al}_2O_3}{102 \text{ g Al}_2O_3} \times \frac{4 \text{ mol Al}}{2 \text{ mol Al}_2O_3} \times \frac{27.0 \text{ g Al}}{1 \text{ mol Al}} = 1060 \text{ g Al}$$

The *percent yield* can then be found:

$$\text{percent yield} = \frac{870 \text{ g Al (actual yield)}}{1060 \text{ g Al (theoretical yield)}} \times 100\% = 82.1\%$$

Chapter Summary

The following terms summarize all the concepts and ideas that were introduced in this chapter. You should be able to explain their meaning and how you would use them in chemistry. They appear in boldface type in this chapter to draw your attention to them. The boldface type also makes the terms easier for you to look up, if you need to. You could also use a search engine on a computer to get a quick and expanded explanation of them.

TERMS YOU SHOULD KNOW

Avogadro's Law

Avogadro's Hypothesis

Avogadro's Number

Gay-Lussac's Law of
  Combining Gases

mass-mass problems

molar mass

molar volume

mole

mole-ratios

percent yield

STP

Chapter 6 Review Exercises

1. What is the molar mass of carbon monoxide?

    (A) 44.0 amu
    (B) 44.0 g
    (C) 28.0 amu
    (D) 28.0 g

2. Describe the particles of neon in 40.4 grams of elemental neon by both number and type.

    (A) 2 mol of neon atoms
    (B) 2 mol of neon molecules
    (C) 6.02×10^{23} neon molecules
    (D) 6.02×10^{23} neon atoms

3. A sample of nitrogen gas with a volume of 11.2 L at STP would contain how many particles of that gas?

    (A) 6.02×10^{23} molecules
    (B) 3.01×10^{23} molecules
    (C) 1.12×10^{23} atoms
    (D) 3.01×10^{23} atoms

4. The density of gases in g/L

   (A) is independent of the identity of the gas
   (B) can be found by dividing its molar mass by 22.4 as long as it is at STP
   (C) can be found by multiplying its molar mass by 22.4 as long as it is at STP
   (D) can be found by dividing its molar mass by 22.4 as long as it is not at STP

5. In the reaction between calcium and sulfur to produce calcium sulfide, the number of moles of sulfur needed to react with 4 moles of calcium is

   (A) 1 mole
   (B) 2 moles
   (C) 3 moles
   (D) 4 moles

6. In the reaction

$$H_2(g) + Cl_2(g) \rightarrow 2HCl(g)$$

   what volume of HCl could be produced when 6 L of H_2 reacts with 4 L of Cl_2?

   (A) 4 L
   (B) 5 L
   (C) 6 L
   (D) 8 L

7. Consider the *unbalanced* reaction equation describing the combustion of ethane.

$$C_2H_6(g) + O_2(g) \rightarrow CO_2(g) + H_2O(g)$$

   What number of moles of carbon dioxide could be produced if 15.0 grams of ethane reacted with an excess of O_2?

   (A) 1.00 mol
   (B) 2.00 mol
   (C) 4.00 mol
   (D) 8.00 mol

8. What volume of hydrogen would be produced in the reaction

$$2Na(s) + 2H_2O(\ell) \rightarrow 2NaOH(aq) + H_2(g)$$

   if 46.0 g of sodium were completely reacted and the hydrogen gas was collected and then stored at STP?

   (A) 11.2 L
   (B) 22.4 L
   (C) 44.8 L
   (D) 60.2 L

9. After a reaction between a silver nitrate solution and copper in the lab, 47.0 grams of silver were retrieved from the reaction container. Stoichiometry concerning the reaction shows that 50.0 grams could be produced. The percent yield for this reaction is

 (A) 100.0%
 (B) 97.0%
 (C) 94.0%
 (D) 47.0%

Answers and Explanations

1. **(D)** The molar mass of a compound is found by adding the masses of the individual elements in the compound and by considering the subscripts associated with that element in the formula. In this case, the formula for carbon monoxide is CO. The molar mass of carbon is 12.0 grams and that of oxygen is 16.0 g. Therefore, the molar mass of carbon monoxide is 28.0 g. The masses in the answers expressed in amu would be appropriate for one molecule of a substance, not for one mole.

2. **(A)** The molar mass of neon is 20.2 g. Dimensional analysis shows that 40.4 g of neon represent 2.00 mol.

$$\frac{40.4 \text{ g Ne}}{1} \times \frac{1 \text{ mol Ne}}{20.2 \text{ g Ne}} = 2.00 \text{ mol Ne}$$

Since neon is a monatomic element, the 2.00 mol of Ne is in reference to moles of atoms.

3. **(B)** Since 1 mole of *any gas* at STP has a volume of 22.4 L, then 11.2 L of oxygen would contain 0.500 moles of nitrogen or 3.01×10^{23} molecules of the gas. This can be shown by using dimensional analysis as

$$\frac{11.2 \text{ L N}_2}{1} \times \frac{1 \text{ mol N}_2}{22.4 \text{ L N}_2} \times \frac{6.02 \times 10^{23} \text{ mol N}_2}{1 \text{ mol N}_2} = 3.01 \times 10^{23} \text{ mol N}_2$$

4. **(B)** The molar mass of a gas divided by 22.4 gives the density of any gas at STP because the molar volume of a gas at STP is 22.4 L. Using dimensional analysis,

$$\frac{\text{grams}}{\text{mole}} \times \frac{1 \text{ mole}}{22.4 \text{ liters}} = \frac{\text{grams}}{\text{liter}}$$

5. **(D)** The balanced reaction describing the reaction between calcium and sulfur to produce calcium sulfide is

$$Ca(s) + S(s) \rightarrow CaS(s)$$

Dimensional analysis shows

$$\frac{4 \text{ mol Ca}}{1} \times \frac{1 \text{ mol S}}{1 \text{ mol Ca}} = 4 \text{ mol S}$$

6. **(D)** Since the volumes for both reactants were given, this is a limiting reactant problem. Using dimensional analysis, the theoretical yield can be found by choosing the smaller volume of HCl that could be made by each. Also, the coefficients in the equation can be construed as volume ratios since both reactants are gases (Gay-Lussac's Law).

$$\frac{6 \text{ L } H_2}{1} \times \frac{2 \text{ L HCl}}{1 \text{ L } H_2} = 12 \text{ L HCl}$$

$$\frac{4 \text{ L } Cl_2}{1} \times \frac{2 \text{ L HCl}}{1 \text{ L } Cl_2} = 8 \text{ L HCl (theoretical yield)}$$

7. **(A)** The balanced reaction equation for the combustion of ethane is

$$2C_2H_6(g) + 7O_2(g) \rightarrow 4CO_2(g) + 6H_2O(g)$$

Using dimensional analysis,

$$\frac{15.0 \text{ g } C_2H_6}{1} \times \frac{1 \text{ mol } C_4H_{10}}{30.0 \text{ g } C_4H_{10}} \times \frac{4 \text{ mol } CO_2}{2 \text{ mol } C_4H_{10}} = 1.00 \text{ mol } CO_2$$

8. **(B)** Using dimensional analysis,

$$\frac{46.0 \text{ g Na}}{1} \times \frac{1 \text{ mol Na}}{23.0 \text{ g Na}} \times \frac{1 \text{ mol } H_2}{2 \text{ mol Na}} \times \frac{22.4 \text{ L } H_2}{1 \text{ mol } H_2} = 22.4 \text{ L } H_2$$

9. **(C)** 47.0 g is the actual yield of silver, while 50.0 g is the theoretical yield of silver. Percent yield can be found as

$$\text{percent yield} = \frac{47.0 \text{ g Ag}}{50.0 \text{ g Ag}} \times 100\% = 94.0\%$$

LIQUIDS, SOLIDS, AND PHASE CHANGES

CHAPTER OBJECTIVES

Upon completing this chapter, you will be able to:

- Explain, using a graph, the distribution of the kinetic energy of molecules of a liquid at different temperatures
- Describe the states of matter and what occurs when a substance changes state
- Define critical temperature and pressure
- Analyze a phase diagram and determine the triple point for a given substance
- Solve water calorimetry problems that include changes of state
- Explain the polarity of the water molecule and hydrogen bonding
- Solve solubility problems, concentration problems, and changes in boiling point/ freezing point problems
- Describe the continuum of water mixtures, including solutions, colloids, and suspensions

Liquids

IMPORTANCE OF INTERMOLECULAR INTERACTION

In a liquid, the volume of the molecules and the intermolecular forces between them are much more important than in a gas. When you consider that in a gas the molecules constitute far less than 1% of the total volume, while in the liquid state the molecules constitute 70% of the total volume, it is clear that in a liquid the forces between molecules are more important. Because of this decreased volume and increased intermolecular interaction, a liquid expands and contracts only very slightly with a change in temperature and lacks the compressibility typical of gases.

KINETICS OF LIQUIDS

Even though the volume of space between molecules has decreased in a liquid and the mutual attraction forces between neighboring molecules can have great effects on the molecules, they are still in motion. This motion can be verified under a microscope when colloidal particles are suspended in a liquid. The particles' zigzag path, called **Brownian movement**, indicates molecular motion and supports the *Kinetic Molecular Theory*.

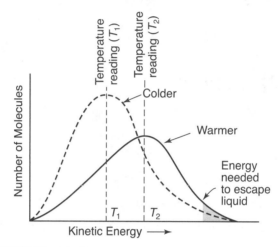

FIGURE 22. Distribution of the kinetic energy of molecules.

VISCOSITY

Viscosity is the friction or resistance to motion that exists between the molecules of a liquid when they move past each other. It is logical that the stronger the attraction between the molecules of a liquid, the greater its resistance to flow—and thus the greater its viscosity. The viscosity of a liquid depends on its intermolecular forces. Because hydrogen bonds are such strong intermolecular forces, liquids with hydrogen bonds tend to have high viscosities. Water is strongly hydrogen-bonded and has a relatively high viscosity. You may have noticed how fast liquids without this trait, such as alcohol and gasoline, flow.

SURFACE TENSION

Molecules at the surface of a liquid experience attractive forces downward, toward the inside of the liquid, and sideways, along the surface of the liquid. This is unlike the uniformly distributed attractive forces that molecules in the center of the liquid experience. This imbalance of forces at the surface of a liquid results in a property called **surface tension**. The uneven forces make the surface behave as if it had a tight film stretched across it. Depending on the magnitude of the surface tension of the liquid, the film is able to support the weight of a small object such as a razor blade or a needle. Surface tension also explains the beading of raindrops on the shiny surface of a car.

Increases in temperature increase the average kinetic energy of molecules and the rapidity of their movement. This is shown graphically in Figure 22. The molecules in the sample of cold liquid have, on the average, less kinetic energy than those in the warmer sample. Hence, the temperature reading T_1 will be less than the temperature reading T_2. If a particular molecule gains enough kinetic energy when it is near the surface of a liquid, it can overcome the attractive forces of the liquid phase and escape into the gaseous phase. This is called a *change of phase*. When fast-moving molecules with high kinetic energy escape, the average energy of the remaining molecules is lower; hence, the temperature is lowered.

Phase Equilibrium

Figure 23 shows water in a container enclosed by a bell jar. Observation of this closed system would show an initial small drop in the water level, but after some time the level would become constant. The explanation is that, at first, more energetic molecules near the surface are escaping into the gaseous phase faster than some of the gaseous water molecules are returning to the surface and possibly being caught by the attractive forces that will retain them in the liquid phase. After some, time the rates of evaporation and condensation equalize. This is known as **phase equilibrium**.

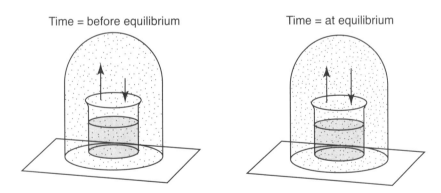

Time = before equilibrium   Time = at equilibrium

FIGURE 23. Closed system in dynamic equilibrium.
Note: Arrows show extent of evaporation (↑) and condensation (↓).

In a closed system like this, when opposing changes are taking place at equal rates, the system is said to have **dynamic equilibrium**. At higher temperatures, since the number of molecules at higher energies increases, the number of molecules in the liquid phase will be reduced and the number of molecules in the gaseous phase will be increased. The rates of evaporation and condensation, however, will again become equal.

The behavior of the system described above illustrates what is known as **Le Châtelier's Principle**. It is stated as follows: When a system at equilibrium is disturbed by the application of a stress (a change in temperature, pressure, or concentration), it reacts so as to minimize the stress and attain a new equilibrium position.

In the discussion above, if the 20°C system is heated to 30°C, the number of gas molecules will be increased while the number of liquid molecules will be decreased

$$\text{Heat} + H_2O(\ell) \rightarrow H_2O(g)$$

The equation shifts to the right (any similar system that is endothermic shifts to the right when temperature is increased) until equilibrium is reestablished at the new temperature.

The molecules in the vapor that are in equilibrium with the liquid at a given temperature exert a constant pressure. This is called the *equilibrium vapor pressure* at that temperature.

Table Ⓜ in the Equations and Tables for Reference section at the back of this book gives the vapor pressure of water at various temperatures.

Boiling Point

The vapor pressure-temperature relation can be plotted on a graph for a closed system. (See Figure 24.) When a liquid is heated in an open container, the liquid and vapor are not in equilibrium and the vapor pressure increases until it becomes equal to the pressure above the liquid. At this point the average kinetic energy of the molecules is such that they are rapidly converted from the liquid to the vapor phase within the liquid as well as at the surface. The temperature at which this occurs is known as the **boiling point**.

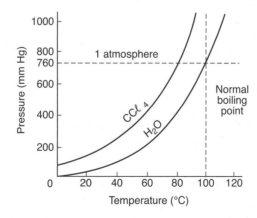

FIGURE 24. Vapor pressure-temperature relationship for carbon tetrachloride and water.

Critical Temperature and Pressure

There are conditions for particular substances when it is impossible for the liquid or solid phase to exist. Since the kinetic energy of a molecular system is directly proportional to the Kelvin temperature, it is logical to assume that there is a temperature at which the kinetic energy of the molecules is so great that the attractive forces between molecules are insufficient for the liquid phase to remain. The temperature above which the liquid phase of a substance cannot exist is called its **critical temperature**. Above its critical temperature, no gas can be liquefied regardless of the pressure applied. The minimum pressure required to liquefy a gas at its critical temperature is called its **critical pressure**.

Solids

Whereas particles in gases have the highest degree of disorder, the solid state has the most ordered system. Particles are fixed in a rather definite position and maintain a definite shape. Particles in solids do vibrate in position, however, and may even diffuse through the solid. (Example: Gold clamped to lead shows diffusion of some gold atoms into the lead over long periods of time.) Other solids do not show diffusion because of strong ionic or covalent bonds in network solids. (Examples: NaCl and diamond, respectively.)

When heated at certain pressures, some solids vaporize directly without passing through the liquid phase. This is called **sublimation**. Solids like solid carbon dioxide and solid iodine exhibit this property because of unusually high vapor pressure.

The temperature at which atomic or molecular vibrations of a solid become so great that the particles break free from fixed positions and begin to slide freely over each other in a liquid state is called the **melting point**. The amount of energy required at the melting point temperature to cause the change of phase to occur is called the *heat of fusion*. The amount of this energy depends on the nature of the solid and the type of bonds present.

Phase Diagrams

The simplest way to discuss a phase diagram is by an example, such as Figure 25.

A phase diagram ties together the effects of temperature and also pressure on the phase changes of a substance. In Figure 25, the line *BD* is essentially the vapor pressure curve for the liquid phase. Notice that at 760 mm (the pressure for 1 atm), the water will boil (change to the vapor phase) at 100°C (point *F*). However, if the pressure is raised, the boiling point temperature increases; and if the pressure is less than 760 mm, the boiling point decreases along the *BD* curve down to point *B*.

At 0°C the freezing point of water is found along the line *BC* at point *E* for pressure at 1 atm or 760 mm. Again, this point is affected by pressure along the line *BC*, so that if the pressure is decreased, the freezing point is slightly higher up to point *B* or 0.01°C.

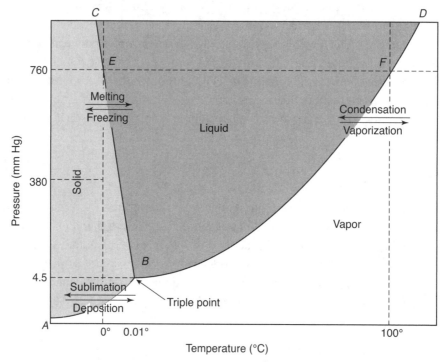

FIGURE 25. Partial phase diagram for water (distorted somewhat to distinguish the triple point from the freezing point).

Point *B* represents the point at which the solid, liquid, and vapor phases can all exist at equilibrium. This point is known as the *triple point*. It is the only temperature and pressure at which three phases of a pure substance can exist in equilibrium with one another in a system containing only the pure substance.

Water

HISTORY OF WATER

Ancient philosophers like Aristotle regarded water as a basic element that typified all liquid substances. Scientists did not discard this view until the latter half of the eighteenth century. In 1781, the British chemist Henry Cavendish synthesized water by detonating a mixture of hydrogen and air. However, the results of his experiments were not clearly interpreted until two years later when the French chemist Antoine Laurent Lavoisier proved that water was not an element but a compound of oxygen and hydrogen. In a scientific paper presented in 1804, the French chemist Joseph Gay-Lussac and the German naturalist Alexander von Humboldt demonstrated jointly that water consisted of two volumes of hydrogen to one of oxygen, as expressed by the well-known formula H_2O.

It was not until 1932 that the American chemist Harold Clayton Urey discovered the presence of a small amount (1 part in 6000 parts) of so-called **heavy water**, or deuterium oxide (D_2O). Deuterium is the hydrogen isotope with a mass number of 2. Therefore, it has a nucleus comprised of one proton and one neutron rather than just one proton, as in the most common isotope of hydrogen—protium. In 1951, the American chemist Aristid Gross discovered that naturally occurring water also contained minute traces of tritium oxide (T_2O). Tritium is the hydrogen isotope with a mass number of 3. Its nucleus is composed of two neutrons and one proton. Both of the isotopes of hydrogen have had use in the nuclear energy field.

PURIFICATION OF WATER

Water is so often involved in chemistry that it is important to have a rather complete understanding of this compound and its properties. Pure water has become a matter of national concern. Both the commercial methods of purification and the usual laboratory method of obtaining pure water, distillation, will be covered.

The process of distillation involves the evaporation and condensation of the water molecules. The usual apparatus for the distillation of any liquid is shown in Figure 26.

This method of purification will remove any substance that has a boiling point higher than that of water. It cannot remove dissolved gases or liquids that boil off before water. These substances will be carried over into the condenser and subsequently into the distillate.

Often naturally occurring water contains suspended and dissolved impurities that make it unsuitable for many purposes. Objectionable organic and inorganic materials are removed by such methods as screening and sedimentation to eliminate suspended materials. This treatment often involves the use of activated carbon to remove tastes and odors, filtration to remove insoluble solids, and chlorination or irradiation to kill infective microorganisms.

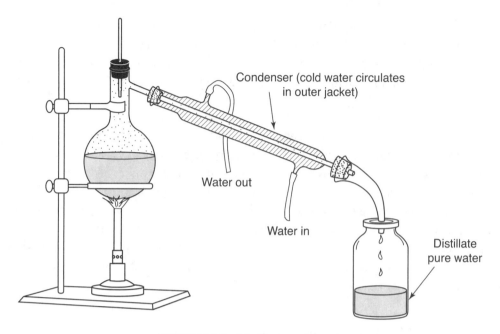

FIGURE 26. Distillation of water.

Another method of water purification is aeration, or the saturation of water with air. In this method, water is brought into contact with air in such a manner as to produce maximum diffusion, usually by spraying the water into the air in fountains. Aeration removes odors and taste caused by decomposing organic matter, as well as industrial wastes such as phenols and volatile gases such as chlorine. It also converts dissolved iron and manganese compounds to insoluble hydrated oxides of the metals, which then settle out as precipitates.

When water contains bicarbonates of Ca, Mg, or Fe it is said to have carbonate hardness, formerly called "temporary hardness." Such hardness may be removed by boiling since heat causes the bicarbonate to decompose to a carbonate precipitate. This precipitate can cause detrimental scales in industrial boilers if not removed.

Natural waters are also apt to contain varying amounts of calcium sulfate, magnesium sulfate, calcium chloride, and magnesium chloride. Water in which these chemicals are dissolved is said to have noncarbonate hardness, formerly known as "permanent hardness." One way of softening water is to pass it through an artificial zeolite that replaces the "hard" positive ions (called cations) with "soft" ions of sodium. It is the "hard" cations that are responsible for "bathtub ring" by reacting with the soap to form an insoluble compound. Another effective method that gives pure water like that obtained by distillation is the use of organic deionizers (e.g., Dowex, Amberlite, and Zeolite). These organic resins exchange H^+ ions for the cations, and OH^- ions for the anions.

With the ever-increasing demands for freshwater, especially in arid and semiarid areas or very densely populated areas, much research has gone into finding efficient methods of removing salt from seawater and brackish water. The state of California is seriously considering **desalinization** as a possible solution to its growing water problems. In the United States, desalinization research is directed by the Bureau of Reclamation, Department of the Interior. Several processes are being developed to produce freshwater cheaply.

Three of the processes involve evaporation followed by condensation of the resultant steam and are known as multiple-effect evaporation, vapor-compression distillation, and flash evaporation. The last-named method, the most widely used, involves heating seawater and pumping it into lower-pressure tanks where it abruptly vaporizes into steam. The steam then condenses and is drawn off as pure water. In 1967, Key West, Florida, opened this type of plant and became the first city in the United States to draw its freshwater from the sea.

Other alternatives are freezing, reverse osmosis, and electrodialysis. The major obstacle to the use of any of these methods is cost. Using conventional fuels, plants with a capacity of 1 million gallons per day or less produce water at a cost of $3.00 or more per 1000 gallons. Nearly 20,000 such plants are in operation today. However, water from conventional sources, such as wells and reservoirs, is sold for less than 30 cents per 1000 gallons delivered to the home.

COMPOSITION OF WATER

Water can be analyzed, that is, broken into its components, by electrolysis. This process is discussed on page 272 and shows that its composition by volume is 2 parts of hydrogen to 1 part of oxygen. Water composition can also be arrived at by synthesis. Synthesis is the formation of a compound by uniting its components. Water can be made by mixing hydrogen and oxygen in a eudiometer over mercury and passing an igniting spark through the mixture. Again the ratio of combination is found to be 2 parts hydrogen to 1 part oxygen. In a steam-jacketed eudiometer, which keeps the water formed in the gas phase, 2 vol of hydrogen combine with 1 vol of oxygen to form 2 vol of steam. Another interesting method is the Dumas experiment pictured in Figure 27.

Data obtained show that the ratio of hydrogen to oxygen combined to form water is 1 : 8 by mass. This means that 1 g of hydrogen combines with 8 g of oxygen to form 9 g of water.

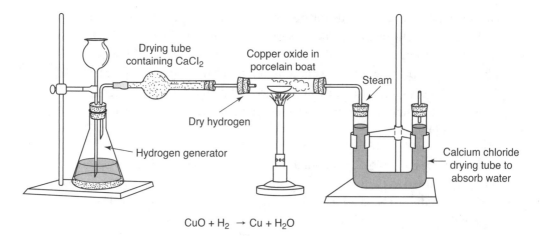

CuO + H₂ → Cu + H₂O

FIGURE 27. Synthesis of water.

Some sample problems involving the composition of water are shown below.

EXAMPLE 1 *(By Mass)*

An electric spark is passed through a mixture of 12 g of hydrogen and 24 g of oxygen in the eudiometer setup shown. Find the number of grams of water formed and the number of grams of gas left uncombined.

SOLUTION

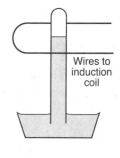

Since water forms in a ratio of 1 : 8 by mass, to use up the oxygen (which by inspection will be the limiting factor since it has enough hydrogen present to react completely), we need only 3 g of hydrogen.

3 g of hydrogen + 24 g of oxygen = 27 g of water

This leaves 12 g − 3 g = 9 g of hydrogen uncombined.

EXAMPLE 2 *(By Volume)*

A mixture of 8 mL of hydrogen and 200 mL of air is placed in a steam-jacketed eudiometer, and a spark is passed through the mixture. What will be the total volume of gases in the eudiometer?

SOLUTION

Since this is a combination by volume, 8 mL of hydrogen require 4 mL of oxygen. (Ratio H_2 : O_2 by volume = 2 : 1.)

The 200 mL of air is approximately 21% oxygen. This will more than supply the needed oxygen and leave 196 mL of the air uncombined.

The 8 mL of hydrogen and 4 mL of oxygen will form 8 mL of steam since the eudiometer is steam-jacketed and keeps the water formed in the gaseous state.

(Ratio by volume of hydrogen : oxygen : steam = 2 : 1 : 2)

TOTAL VOLUME = 196 mL of air + 8 mL of steam = 204 mL

Hydrogen Peroxide

The prefix *per-* indicates that this compound contains more than the usual oxide. Its formula is H_2O_2. It is a well-known bleaching and oxidizing agent. Its electron-dot formula is shown in Figure 28.

FIGURE 28. Hydrogen peroxide.

PROPERTIES AND USES OF WATER

Water has been used in the definition of various standards.

1. For mass—1 mL (or 1 cc) of water at 4°C is 1 g

2. For heat— (a) the heat needed to raise 1 g of water 1°C on the Celsius scale = 1 cal; 1000 cal = 1 kcal

    (b) the heat needed to raise 1 lb of water 1°C on the Fahrenheit scale = 1 British thermal unit (Btu)

3. Degree of heat—the freezing point of water = 0°C, 32°F, 273 K
    —the boiling point of water = 100°C, 212°F, 373 K

WATER CALORIMETRY PROBLEMS

A calorimeter is a container well insulated from outside sources of heat or cold so that most of its heat is contained in the vessel. If a very hot object is placed in a calorimeter containing some ice crystals, we can find the final temperature of the mixture mathematically and check it experimentally. To do this, certain behaviors must be understood. Ice changing to water and then to steam is not a continuous and constant change of temperature as time progresses. In fact, the chart would look as shown in Figure 29.

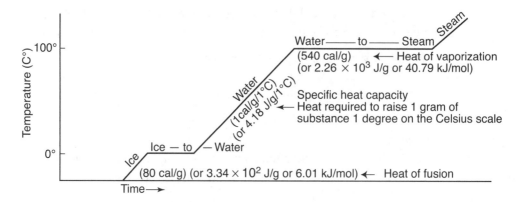

FIGURE 29. Changing ice to steam.

From this graph, you see that heat is being used at 0°C and 100°C to change the state of water but not its temperature. One gram of ice at 0°C needs 80 cal or 3.34×10^2 J to change to water at 0°C. This is called its **heat of fusion**. Likewise, energy is being used at 100°C to change water to steam, not to change the temperature. One gram of water at 100°C absorbs 540 cal or 2.26×10^3 J of heat to change to 1 g of steam at 100°C. This is called its **heat of vaporization**. This energy being absorbed at the plateaus in the curve is being used to break up the bonding forces between molecules by increasing their potential energy content so that a specific change of state can occur.

EXAMPLE 1 What quantity of ice at 0°C can be melted by 100 J of heat?

SOLUTION

Heat to fuse (melt) a substance =
heat of fusion of the substance × mass of the substance.

This can be expressed by the following formula, where q is used to denote heat measurement made in a calorimeter:

$$q = m \text{ (mass)} \times \Delta H_{\text{fus}} \text{ (heat of fusion)}$$

Solving for m, we get

$$m = q/\Delta H_{\text{fus}}$$
$$= 100 \text{ J}/3.34 \times 10^2 \text{ J/g}$$
$$= 29.9 \times 10^{-2} \text{ g or } 0.299 \text{ g of ice melted}$$

Because heat is absorbed in this melting, this is an endothermic action.

EXAMPLE 2

How much heat is needed to change 100 g of ice at 0°C to steam at 100°C?

To melt 100 g of ice at 0°C:

SOLUTION

Use m (mass) $\times \Delta H_{\text{fus}}$ (heat of fusion) $= q$ (quantity of heat).

$$100 \ \cancel{g} \times \frac{80 \text{ cal}}{1 \ \cancel{g}} = 8000 \text{ cal} = 8 \text{ kcal}$$

To heat 100 g of water from 0°C to 100°C:

Use m (mass) $\times \Delta T \times$ specific heat $= q$ (quantity of heat).

$$100 \ \cancel{g} \times \overbrace{(100° - 0°)° \cancel{C}}^{\text{temperature change}} \times \frac{1 \text{ cal}}{1 \ \cancel{g} \times 1° \cancel{C}} = 10{,}000 \text{ cal} = 10 \text{ kcal}$$

To vaporize 100 g of water at 100°C to steam at 100°C:

Use m (mass) $\times \Delta H_{\text{vap}}$ (heat of vaporization) $= q$ (quantity of heat).

$$100 \ \cancel{g} \times \frac{540 \text{ cal}}{1 \ \cancel{g}} = 54{,}000 \text{ cal} = 54 \text{ kcal}$$

$$\text{Total heat} = 8 + 10 + 54 = 72 \text{ kcal}$$

Using dimensional analysis to express the answer in joules, we have

$$72 \ \cancel{\text{kcal}} \times \frac{1000 \ \cancel{\text{cal}}}{1 \ \cancel{\text{kcal}}} \times \frac{4.18 \text{ J}}{1 \ \cancel{\text{cal}}} = 3.01 \times 10^5 \text{ J}$$

REACTIONS OF WATER WITH ANHYDRIDES

Anhydrides are certain oxides that react with water to form two classes of compounds—acids and bases.

Many metal oxides react with water to form bases such as sodium hydroxide, potassium hydroxide, and calcium hydroxide. For this reason, they are called *basic anhydrides*. The common bases are water solutions that contain the hydroxide (OH^-) ion. Some common examples are

$$Na_2O + H_2O \rightarrow 2NaOH, \text{ sodium hydroxide}$$
$$CaO + H_2O \rightarrow Ca(OH)_2, \text{ calcium hydroxide}$$
In general, then: Metal oxide $+ H_2O \rightarrow$ Metal hydroxide

In a similar manner, nonmetallic oxides react with water to form an acid such as sulfuric acid, carbonic acid, or phosphoric acid. For this reason, they are referred to as *acid anhydrides*. The common acids are water solutions containing hydrogen ions (H^+). Some common examples are

$$CO_2 + H_2O \rightarrow H_2CO_3, \text{ carbonic acid}$$
$$SO_3 + H_2O \rightarrow H_2SO_4, \text{ sulfuric acid}$$
$$P_2O_5 + 3H_2O \rightarrow 2H_3PO_4, \text{ phosphoric acid}$$
In general, then: Nonmetallic oxide + $H_2O \rightarrow$ Acid

Polarity and Hydrogen Bonding

Water is different from most liquids in that it reaches its greatest density at 4°C and then begins to expand from 4°C to 0°C (which is its freezing point). When it freezes at 0°C, its volume expands by about 9%. Most liquids contract as they cool and change state to a solid because their molecules have less energy, move more slowly, and are closer together. This abnormal behavior can be explained as follows. X-ray studies of ice crystals show that H_2O molecules are bound into large molecules in which each oxygen atom is connected through *hydrogen bonds* to four other oxygen atoms as shown in Figure 30.

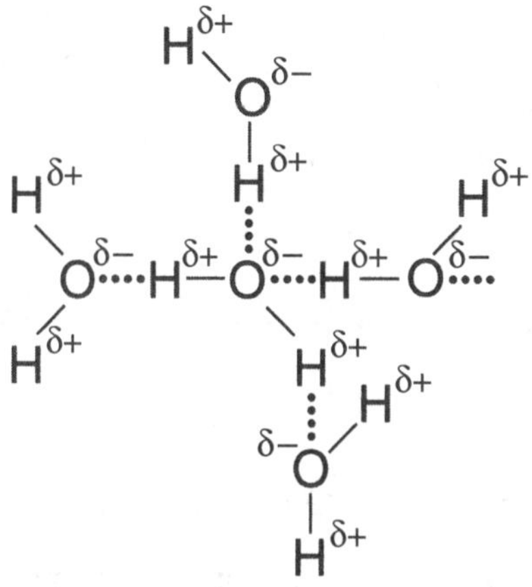

FIGURE 30. Study of ice crystal.
Note: δ indicates partial charge.

This is a rather wide open structure, which accounts for the low density of ice. As heat is applied and melting begins, this structure begins to collapse, but not all the hydrogen bonds are broken. The collapsing increases the density, but the remaining bonds keep the structure from completely collapsing. As the heat is absorbed, the kinetic energy of the molecules breaks more of these bonds as the temperature goes from 0°C to 4°C. At the same time, this added kinetic energy tends to distribute the molecules farther apart. At 4°C, these opposing forces are in balance—thus the greatest density. Above 4°C, the increasing molecular motion again causes a decrease in density since it is the dominant force and offsets the breaking of any more hydrogen bonds.

This behavior of water can be explained by studying the water molecule itself. The water molecule is composed of two hydrogen atoms bonded by a polar covalent bond to one oxygen atom.

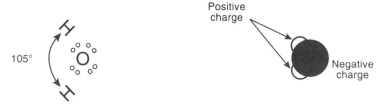

Because of the polar nature of the bond, the molecule exhibits the charges shown in the above drawing. It is this polar charge that causes the polar bonding discussed in Chapter 3 as the hydrogen bond. This bonding is stronger than the usual molecular attraction called **van der Waals forces**.

Solubility

Water is often referred to as "the universal solvent" because of the number of common substances that dissolve in it. When substances are dissolved in water to the extent that no more will dissolve at that temperature, the solution is said to be *saturated*. The substance dissolved is called the **solute**, and the dissolving medium is called a **solvent**. To give an accurate statement of a substance's solubility, three conditions are mentioned: the amount of solute, the amount of solvent, and the temperature of the solution. Since the solubility varies for each substance and for different temperatures, a student must be acquainted with the use of solubility curves such as those shown in Figure 31. A more complete table is given as Table Ⓑ in the Equations and Tables for Reference section.

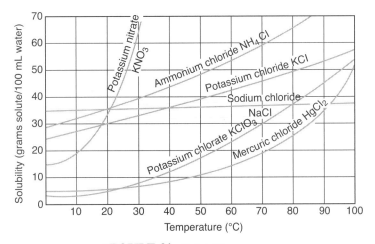

FIGURE 31. Solubility curves.

These curves show the number of grams of solute that will dissolve in 100 g of water over a temperature of 0°C to 100°C. Take, for example, the very lowest curve at 0°C. This curve shows the number of grams of $KClO_3$ that will dissolve in 100 g of water over a temperature range of 0°C to 100°C. To find the solubility at any particular temperature, for example, at 50°C, follow the vertical line up from 50°C until it crosses the curve. At that point, place a ruler so that it is horizontal across the page and take the reading on the vertical axis. This point happens to be slightly below the 20-g mark, or 18 g. This means that 18 g of $KClO_3$ will dissolve in 100 g of water at 50°C.

TYPE OF PROBLEM USING THE SOLUBILITY CURVE

A solution contains 20 g of $KClO_3$ in 200 g of H_2O at 80°C. How many more grams of $KClO_3$ can be dissolved to saturate the solution at 90°C?

Reading the graph at 90°C and up to the graph line for $KClO_3$, we find that 100 g of H_2O can dissolve 48 g. So 200 g can hold (2×48) g or 96 g. Therefore, 96 g – 20 g = 76 g $KClO_3$ can be added to the solution.

GENERAL RULES OF SOLUBILITY

All nitrates, acetates, and chlorates are soluble.

All common compounds of sodium, potassium, and ammonium are soluble.

All chlorides are soluble except those of silver, mercury (I), and lead (II). (Lead (II) chloride is noticeably soluble in hot water.)

All sulfates are soluble except those of lead, barium, strontium, and calcium. (Calcium sulfate is slightly soluble.)

The normal carbonates, phosphates, silicates, and sulfides are insoluble except those of sodium, potassium, and ammonium.

All hydroxides are insoluble except those of sodium, potassium, ammonium, calcium, barium, and strontium.

Some general trends of solubility are shown in the chart below.

	Temperature Effect	Pressure Effect
Solid	Solubility usually increases with temperature increase	Little effect
Gas	Solubility usually decreases with temperature increase	Solubility varies in direct proportion to the pressure applied to it. *Henry's Law*

FACTORS THAT AFFECT RATE OF SOLUBILITY

The following procedures increase the rate of solubility:

Pulverizing—Increases surface exposed to solvent

Stirring—Brings more solvent that is unsaturated into contact with solute

Heating—Increases molecular action and gives rise to mixing by convection currents. (This heating affects the solubility as well as the rate of solubility.)

SUMMARY OF TYPES OF SOLUTES AND RELATIONSHIP OF TYPE TO SOLUBILITY

Generally speaking, solutes are most likely to dissolve in solvents with similar characteristics; that is, ionic and polar solutes dissolve in polar solvents, and nonpolar solutes dissolve in nonpolar solvents.

It should also be mentioned that polar molecules that do not ionize in aqueous solution (e.g., sugar, alcohol, glycerol) have molecules as solute particles; polar molecules that partially ionize in aqueous solution (e.g., ammonia, acetic acid) have a mixture of molecules and ions as solute particles; and polar molecules that completely ionize in aqueous solution (e.g., hydrogen chloride, hydrogen iodide) have ions as solute particles.

Water Solutions

In order to make molecules or ions of another substance go into solution, water molecules must overcome the forces that hold these molecules or ions together. The mechanism of the actual process is complex. To make sugar molecules go into solution, the water molecules cluster around the sugar molecules, pull them off, and disperse, forming the solution.

For an ionic crystal such as salt, the water molecules orient themselves around the ions (which are charged particles) and again must overcome the forces holding the ions together. Since the water molecule is polar, this orientation around the ion is an attraction of the polar ends of the water molecule. For example,

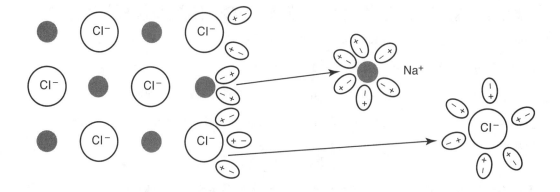

Once surrounded, the ion is insulated to an extent from other ions in solution because of the *dipole* property of water. The water molecules that surround the ion differ in number for various ions, and the whole group is called a *hydrated ion*. In general, as stated in the preceding section, polar substances and ions dissolve in polar solvents, and nonpolar substances such as fats dissolve in nonpolar solvents such as gasoline. The process of going into solution is **exothermic** if energy is released in the process, and **endothermic** if energy from the water is used up to a greater extent than energy is released in freeing the particle.

When two liquids are mixed and they dissolve in each other in all portions, they are said to be *miscible*. If they separate and do not mix, they are said to be *immiscible*.

Two molten metals may be mixed and allowed to cool. This gives a "solid solution" called an *alloy*.

Continuum of Water Mixtures

Figure 32 shows the general sizes of the particles found in a water mixture.

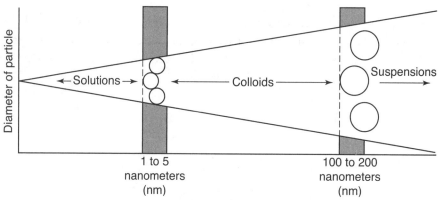

FIGURE 32. Size of particles in water mixture.

The basic difference between a colloid and a suspension is in the diameter of the particles dispersed. All the boundaries marked are only the general ranges in which the distinctions among solutions, colloids, and suspensions are usually made.

The characteristics of water mixtures are as follows:

Solutions	Colloids	Suspensions
.... 1 nm .... 1000 nm		
Clear; may have color		Cloudy; opaque color
Particles do not settle		Particles settle on standing
Particles pass through ordinary filter paper		
Particles pass through membranes	Do not pass through semipermeable membranes like animal bladders, cellophane, and parchment, which have very small pores*	Do not pass through ordinary filter paper
Particles are not visible	Visible in ultramicroscope	Visible with microscope or naked eye
	Show Brownian movement	No Brownian movement

*Separation of a solution from a colloidal dispersion through a semipermeable membrane is called *dialysis*.

When a bright light is directed at right angles to the stage of an ultramicroscope, the individual reflections of colloidal particles can be seen. They can be observed to be following a random zigzag path. This is explained as follows: The molecules in the dispersing medium are in motion and continuously bumping into the colloidal particles, causing them to change direction in a random fashion. This motion is called **Brownian movement** after the scientist who first observed it.

Expressions of Concentration

There are general terms and very specific terms used to express the concentration of a solution. The general terms and their definitions are

Dilute—Small amount of solute is dispersed in the solvent.

Concentrated—Large amount of solute is dissolved in the solvent.

Saturated—The solution is holding all the solute possible at that temperature. This is not a static condition; that is, some solute particles are exchanging places with some of the undissolved particles, but the total solute in solution remains the same. This is an example of equilibrium.

Unsaturated—More solute can go into solution at that temperature. The solvent has further capacity to hold more solute.

Supersaturated—Sometimes a saturated solution at a higher temperature can be carefully cooled so that the solute does not have a chance to come out of solution. At a lower temperature, then, the solution will be holding more solute in solution than it should for saturation and is said to be supersaturated. As soon as the solute particles are jarred or a "seed" particle is added to the solution to act as a nucleus, they rapidly come out of solution so that the solution reverts to the saturated state.

It is interesting to note that the words *saturated* and *concentrated* are NOT synonymous. In fact, it is possible to have a saturated dilute solution when the solute is only slightly soluble and a small amount of it makes the solution saturated, but its concentration is still dilute.

The more specific terms used to describe concentration are mathematically calculated.

Percentage concentration is based on the percent of solute in the solution by mass. The general formula is

$$\frac{\textbf{No. of grams of solute}}{\textbf{No. of grams of solution}} \times \textbf{100\%} = \textbf{\% concentration}$$

EXAMPLE How many grams of NaCl are needed to prepare 200 g of a 10% salt solution?

SOLUTION

$$10\% \text{ of } 200 \text{ g} = 20 \text{ g of salt}$$

You could also solve the problem using the above formula and solving for the unknown quantity.

$$\frac{x \text{ grams solute}}{200 \text{ g solution}} \times 100\% = 10\%$$

$$x \text{ grams solute} = \frac{10}{100} \times 200 = 20 \text{ g of solute}$$

Using Specific Gravity in Solutions

Specific gravity is the ratio of the mass of a substance to the mass of an equal volume of water. Often chemistry handbooks give the specific gravity of solutions. These in turn can be used to determine the mass of a solute in solution. Specific gravity has no units, but since the density of water is 1.0 g/mL, specific gravity is numerically equal to the density in grams per milliliter. For example, 7.9 is the specific gravity of iron. This means that the density of iron is 7.9 g/mL.

This relationship can be used in solving problems of concentration that use percentage composition and a known specific gravity.

EXAMPLE Suppose that a solution of hydrogen chloride gas dissolved in water (hydrochloric acid) has a concentration of 30% HCl by mass. How much solute is there in 100 mL of this solution if the specific gravity of the solution is 1.15? (This means that its density is 1.15 g/mL.)

SOLUTION

First, determine the mass of 100 mL of the solution:

$$\text{Mass} = \text{Density} \times \text{volume}$$
$$\text{Mass} = 1.15 \text{ g/mL} \times 100 \text{ mL} = 115 \text{ g}$$

Then, determine what mass of HCl is dissolved in 115 g of the solution:

$$\text{Mass HCl} = 30\% \times \text{mass of solution}$$
$$\text{Mass HCl} = 0.30 \times 115 \text{ g} = 34.5 \text{ g}$$

So, there are 34.5 g of HCl gas dissolved in 100 mL of this solution.

The next two expressions depend on the fact that if the formula mass of a substance is expressed in grams, it is called a *gram-formula mass* (gfm), *molar mass*, or 1 *mol*. *Gram-molecular mass* can be used in place of gram-formula mass when the substance is really of molecular composition and not ionic like NaCl or NaOH. The definitions and examples are:

Molarity (abbreviated M) is defined as the number of moles of a substance dissolved in 1 L of solution.

EXAMPLE 1 A 1 M H_2SO_4 solution has 98 g of H_2SO_4 (its gram-formula mass) in 1 L of the solution.

This can be expressed as a formula, so that

$$\text{Molarity} = \frac{\text{No. of moles of solute}}{1 \text{ L of solution}}$$

If the molarity (M) and the volume of a solution are known, the mass of the solute can be determined by using the above equations and solving for the number of moles of solute and then multiplying this number by the molar mass.

EXAMPLE
2

How many grams of NaOH are dissolved in 200 mL of solution if its concentration is 1.50 *M*?

SOLUTION

$$M = \frac{\text{No. of moles of solute}}{1\,\text{L of solution}}$$

Solving for number of moles, we have

No. of moles of solute $= M \times$ volume in liters of solution
No. of moles of NaOH $= 1.50\,M \times$ volume in liters of solution
No. of moles of NaOH $= 1.5\,\text{mol/L} \times 0.200\,\text{L} = 0.30\,\text{mol of NaOH}$

The molar mass of NaOH $= 23 + 16 + 1 = 40$ g of NaOH.

$$0.30\ \cancel{\text{mol NaOH}} \times \frac{40\text{ g of NaOH}}{1\ \cancel{\text{mol NaOH}}} = 12\text{ g of NaOH}$$

Molality (abbreviated *m*) is defined as the number of moles of the solute dissolved in 1000 g of solvent.

EXAMPLE
1

A 1 *m* solution of H_2SO_4 has 98 g of H_2SO_4 dissolved in 1000 g of water. This, you will notice, gives a total volume greater than 1 L, whereas the molar solution had 98 g in 1 L of solution since

$$\text{Molality } (m) = \frac{\text{Moles of solute}}{1000\text{ g of solvent}}$$

EXAMPLE
2

Suppose that 0.25 mol of sugar are dissolved in 500 g of water. What is the molality of this solution?

SOLUTION

$$m = \frac{\text{Moles of solute}}{1000\text{ g of solvent}}$$

If there are 0.25 mol in 500 g of H_2O, then there are 0.50 mol in 1000 g of H_2O. So,

$$m = \frac{0.50\text{ mol of sugar}}{1000\text{ g of }H_2O} \quad \text{or} \quad 0.50$$

Normality (abbreviated *N*) is used less frequently to express concentration and depends on knowledge of what a *gram-equivalent mass* is. This can be defined as the amount of a substance that reacts with or displaces 1 g of hydrogen or 8 g of oxygen. A simple method of determining the number of equivalents in a formula is to count the number of hydrogens or find the total positive oxidation numbers since each 1+ charge can be replaced by a hydrogen.

In H_2SO_4 (gram-formula mass = 98 g), there are 2 hydrogens, and so the gram-equivalent mass will be

$$\frac{98 \text{ g}}{2} = 49 \text{ g}$$

In Al_2SO_4 (molar mass = 150 g), there are 2 aluminums, each of which has a +3 oxidation number, making a total of +6. The gram-equivalent mass will be

$$\frac{150 \text{ g}}{6} = 49 \text{ g}$$

This idea of gram-equivalent mass is used in the definition of this means of expressing concentration, namely, the normality of a solution.

Normality is defined as the number of gram-equivalent masses of solute in 1 L of solution.

EXAMPLE If 49 g of H_2SO_4 is mixed with enough H_2O to make 500 mL of solution, what is the normality?

SOLUTION

Since normality is expressed for a liter, we must double the expression to 98 g of H_2SO_4 in 1000 mL of solution. To find the normality we find the number of gram-equivalents in a liter of solution. So

$$\frac{98 \text{ g (no. of grams 1 L of solution)}}{49 \text{ g (gram-equivalent mass)}} = 2N \text{ (normal)}$$

Mole fraction is another way of indicating the concentration of a component in a solution. It simply is the number of moles of that component divided by the total moles of all solution parts. The mole fraction of component *i* is written as X_i. For a solution consisting of n_A moles of component A, n_B moles of component B, n_C moles of component C, and so on, the mole fraction of component A is given by

$$X_A = \frac{n_A}{n_A + n_B + n_C + \cdots}$$

As an example, if a mixture is obtained by dissolving 10 moles of NaCl in 90 moles of water, the mole fraction of NaCl in the mixture is 10 (moles of NaCl) divided by 100 (or 10 + 90), giving an answer of 0.1, the mole fraction of NaCl in the solution.

Dilution

Since the expression of molarity gives the quantity of solute per volume of solution, the amount of solute dissolved in a given volume of solution is equal to the product of the concentration and the volume. Hence, 0.5 L of 2 M solution contains

$$M \times V = \text{amount of solute (in moles)}$$

$$\frac{2 \text{ mol}}{\cancel{L}} \times 0.5 \cancel{L} = 1 \text{ mol (of solute in 0.5 L)}$$

Notice that volume units must be identical.

If you dilute a solution with water, the amount or number of moles of solute present remains the same, but the concentration changes. So you can use the expression

$$\overset{\text{Before}}{M_1 V_1} = \overset{\text{After}}{M_2 V_2}$$

This expression is useful in solving problems involving dilution.

EXAMPLE If you wish to make 1 L of solution that is 6 M into a 3 M solution, how much water must be added?

SOLUTION

$$M_1 V_1 = M_2 V_2$$
$$6\,M \times 1\,\text{L} = 3\,M \times ?\,\text{L}$$

Solving this expression:

? liters = 2 L. This is the total volume of the solution after dilution and means that 1 L of water has to be added to the original volume of 1 L to get a total of 2 L for the dilute solution volume.

An important use of the molarity concept is in the solution of *titration* problems, which are covered in Chapter 11 along with pH expressions of concentration for acids.

Colligative Properties of Solutions

Colligative properties are properties that depend primarily on the concentration of particles and not the type of particle. There is usually a direct relationship between the concentration of particles and the effect recorded.

The vapor pressure of an aqueous solution is always lowered by the addition of more solute. From the molecular standpoint, it is easy to see that there are fewer molecules of water per unit volume in the liquid, and therefore fewer molecules of water in the vapor phase are required to maintain equilibrium. The concentration in the vapor drops and so does the pressure that molecules exert. This is shown graphically in Figure 33.

Notice that the effects of this change in vapor pressure are registered in the freezing point and the boiling point. The freezing point is lowered, and the boiling point is raised, in direct proportion to the number of particles of solute present. For water solutions, the concentration expression that expresses this relationship is molality (m), that is, the number of moles of solute per kilogram of solvent. For molecules that do not dissociate, it has been found that a 1 m solution freezes at −1.86°C and boils at 100.51°C. A 2 m solution would then freeze at twice this lowering, or −3.72°C, and boil at twice the 1 m increase of 0.51°C, or 101.02°C.

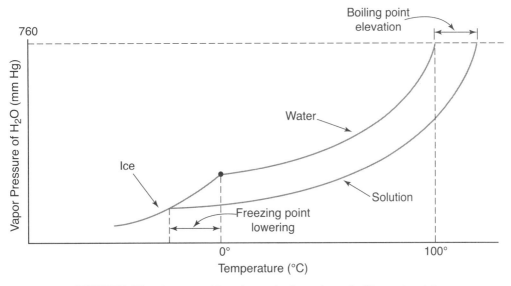

FIGURE 33. Diagram of freezing point lowering—boiling point rising.

The table below summarizes the colligative effect for aqueous solutions.

Type	Concentration (m)	Examples	Moles of Particles	Freezing Point (°C)	Boiling Point (°C)
Molecular, nonionizing	1	Sugar, urea	1	−1.86	100.51
Molecular, completely ionized	1	HCl	2	−3.72	101.02
Ionic, completely dissociated	1	NaCl	2	−3.72	101.02
Ionic, completely dissociated	1	$CaCl_2$ $Cu(NO_3)_2$	3	−5.58	101.53

EXAMPLE
1

A 1.50 g sample of urea is dissolved in 105.0 g of water and produces a solution that boils at 100.12°C. From the data, what is the molecular mass of urea?

SOLUTION

Since this property is related to the molality, then

$$\frac{1.5\ g}{105.0\ g} = \frac{x\ grams}{1000\ g}$$

$$x = 14.28\ g\ in\ 1000\ g\ of\ water$$

The boiling point change is 0.12°C, and since each mole of particles causes a 0.51° increase, then

$$0.12°C \div \frac{0.51°C}{mol} = 0.235\ mol$$

Then 14.28 g = 0.235 mol and

$$\frac{14.28\ g}{0.235\ mol} = \frac{x\ grams}{1\ mol}$$

$$x = 60.8\ g/mol$$

EXAMPLE
2

Suppose that there are two water solutions, one of glucose (molar mass = 180), and the other of sucrose (molar mass = 342). Each contains 50 g of solute in 1000 g (1 kg) of water. Which has the higher boiling point? The lower freezing point?

SOLUTION

The molality of each of these nonionizing substances is found by dividing the number of grams of solute by the molecular mass.

$$Glucose: \frac{50\ g}{180\ g/mol} = 0.278\ mol$$

$$Sucrose: \frac{50\ g}{342\ g/mol} = 0.146\ mol$$

Both are in 1000 g of water, and so their respective molalities are 0.278 m and 0.146 m. Since the freezing point and boiling point are colligative properties, the effect depends only on the concentration. Because the glucose has a higher concentration, it will have a higher boiling point and a lower freezing point. The respective boiling points are

$$0.278\ \cancel{m} \times \frac{0.51°C\ rise}{1\ \cancel{m}} = 0.14°C\ rise\ or\ 100.14°C\ for\ glucose\ and$$

$$0.146\ \cancel{m} \times \frac{0.51°C\ rise}{1\ \cancel{m}} = 0.07°C\ rise\ in\ boiling\ point\ or\ 100.07°C\ for\ sucrose.$$

The lowering of the freezing point is

$$0.278 \, \cancel{m} \times \frac{-1.86°\text{C drop}}{1 \, \cancel{m}} = -0.52°\text{C drop below } 0°\text{C for glucose}$$

and

$$0.146 \, \cancel{m} \times \frac{-1.86°\text{C drop}}{1 \, \cancel{m}} = -0.27°\text{C drop below } 0°\text{C for sucrose}$$

Using a solute that is an ionic solid that completely ionizes in an aqueous solution introduces a consideration of the number of particles present in the solution. Notice in the previous chart that a 1 m solution of NaCl yields a solution with 2 mol of particles.

This is because

NaCl(aq)	→ Na$^+$	+ Cl$^-$
1 mol of ionic sodium chloride salt	1 mol of Na$^+$ ions	1 mol of Cl$^-$ ions

So a 1 m solution of salt has 2 mol of ion particles in 1000 g of solvent. The colligative property of lowering the freezing point and raising the boiling point depends primarily on the concentration of particles and not the type of particles.

In a 1 m solution of NaCl there is 1 mol of sodium chloride salt dissolved in 1000 g of water, and so there is a total of 2 mol of ions released—1 mol of Na$^+$ ions and 1 mol of Cl$^-$ ions. Because this provides 2 mol of particles, it will cause a $2 \times -1.86°$C (drop caused by 1 mol) $= -3.72°$C drop in the freezing point and likewise a $2 \times 0.51°$C (rise caused by 1 mol) $= 1.02°$C rise in the boiling point or a boiling point of 101.02°C.

The chart also shows that CaCl$_2$ releases 3 mol of particles from 1 mol of CaCl$_2$ dissolved in 1000 g of water. Note that its effect is to lower the freezing point $3 \times -1.86°$C or to $-5.58°$C. The boiling point rise is also three times the molal rise of 0.51°C and gives a boiling point of 101.53°C.

This explains the use of salt on icy roads in the winter and the increased effectiveness of calcium chloride per mole of solute. The use of glycols in antifreeze solutions in automobile radiators is based on this same concept.

Crystallization

Many substances form a repeated pattern structure as they come out of solution. The structure is bounded by plane surfaces that make definite angles with each other to produce a geometric form called a **crystal**. The smallest portion of the crystal lattice that is repeated throughout the crystal is called the *unit cell*. Some kinds of unit cells are shown in Figure 34. For all the cubic unit cells shown in the following diagram, the

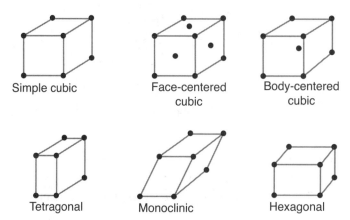

Simple cubic   Face-centered cubic   Body-centered cubic

Tetragonal   Monoclinic   Hexagonal

FIGURE 34. Kinds of unit cells.

edge lengths of each are identical and the angles between the edges are all 90°. For the tetragonal unit cell, the width is different in length than the height and depth, which are the same length. All angles between them are 90°. For the monoclinic, none of the edge lengths are equal and one angle between two of the edges is not 90°. The hexagonal unit cell exhibits two edges with identical length and an angle of 120° between them and a third edge different in length and at 90° from the other two edges. The crystal structure can also be classified by its internal axes, as shown in Figure 35.

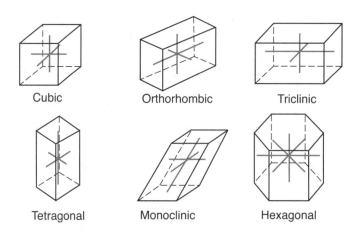

Cubic   Orthorhombic   Triclinic

Tetragonal   Monoclinic   Hexagonal

FIGURE 35. Crystal structures classified by internal axes.

A substance that holds a definite proportion of water in its crystal structure is called a **hydrate**. The formulas of hydrates show this water in the following manner: $CuSO_4 \cdot 5H_2O$; $CaSO_4 \cdot 2H_2O$; and $Na_2CO_3 \cdot 10H_2O$. (The • is read as "with.") When these crystals are heated gently, the water of hydration can be forced out of the crystal and the structure collapses into an anhydrous (without water) powder. The dehydration of hydrated $CuSO_4$ serves as a good example since the hydrated crystals are deep blue because of the water molecules present with the copper ions. When this water is removed, the structure crumbles into the anhydrous white powder. Some hydrated crystals, such as magnesium sulfate (Epsom salt), lose the water of hydration on exposure to air at ordinary temperatures. They are said to be *efflorescent*. Other hydrates, such as magnesium chloride and calcium chloride, absorb water from the air and become wet. They are said to be *deliquescent*, or *hygroscopic*. This property explains why calcium chloride is often used as a drying agent in laboratory experiments.

Chapter Summary

The following terms summarize all the concepts and ideas that were introduced in this chapter. You should be able to explain their meaning and how you would use them in chemistry. They appear in boldface type in this chapter to draw your attention to them. The boldface type also makes the terms easier for you to look up, if you need to. You could also use a search engine on a computer to get a quick and expanded explanation of them.

TERMS YOU SHOULD KNOW

anhydride
boiling point
Brownian movement
colligative property
critical pressure
critical temperature
crystal
desalination
dynamic equilibrium
endothermic
exothermic sublimation
heat of fusion
heat of vaporization
heavy water
hydrate

Le Châtelier's Principle
melting point
molality
molarity
mole fraction
normality
phase equilibrium
polarity
solute
solvent
specific gravity
surface tension
van der Waals forces
viscosity

Chapter 7 Review Exercises

1. Distillation for use in purification of water cannot remove

   (A) volatile liquids like alcohol
   (B) dissolved salts
   (C) suspensions
   (D) precipitates

2. The ratio in water of hydrogen to oxygen by mass is

   (A) $1:9$
   (B) $2:1$
   (C) $1:2$
   (D) $1:8$

3. Decomposing water by an electric current will give a ratio of hydrogen to oxygen by volume equal to

   (A) $1:9$
   (B) $2:1$
   (C) $1:2$
   (D) $1:8$

4. If 10.0 g of ice melts at 0°C, the total quantity of heat absorbed is

   (A) 10 J
   (B) 334 J
   (C) 3340 J
   (D) 33,400 J

5. To heat 10.0 g of water from 4°C to 14°C requires

   (A) 10 J
   (B) 4.18 J
   (C) 418 J
   (D) 4180 J

6. The abnormally high boiling point of water in comparison to similar compounds is due primarily to

   (A) van der Waals forces
   (B) polar covalent bonding
   (C) dipole insulation
   (D) hydrogen bonding

7. A metallic oxide placed in water would most likely yield a(n)

   (A) acid
   (B) base
   (C) metallic anhydride
   (D) basic anhydride

8. A solution can be *both*

   (A) dilute and concentrated
   (B) saturated and dilute
   (C) saturated and unsaturated
   (D) supersaturated and saturated

9. The solubility of a solute must indicate

   (A) the temperature of the solution
   (B) the quantity of solute
   (C) the quantity of solvent
   (D) all of these

10. A foam is an example of

    (A) a gas dispersed in a liquid
    (B) a liquid dispersed in a gas
    (C) a solid dispersed in a liquid
    (D) a liquid dispersed in a liquid

11. When another crystal was added to a water solution of the same substance, the crystal seemed to remain unchanged. Its particles were

    (A) going into an unsaturated solution
    (B) exchanging places with others in the solution
    (C) causing the solution to become supersaturated
    (D) not going into solution in this static condition

12. A 10% solution of NaCl means that in 100 g of solution there is

    (A) 5.85 g NaCl
    (B) 58.5 g NaCl
    (C) 10 g NaCl
    (D) 94 g of H_2O

13. The gram-equivalent mass of $AlCl_3$ is its gram-formula mass divided by

 (A) 1
 (B) 3
 (C) 4
 (D) 6

14. The molarity of a solution made by placing 98 g of H_2SO_4 in sufficient water to make 500 mL of solution is

 (A) 0.5
 (B) 1
 (C) 2
 (D) 3
 (E) 4

15. If 684 g of sucrose (molecular mass = 342 g) is dissolved in 2000 g of water (essentially 2 L), what will be the freezing point of this solution?

 (A) −0.51°C
 (B) −1.86°C
 (C) −3.72°C
 (D) −6.58°C

Answers and Explanations

1. **(A)** Volatile liquids cannot be removed by distillation because they would come over with the steam in this process.

2. **(D)** In the molar mass of water, there are two hydrogen = 2 g and one oxygen = 16 g. So the H : O ratio by mass is 2 g : 16 g or 1 : 8.

3. **(B)** In the balanced reaction, $2\,H_2O \rightarrow 2\,H_2\,(g) + O_2\,(g)$, 2 volumes of hydrogen are released to 1 volume of oxygen, so the ratio is 2 : 1.

4. **(C)** It takes 334 J to melt 1 g (heat of fusion for 1 gram), so $10\,g \times 334\,J/g = 3340\,J$ to melt the ice.

5. **(C)** m (mass) $\times \Delta T$ (change of temperature) $\times$ specific heat = q (quantity of heat). So $10\,g \times (14°C - 4°C = 10°C) \times 4.184\,J/g°C = 418\,J$.

6. **(D)** Because hydrogen bonding is so strong in water, water takes more heat energy to reach its boiling point.

7. **(B)** Metal oxides are basic anhydrides because they react with water to form bases. Example: $CaO + H_2O \rightarrow Ca(OH)_2$

8. **(B)** Saturated means that the solution is holding all the solute it can at that temperature. Dilute means that there is a small amount of solute in solution compared with the amount of solvent. So a substance that is only slightly soluble can form a saturated dilute solution.

9. **(D)** Solubility of a solute must give the amount of solute dissolved in a given amount of solvent at a particular temperature.

10. **(A)** A foam is an example of a colloidal dispersion of a gas in a liquid.

11. **(B)** This situation describes a saturated solution where an equilibrium exists between the undissolved solute and the solute particles in solution.

12. **(C)** A 10% solution contains 10 grams of solute per 100 g of solution because the percentage is mass per 100 grams of solution.

13. **(B)** The gram-equivalent mass is equal to gram-formula mass divided by 3 because Al has a +3 oxidation number that is equivalent to 3 hydrogens each with a +1 oxidation number.

14. **(C)** 98 g H_2SO_4 = 1 mol H_2SO_4

    So, if there are 500 mL (which is 0.5 L), then 1 mol in 0.5 L would be equivalent to 2 mol in 1 L or a 2-molar solution.

15. **(B)**

$$684 \text{ g of sucrose} = \frac{684 \text{ g}}{342 \text{ g/mol}} = 2 \text{ mol of sucrose}$$

$$\text{So } \frac{2 \text{ mol}}{2000 \text{ g } H_2O} = 1 \text{ molal solution}$$

The freezing point is lowered to $-1.86°C$.

PART V

CHEMICAL REACTIONS

CHEMICAL REACTIONS AND THERMOCHEMISTRY

CHAPTER OBJECTIVES

Upon completing this chapter, you will be able to:

- Identify the driving force for these four major types of chemical reactions and write balanced equations for each: combination (or synthesis), decomposition (or analysis), single replacement, and double replacement

- Explain hydrolysis using a balanced equation

- Identify and explain graphically enthalpy changes in exothermic and endothermic reactions

- Use Hess's Law to show the additivity of heats of reactions

- Calculate enthalpy from bond energies

Types of Reactions

The many kinds of reactions you may encounter can be placed in four basic categories: combination, decomposition, single replacement, and double replacement.

The first type, *combination*, can also be called *synthesis*. This means the formation of a compound from the union of its elements. Some examples of this type are

$$Zn + S \rightarrow ZnS$$
$$2H_2 + O_2 \rightarrow 2H_2O$$
$$C + O_2 \rightarrow CO_2$$

The second type of reaction, *decomposition*, can also be referred to as *analysis*. This means the breakdown of a compound to release its components as individual elements or other compounds. Some examples of this type are

$$2H_2O \rightarrow 2H_2 + O_2 \text{ (electrolysis of water)}$$
$$C_{12}H_{22}O_{11} \rightarrow 12C + 11H_2O$$
$$2HgO \rightarrow 2Hg + O_2$$

The third type of reaction is called *single replacement* or *single displacement*. This type can best be shown by examples in which one substance is displacing another. Some examples are

$$Fe + CuSO_4 \rightarrow FeSO_4 + Cu$$
$$Zn + H_2SO_4 \rightarrow ZnSO_4 + H_2(g)$$
$$Cu + 2AgNO_3 \rightarrow Cu(NO_3)_2 + 2Ag$$

The last type of reaction is called *double replacement* or *double displacement* because there is an actual exchange of "partners" to form new compounds. Some examples of this are

$$AgNO_3 + NaCl \rightarrow AgCl + NaNO_3$$
$$H_2SO_4 + 2NaOH \rightarrow Na_2SO_4 + 2H_2O \text{ (neutralization)}$$
$$CaCO_3 + 2HCl \rightarrow H_2CO_3 + CaCl_2$$
$$\qquad\qquad\qquad\quad \big\downarrow \text{ (unstable)}$$
$$\qquad\qquad\qquad\quad \hookrightarrow H_2O + CO_2(g)$$

Predicting Reactions

One of the most important topics of chemistry deals with the reasons why reactions take place. Taking each of the above types of reactions, let us see how a prediction can be made concerning how the reaction gets the driving force to make it occur.

COMBINATION (SYNTHESIS)

The best source of information in predicting a chemical combination is the heat of formation table. A **heat of formation** table gives the number of calories evolved or absorbed when a mole (gram-formula mass) of the compound in question is formed by the direct union of its elements. In this book, a positive number indicates that heat is absorbed, and a negative number that heat is evolved. It makes some difference whether the compounds formed are in the solid, liquid, or gaseous state. Unless otherwise indicated (g = gas, ℓ = liquid), the compounds are in the solid state. The values given are in kilojoules, 4.18 joules is the amount of heat needed to raise the temperature of 1 g of water one degree on the Kelvin scale. The symbol ΔH is used to indicate the heat of formation.

If the heat of formation is a large number preceded by a minus sign, the combination is likely to occur spontaneously and the reaction is exothermic. If, on the other hand, the number is small and negative or is positive, heat will be needed to get the reaction to go at any noticeable rate.

EXAMPLE 1

$$Zn + S \rightarrow ZnS + 202.7 \text{ kJ} \qquad \Delta H = -202.7 \text{ kJ}$$

SOLUTION

This means that 1 mol of zinc (65 g) reacts with 1 mol of sulfur (32 g) to form 1 mol of zinc sulfide (97 g) and releases 202.7 kJ of heat.

EXAMPLE 2

$$Mg + \tfrac{1}{2}O_2 \rightarrow MgO + 601.6 \text{ kJ} \qquad \Delta H = -601.6 \text{ kJ}$$

SOLUTION

This indicates that the formation of 1 mol of magnesium oxide requires 1 mol of magnesium and $\tfrac{1}{2}$ mol of oxygen with the release of 601.6 kJ of heat. Notice the use of the fractional coefficient for oxygen. If the equation had been written with the usual whole-number coefficients, 2 mol of magnesium oxide would have been released.

$$2Mg + O_2 \rightarrow 2MgO + 2(+601.6) \text{ kJ}$$

Since, by definition, the heat of formation is given for the formation of *one* mole, this latter thermal equation shows 2(+601.6) kJ released.

EXAMPLE 3

$$H_2(g) + \tfrac{1}{2}O_2(g) \rightarrow H_2O(\ell) + 285.8 \text{ kJ} \qquad \Delta H = -285.8 \text{ kJ}$$

SOLUTION

In combustion reactions, the heat evolved when 1 mol of a substance is completely oxidized is called the **heat of combustion** of the substance. So, in the equation

$$C(g) + O_2(g) \rightarrow CO_2(g) + 393.5 \text{ kJ} \qquad \Delta H = -393.5 \text{ kJ}$$

ΔH is the heat of combustion of carbon. Because the energy of a system is conserved during chemical activity, the same equation could be arrived at by adding the following equations:

$$
\begin{array}{llll}
C(s) & + \tfrac{1}{2}O_2(g) \rightarrow & \cancel{CO}(g) & \Delta H = -110.5 \text{ kJ} \\
\cancel{CO}(g) & + \tfrac{1}{2}O_2(g) \rightarrow & CO_2(g) & \Delta H = -283.0 \text{ kJ} \\
\hline
C(s) & + \phantom{\tfrac{1}{2}}O_2(g) \rightarrow & CO_2(g) & \Delta H = -393.5 \text{ kJ}
\end{array}
$$

DECOMPOSITION (ANALYSIS)

The prediction of **decomposition reactions** uses the same source of information, the heat of formation table. If the heat of formation is a high exothermic (ΔH is negative) value, the compound will be difficult to decompose since this same quantity of energy must be returned to the compound. A low heat of formation indicates decomposition would not be difficult, such as the decomposition of mercuric oxide with $\Delta H = -90.8$ kJ/mol

$$2HgO \rightarrow 2Hg + O_2 \quad \text{(Priestley's method of preparation)}$$

A high positive heat of formation indicates extreme instability of a compound, and it can decompose explosively.

SINGLE REPLACEMENT

A prediction of the feasibility of this type of reaction can be based on a comparison of the heat of formation of the original compound and that of the compound to be formed. For example, in a reaction of zinc with hydrochloric acid, the 2 mol of HCl have $\Delta H = 2 \times -92.3$ kJ and the zinc chloride has $\Delta H = -415.5$ kJ.

$$Zn + 2HCl \rightarrow ZnCl_2 + H_2(g) \quad \textit{Note: } \Delta H = 0 \text{ for elements}$$
$$2 \times -92.3 \text{ kJ}$$
$$-184.6 \text{ kJ} \neq -415.5 \text{ kJ}$$

This comparison leaves an excess of 230.9 kJ of heat given off, and so the reaction would occur.

In the next example, $-928.4 - (-771.4) = -157.0$ excess kilojoules to be given off as the reaction occurs

$$Fe + CuSO_4 \rightarrow FeSO_4 + Cu$$
$$-771.4 \text{ kJ} \qquad -928.4 \text{ kJ}$$

Another simple way of predicting single replacement reactions is to check the relative positions of the two elements in the following activity series. If the element that is to replace the other in the compound is higher on the chart, the reaction will occur. If it is below, there will be no reaction.

Some simple examples of this are the following reactions:

In predicting the replacement of hydrogen by zinc in hydrochloric acid, reference to the electromotive chart shows that zinc is above hydrogen. This reaction will occur

$$Zn + 2HCl \rightarrow ZnCl_2 + H_2(g)$$

In fact, all the metals above hydrogen in the chart on page 265 will replace hydrogen in an acid solution. If a metal such as copper is chosen, there will be no reaction.

$$Cu + HCl \rightarrow \text{No reaction}$$

The determination of these replacements using a quantitative method is covered in Chapter 12.

ACTIVITY SERIES OF COMMON ELEMENTS

	Activity of Metals		Activity of Halogen Nonmetals
Most active	Li Rb K Ba Sr Ca Na	Reacts with cold water and acids, replacing hydrogen. React with oxygen, forming oxides.	F_2 Cl_2 Br_2 I_2
	Mg Al Mn Zn Cr Fe Cd	React with steam (but not cold water) and acids, replacing hydrogen. Reacts with oxygen, forming oxides.	
	Cu Ni Sn Pb H_2 Sb Bi Cu Hg	Do not react with water. React with acids, replacing hydrogen. React with oxygen, forming oxides. React with oxygen, forming oxides.	
Least active	Ag Pt Au	Fairly unreactive, forming oxides only indirectly.	

DOUBLE REPLACEMENT

For **double replacement reactions** to go to completion, that is, proceed until the supply of one of the reactants is exhausted, one of the following conditions must be present: an insoluble precipitate is formed, a nonionizing substance is formed, or a gaseous product is given off.

To predict the formation of an insoluble precipitate, you should have some knowledge of the solubilities of compounds. Table 9 gives some general solubility rules. (A table of solubilities could also be used as reference.)

An example of this type of reaction is given in its complete ionic form.

$$(K^+ + Cl^-) + (Ag^+ + NO_3^-) \rightarrow AgCl(s) + (K^+ + NO_3^-)$$

The silver ions combine with the chloride ions to form an insoluble precipitate, silver chloride. If the reaction had been like

$$(K^+ + Cl^-) + (Na^+ + NO_3^-) \rightarrow K^+ + NO_3^- + Na^+ + Cl^-$$

only a mixture of the ions would have been shown in the final solution.

TABLE 9

SOLUBILITIES OF COMPOUNDS

Soluble	Except
Na^+ NH_4^+ } compounds K^+	
Acetates	
Bicarbonates	
Chlorates	
Chlorides	Ag^+, Hg^+, Pb ($PbCl_2$, sol. in hot water)
Nitrates	
Sulfates	Ba, Ca (slight), Pb
Insoluble	
Carbonates, phosphates	Na, NH_4, K compounds
Sulfides, hydroxides.	Na, NH_4, K, Ba, Ca

Another reason for a reaction of this type to go to completion is the formation of a nonionizing product such as water. This weak electrolyte keeps its component ions in molecular form and thus removes the possibility of reversing the reaction. All neutralization reactions are of this type.

$$(H^+ + Cl^-) + (Na^+ + OH^-) \rightarrow H_2O + Na^+ + Cl^-$$

This example shows the ions of the reactants, hydrochloric acid and sodium hydroxide, and the nonelectrolyte product, water with sodium and chloride ions in solution. Since the water does not ionize to any extent, the reverse reaction cannot occur.

The third reason for double displacement to occur is the evolution of a gaseous product. An example of this is calcium carbonate reacting with hydrochloric acid

$$CaCO_3 + 2HCl \rightarrow CaCl_2 + H_2O + CO_2(g)$$

Another example of a compound that evolves a gas in sodium sulfite with an acid is

$$Na_2SO_3 + 2HCl \rightarrow 2NaCl + H_2O + SO_2(g)$$

In general, acids with carbonates or sulfites are good examples of this type of equation.

HYDROLYSIS REACTIONS

Hydrolysis reactions are the opposite of neutralization reactions. In hydrolysis, the salt and water react to form an acid and a base. For example, if sodium chloride is placed in solution, this reaction occurs to some degree:

$$(Na^+ + Cl^-) + H_2O \rightarrow (Na^+ + OH^-) + (H^+ + Cl^-)$$

In this hydrolysis reaction, the same number of hydrogen ions and of hydroxide ions is released, and so the solution is neutral. This is because sodium hydroxide is a strong base and hydrochloric acid is a strong acid. (There is a chart of acid and base strengths on page 489 to use as a reference.) Because they are both classified as strong, it means that they essentially exist as ions in solutions. Therefore, there is an excess of neither hydrogen nor hydroxide ions in the solution and it will test neutral. So, the salt of a strong acid and strong base forms a neutral solution when dissolved in water. However, if Na_2CO_3 is dissolved, we have

$$(2Na^+ + CO_3^{2-}) + 2H_2O \rightarrow (2Na^+ + 2OH^-) + H_2CO_3$$

The H_2CO_3 is written together because it is a slightly ionized acid or, in other words, a weak acid. Since the hydroxide ions are free in the solution, the solution is basic. Notice that here it is the salt of a strong base and a weak acid that forms a basic solution. This generalization is true for this type of salt.

If we use the salt of a strong acid and a weak base, the reaction will be

$$(Zn^{2+} + 2Cl^-) + H_2O \rightarrow (2H^+ + 2Cl^-) + Zn(OH)_2$$

In this case the hydroxide ions are held in the weakly ionizing compound while the hydrogen ions are free to make the solution acidic. In general, then, the salt of a strong acid and a weak base forms an acid solution by hydrolysis.

The fourth possibility is that of a salt of a weak acid and a weak base dissolving in water. An example is ammonium carbonate, $(NH_4)_2CO_3$, which is the salt of a weak base and a weak acid. The hydrolysis reaction is

$$(NH_4)_2CO_3 + 2H_2O \rightarrow 2NH_4OH + 2H_2CO_3$$

Both the ammonium hydroxide, NH_4OH, and the carbonic acid, H_2CO_3, are written as nonionized compounds because they are classified as a weak base and a weak acid, respectively. Therefore, a salt of a weak acid and a weak base forms a neutral solution since neither hydrogen ion nor hydroxide ion will be present in excess.

ENTROPY

In many of the preceding predictions of reactions, we used the concept that reactions will occur when they result in the lowest possible energy state.

There is, however, another driving force for reactions that relates to their state of disorder or of randomness. This state of disorder is called **entropy**. A reaction is also driven, then, by a need for a greater degree of disorder. An example is the intermixing of gases in two connected flasks when a valve is opened to allow the two previously isolated gases to travel between the two flasks. Since temperature remains constant throughout the process, the total heat content cannot have changed to a lower energy level, and yet the gases will become evenly distributed in the two flasks. The system has thus reached a higher degree of disorder or entropy. A more substantive treatment of entropy is given on pages 226–227.

Thermochemistry

In general, all chemical reactions either liberate or absorb heat. The origin of chemical energy lies in the position and motion of atoms, molecules, and subatomic particles. The total energy possessed by a molecule is the sum of all the forms of potential and kinetic energy associated with it.

The energy changes in a reaction are due, to a large extent, to the changes in potential energy that accompany the breaking of chemical bonds in reactants to form new bonds in products.

The molecule may also have rotational, vibrational, and translational energy, along with some nuclear energy sources. All these make up the total energy of molecules. In beginning chemistry, the greatest concern in reactions is the electronic energy involved in the making and breaking of chemical bonds.

Since it is virtually impossible to measure the total energy of molecules, the energy change is usually the experimental data that we deal with in reactions. This change in quantity of energy is known as the change in **enthalpy** (heat content) of the chemical system and is symbolized by ΔH.

Changes in Enthalpy

Changes in enthalpy for exothermic and endothermic reactions can be shown graphically.

Notice that the ΔH for an endothermic reaction is positive, while that for an exothermic reaction is negative. It should be noted also that changes in enthalpy are always independent of the path taken to change a system from the initial state to the final state.

Since the quantity of heat absorbed or liberated during a reaction varies with the temperature, scientists have adopted 25°C and 1 atm pressure as the *standard state* condition for reporting heat data. A superscript zero on ΔH (i.e., ΔH^0) indicates that the corresponding process is carried out under standard conditions. The **standard enthalpy**

of formation (ΔH_f^0) of a compound is defined as the change in enthalpy that accompanies the formation of 1 mol of a compound from its elements with all substances in their standard states at 25°C. This is called the **molar heat of reaction**.

To calculate the enthalpy of a reaction, it is necessary to write an equation for the reaction. The standard enthalpy change, ΔH, for a given reaction is usually expressed in kilojoules and depends on how the equation is written. For example, here is the equation for the reaction of hydrogen with oxygen written in two ways

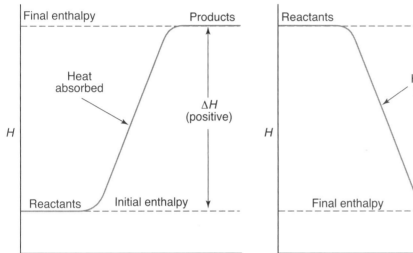

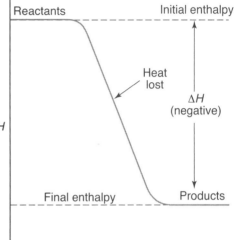

Course of Endothermic Reaction   **Course of Exothermic Reaction**

$$H_2(g) + \tfrac{1}{2}O_2(g) \rightarrow H_2O(g) \qquad \Delta H_f^0 = -241.8 \text{ kJ}$$
$$2H_2(g) + O_2(g) \quad \rightarrow 2H_2O(g) \qquad \Delta H_f^0 = -483.6 \text{ kJ}$$

Experimentally, ΔH_f^0 for the formation of 1 mol of $H_2O(g)$ is –241.8 kJ. Since the second equation represents the formation of 2 mol of $H_2O(g)$, the quantity is twice –241.8 or –483.6 kJ. It is assumed that the initial and final states are measured at 25°C and 1 atm, although the reaction occurs at a higher temperature.

EXAMPLE   How much heat is liberated when 40.0 g of $H_2(g)$ reacts with excess $O_2(g)$?

SOLUTION

The reaction equation is

$$H_2(g) + \tfrac{1}{2}O_2(g) \rightarrow H_2O(g) \qquad \Delta H_f^0 = -241.8 \text{ kJ}$$

This represents 1 mol or 2 g of H(g) forming 1 mol of $H_2O(g)$

$$40 \ \cancel{g} \times \frac{1 \text{ mol}}{2 \ \cancel{g}} = 20 \text{ mol of hydrogen}$$

Since each mole gives off -241.8 kJ, then

$$20 \text{ mol} \times \frac{-241.8 \text{ kJ}}{1 \text{ mol}} = -4{,}836 \text{ kJ}$$

Notice that the physical state of each participant must be given since the phase changes involve energy changes.

Combustion reactions produce a considerable amount of energy in the form of light and heat when a substance is combined with oxygen, The heat released by the complete combustion of one mole of a substance is called the **heat of combustion** of the substance. Heat of combustion is defined in terms of one mole of reactant, whereas heat of formation is defined in terms of one mole of product. All substances are in their standard state. The general enthalpy notation, ΔH, applies to heats of reaction, but with the addition of a subscripted c, ΔH_c, refers specifically to heat of combustion.

Additivity of Reaction Heats and Hess's Law

Chemical equations and ΔH^0 values may be manipulated algebraically. Finding the ΔH for the formation of vapor from liquid water shows how this can be done.

$$H_2(g) + \tfrac{1}{2}O_2(g) \rightarrow H_2O(g) \qquad \Delta H_f^0 = -241.8 \text{ kJ}$$
$$H_2(g) + \tfrac{1}{2}O_2(g) \rightarrow H_2O(\ell) \qquad \Delta H_f^0 = -285.8 \text{ kJ}$$

Since we want the equation for $H_2O(\ell) \rightarrow H_2O(g)$, we can reverse the second equation. This changes the sign of ΔH.

$$H_2O(\ell) \rightarrow H_2O(g) + \tfrac{1}{2}O_2(g) \qquad \Delta H_f^0 = +285.8 \text{ kJ}$$

Adding

$$H_2(g) + \tfrac{1}{2}O_2(g) \rightarrow H_2O(g) \qquad \Delta H_f^0 = -241.8 \text{ kJ}$$

yields

$$H_2O(\ell) + H_2(g) + \tfrac{1}{2}O_2(g) \rightarrow$$
$$H_2(g) + \tfrac{1}{2}O_2(g) + H_2O(g) \qquad \Delta H_f^0 = +44.0 \text{ kJ}$$

Simplification gives a net equation of

$$H_2O(\ell) \rightarrow H_2O(g) \qquad \Delta H_f^0 = +44.0 \text{ kJ}$$

The principle underlying the preceding calculations is known as **Hess's Law of Heat Summation**. This principle states that when a reaction can be expressed as the algebraic sum of two or more other reactions, the heat of the reaction is the algebraic sum of the heats of these reactions. This is based upon the **First Law of Thermodynamics**, which, simply stated, says that the total energy of the universe is constant and cannot be created or destroyed.

These laws allow calculations of ΔH values that cannot be easily determined experimentally. An example is determination of the ΔH of CO from the ΔH_f^0 of CO_2.

$$C(s) + \tfrac{1}{2}O_2(g) \rightarrow CO(g) \qquad \Delta H = ?$$

The calculation of ΔH is done in kJ/mol.

$$C(s) + O_2(g) \rightarrow CO_2(g) \qquad \Delta H_f^0 = -393.5 \text{ kJ/mol}$$
$$CO(g) + \tfrac{1}{2}O_2(g) \rightarrow CO_2(g) \qquad \Delta H_f^0 = -283.0 \text{ kJ/mol}$$

The equation we want is

$$C(s) + \tfrac{1}{2}O_2(g) \rightarrow CO(g)$$

To get this, we reverse the second equation and add it to the first:

$$C(s) + O_2(g) \rightarrow CO_2(g) \qquad \Delta H_f^0 = -393.5 \text{ kJ/mol}$$
$$CO_2(g) \rightarrow \tfrac{1}{2}O_2(g) + CO(g) \qquad \Delta H_f^0 = +283.0 \text{ kJ/mol}$$

Addition yields:

$$C(s) + \tfrac{1}{2}O_2(g) \rightarrow CO(g) \qquad \Delta H_f^0 = -110.5 \text{ kJ/mol}$$

The relationship can be shown schematically as given below.

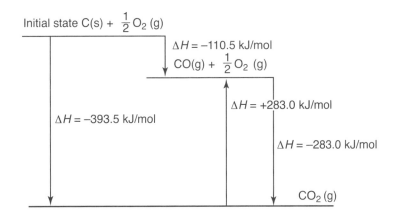

Some commonly used standard heats of formation (enthalpy), designated as ΔH_f^0, are listed in Table Ⓔ in the Equations and Tables for Reference section.

An alternative (and easier) method of calculating enthalpies is based on the concept that $\Delta H^0_{reaction}$ is equal to the difference between the total enthalpy of the reactants and that of the products. This can be expressed as

$$\Delta H^0_{reaction} = \Sigma \text{ (sum of) } \Delta H^0_f \text{ (products)} - \Sigma \text{ (sum of) } \Delta H^0_f \text{(reactants)}$$

EXAMPLE 1

Calculate $\Delta H^0_{reaction}$ for the decomposition of sodium chlorate.

SOLUTION

$$NaClO_3(s) \rightarrow NaCl(s) + \tfrac{3}{2}O_2(g)$$

Step 1: Obtain ΔH^0_f for all substances.

$$
\begin{aligned}
NaClO_3(s) &= -358.2 \text{ kJ} \\
NaCl(s) &= -410.5 \text{ kJ} \\
O_2(g) &= 0 \text{ kJ/mol (all elements} = 0)
\end{aligned}
$$

Step 2: Substitute these values in the equation.

$$
\begin{aligned}
\Delta H^0_{reaction} &= \Sigma \Delta H^0_f \text{ (products)} - \Sigma \Delta H^0_f \text{ (reactants)} \\
\Delta H_{reaction} &= -410.5 - (-358.2) \text{ kJ} \\
\Delta H_{reaction} &= -52.2 \text{ kJ}
\end{aligned}
$$

EXAMPLE 2

Calculate $\Delta H_{reaction}$ for this oxidation of ammonia:

$$4NH_3(g) + 5O_2(g) \rightarrow 6H_2O(g) + 4NO(g)$$

SOLUTION

The individual ΔH^0_f values are

$$
\begin{aligned}
4NH_3 &= 4(-45.9 \text{ kJ}) &= -183.9 \text{ kJ} \\
5O_2(g) &= 0 \\
6H_2O(g) &= 6(-241.8 \text{ kJ}) &= -1450.8 \text{ kJ} \\
4NO(g) &= 4(90.29) &= 361.16 \text{ kJ}
\end{aligned}
$$

Substituting these in the $\Delta H_{reaction}$ equation gives

$$
\begin{aligned}
\Delta H_{reaction} &= [-1450.8 \, (6H_2O) + 361.16 \, (4NO)] - [-183.9 \, (4NH_3) + 0 \, (5O_2)] \\
\Delta H_{reaction} &= -905.7 \text{ kJ}
\end{aligned}
$$

Bond Dissociation Energy

The same principle of additivity applies to bond energies. Experimentation has found average bond energies for particular bonds; some of the more common are shown in the chart.

Bond	Energy (kJ/mol)	Bond	Energy (kJ/mol)
H—H	432	O=O	495
C—H	413	O—H	467
C—C	347	C—O	358
C=C	614	C=O	745
C≡C	839	H—F	565
O—O	146	H—Cl	427
N—N	160	H—Br	363
Cl—Cl	239	H—I	295

EXAMPLE Find the bond energy, H_{be}, for ethane.

SOLUTION

$$\text{Ethane} = \begin{array}{ccc} & \text{H} & \text{H} \\ & | & | \\ \text{H}- & \text{C}-\text{C} & -\text{H} \\ & | & | \\ & \text{H} & \text{H} \end{array} \text{ and has 6 C—H bonds and 1 C—C bond.}$$

$$H_{be} = 6E_{C-H} + 1E_{C-C} = 6(413 \text{ kJ}) + 1(347 \text{ kJ}) = 2825 \text{ kJ}$$

Enthalpy from Bond Energies

A reaction's enthalpy can be approximated through the summation of bond energies.

An example is the reaction

$$H_2(g) + Cl_2(g) = 2HCl(g)$$

Bonds Broken	Heat Energy Absorbed (kJ)	Bonds Formed	Heat Energy Evolved (kJ)
H—H	432 kJ		
Cl—Cl	239 kJ	2H—Cl	2 (427) = 854
Total	671 kJ		854 kJ

The difference between heat evolved and heat absorbed is $854 - 671 = 183.0$ kJ. This is for 2 mol of HCl, and so -183.0 kJ divided by $2 = -91.5$ kJ/mol. This answer is very close to the experimentally determined ΔH_f^0.

Chapter Summary

The following terms summarize all the concepts and ideas that were introduced in this chapter. You should be able to explain their meaning and how you would use them in chemistry. They appear in boldface type in this chapter to draw your attention to them. The boldface type also makes the terms easier for you to look up if you need to. You could also use a search engine on your computer to get a quick and expanded explanation of these terms, laws, and formulas.

TERMS YOU SHOULD KNOW

combination reaction
decomposition reaction
double replacement reaction
enthalpy
entropy
First Law of Thermodynamics
heat of combustion

heat of formation
Hess's Law of Heat Summation
hydrolysis reaction
molar heat of formation
reaction mechanism
single replacement reaction
standard enthalpy of formation

Chapter 8 Review Exercises

1. A synthesis reaction will occur spontaneously if the heat of formation of the product is

    (A) large and negative
    (B) small and negative
    (C) large and positive
    (D) small and positive

2. The reaction of aluminum with dilute H_2SO_4 can be classified as

    (A) synthesis
    (B) decomposition
    (C) single replacement
    (D) double replacement

3. For a metal atom to replace another kind of metallic ion in a solution, the metal atom must be

    (A) a good oxidizing agent
    (B) higher on the electromotive chart than the metal in solution
    (C) lower on the electromotive chart than the metal in solution
    (D) equal in activity to the metal in solution

4. One reason for a double displacement reaction to go to completion is that

   (A) a product is soluble
   (B) a product is given off as a gas
   (C) the products can react with each other
   (D) the products are miscible

5. Hydrolysis will give an acid reaction when which of these is placed in solution with water?

   (A) Na_2SO_4
   (B) K_2SO_4
   (C) $NaNO_3$
   (D) $Cu(NO_3)_2$

6. A salt derived from a strong base and a weak acid will undergo hydrolysis and give a solution that will be

   (A) basic
   (B) acid
   (C) neutral
   (D) volatile

7. Enthalpy is an expression for the

   (A) heat content
   (B) energy state
   (C) reaction rate
   (D) activation energy

8. The ΔH_f^0 of a reaction is recorded for

   (A) 0°C
   (B) 25°C
   (C) 100°C
   (D) 200°C

9. The property of being able to add enthalpies is based on the

   (A) Law of Conservation of Heat
   (B) First Law of Thermodynamics
   (C) Law of Constants
   (D) Law of $E = mc^2$

10. If $\Delta H_{reaction}$ is -100 kcal/mol, it indicates the reaction is

   (A) endothermic
   (B) unstable
   (C) in need of a catalyst
   (D) exothermic

11. The algebraic sign for the expression of the heat of reaction for an endothermic reaction is

   (A) +
   (B) –
   (C) either + or –
   (D) dependent on the amount of activation energy needed

12. The type of reaction exemplified by the burning of carbon with an excess of oxygen is

   (A) synthesis
   (B) decomposition
   (C) single replacement
   (D) double replacement

13. That the heat of reaction can be arrived at by the algebraic summation of two or more thermal reactions is explained by

   (A) entropy theory
   (B) enthalpy theory
   (C) combination theory
   (D) Hess's Law

14. Entropy can be described as

   (A) the state of disorder of a system
   (B) the heat of formation released by a reaction
   (C) the heat of formation absorbed by a system
   (D) the hydrolysis reaction of a system

15. The reaction in which HgO is heated to release O_2 is called

   (A) synthesis theory
   (B) decomposition
   (C) single replacement
   (D) double replacement

Answers and Explanations

1. **(A)** A large negative heat of formation indicates that the reaction will give off a large amount of energy and self-perpetuate itself after getting the activation energy necessary for it to start. An example is burning paper.

2. **(C)** Since the aluminum replaces the hydrogen to form aluminum sulfate, this is classified as a single replacement reaction.

3. **(B)** For a metal to replace another metallic ion in a solution, it must be higher in the activity series of elements than the metal in solution.

4. **(B)** For a double displacement to go to completion, a product or products must either deposit as a precipitate or leave the reaction as a gas.

5. **(D)** The only salt that hydrolyzes to a strong acid and a weak base is $Cu(NO_3)_2$. This would give an acid solution. All the others are salts of strong acids and strong bases, so they would not.

6. **(A)** Salts of strong bases and weak acids react with water to give a basic solution. An example would be Na_2CO_3 because it hydrolyzes into NaOH, a strong base, and H_2CO_3, a weak acid.

7. **(A)** Enthalpy is defined as the heat content of a system.

8. **(B)** The ΔH^0 with the superscript zero indicates that the process was carried out under standard conditions, which are 25°C and 1 atmosphere pressure.

9. **(B)** The property of being able to add enthalpies comes from Hess's Law of Heat Summation and is based upon the First Law of Thermodynamics, which simply says that the total energy of the universe is constant and cannot be created nor destroyed.

10. **(D)** Since the heat of the reaction is negative, energy (heat) is given off when this reaction occurs.

11. **(A)** The convention in chemistry is that a negative heat of formation is associated with an exothermic reaction. This means that energy is released as a product of the reaction.

12. **(A)** The combining of carbon and oxygen to form carbon dioxide exemplifies a combination reaction that is also referred to as a synthesis reaction.

13. **(D)** Hess's Law is the basis for arriving at a heat of formation for a reaction by algebraically adding two or more thermal-reaction equations.

14. **(A)** Entropy is a measure of a system's randomness of disorder. The greater the randomness, the greater the entropy.

15. **(B)** The reaction that shows HgO producing Hg and O_2 is an example of a decomposition (or analysis) reaction. The equation is

$$2\ HgO(s) \rightarrow 2\ HgO(s) + O_2(g)$$

RATES OF CHEMICAL REACTIONS

CHAPTER OBJECTIVES

Upon completing this chapter, you will be able to:

- Explain how each of the following factors affect the rate of a chemical reaction: nature of the reactants, surface area exposed, concentrations, temperature, and the presence of a catalyst

- Describe collision theory

- Draw reaction diagrams with and without a catalyst

- Explain the Law of Mass Action

- Describe the relationship between reaction mechanisms and rates of reaction

Measurements of Reaction Rates

The measurement of reaction rate is based on the rate of appearance of a product or disappearance of a reactant. It is usually expressed in terms of a change in concentration of one of the participants per unit time.

Experiments have shown that for most reactions the concentrations of all participants change most rapidly at the beginning of the reaction, that is, the concentration of the product shows the greatest rate of increase, and the concentrations of the reactants the highest rate of decrease, at this point. This means that the rate of a reaction changes with time. Therefore, a rate must be identified with a specific time.

Factors Affecting Reaction Rates

Five important factors control the rate of a chemical reaction. These are summarized below.

1. ***The nature of the reactants.*** In chemical reactions, some bonds are broken and others are formed. Therefore, the rates of chemical reactions should be affected by the nature of the bonds in the reacting substances. For example, reactions between ions in an aqueous solution may take place in a fraction of a second. Thus, the reaction between silver nitrate and sodium chloride is very fast. The appearance of the white silver chloride precipitate is immediate. In reactions where many covalent bonds must be broken, reactions usually take place slowly at room temperatures. The decomposition of hydrogen peroxide into water and oxygen happens slowly at room temperature. In fact, it takes about 17 minutes (min) for half the peroxide in a 0.50 M solution to decompose.

2. ***The surface area exposed.*** Since most reactions depend on the reactants coming into contact, the surface exposed proportionally affects the rate of the reaction. Consider what happens to the surface area exposed when a cube is sliced in half.

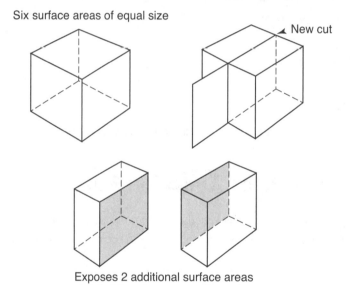

Six surface areas of equal size

New cut

Exposes 2 additional surface areas
or a total of 8 surface areas of equal area.

There is an increase of $33\frac{1}{3}\%$ of surface area that can now become involved in a chemical reaction. This is why reactants that depend on surface reactions are often introduced into reaction vessels as powders. It also explains the danger of a dust explosion if the reactant is flammable.

3. ***The concentrations.*** The reaction rate is usually proportional to the concentrations of the reactants. The usual dependence of the reaction rate on the concentration of the reactants can be simply explained by theorizing that if there are more

molecules or ions of the reactant in the reaction area, then there is a greater chance that more reactions will occur. This idea is further developed in the collision theory described below.

4. *The temperature.* A temperature increase of 10°C above room temperature usually causes the reaction rate to double or triple. The basis for this generality is that as the temperature increases, the average kinetic energy of the particles involved increases. This means that the particles move faster and have a greater probability of hitting another reactant particle and, because they have more energy, causing an effective collision that results in the chemical reaction that forms the product substance.

5. *The catalyst.* A substance that increases the rate of chemical reaction without itself undergoing any permanent chemical change. The catalyst provides an alternative pathway by which the reaction can proceed and in which the activation energy is lower. It thus increases the rate at which the reaction comes to completion or equilibrium. Generally, the term is used for a substance that increases the reaction rate (a positive catalyst). Some reactions can be slowed down by negative catalysts.

Collision Theory of Reaction Rates

This theory makes the assumption that, for a reaction to occur, there must be collisions between the reacting species. This means that the rate of reaction depends on two factors: the number of collisions per unit time, and the fraction of these collisions that are successful because enough energy is involved.

There is a definite relationship between the concentrations of the reactants and the number of collisions. Graphically, the reaction of

$$A + B \rightarrow AB$$

can be examined as the concentrations are changed.

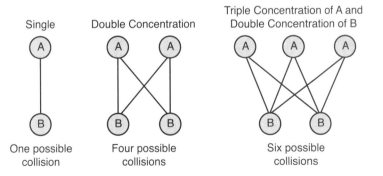

This shows that the number of collisions, and consequently the rate of reaction, are proportional to the product of the concentrations. Simply stated, then, the rate of reaction is directly proportional to the concentrations.

Activation Energy

Often a reaction rate may be increased or decreased by affecting the activation energy, that is, the energy necessary to cause a reaction to occur. This is shown graphically below.

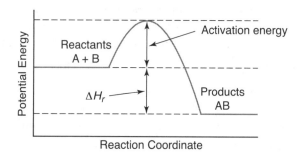

A **catalyst** is a substance that is introduced into a reaction to speed up the reaction. This is accomplished by changing the amount of activation energy needed. The effect of a catalyst used to speed up a reaction can be shown as

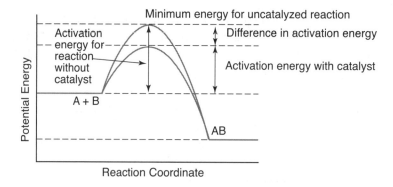

Reaction Rate Law

The relationship between the rate of a reaction and the masses (expressed as concentrations) of the reacting substances is summarized in the **Law of Mass Action**. It states that the rate of a chemical reaction is proportional to the product of the concentrations of the reactants. For a general reaction between A and B, represented by

$$a\text{A} + b\text{B} \rightarrow \cdots$$

the rate law expression is

$$r \propto [\text{A}]^{a}[\text{B}]^{b}$$

or, inserting a constant of proportionality that mathematically changes the expression to an equality, we have

$$r = k[A]^a[B]^b$$

Here k is called the *specific rate* constant for the reaction at the temperature of the reaction.

The exponents a and b may be added to give the total reaction order. For example,

$$H_2(g) + I_2(g) \rightarrow 2HI(g)$$
$$r = k[H_2]^1[I_2]^1$$

The sum of the exponents is $1 + 1 = 2$, and therefore we have a second-order reaction.

REACTION MECHANISM AND RATES OF REACTION

Under the factors listed as affecting the rates of reaction, it was stated that the reaction rate is *usually* proportional to the concentration of the reactants. The reason for this is that some reactions do not occur directly between the reactants but may go through intermediate steps to get to the final product. The series of steps by which the reacting particles rearrange themselves to form the products of a chemical reaction is called the **reaction mechanism**. For example,

Step 1:	$A + B$	$\rightarrow I_1$ (fast)
Step 2:	$A + I_1$	$\rightarrow I_2$ (slow)
Step 3:	$C + I_2$	$\rightarrow D$ (fast)
Net equation:	$2A + B + C \rightarrow D$	

Notice that the reactions in Steps 1 and 3 occur relatively fast compared to the reaction in Step 2. Now suppose that we increase the concentration of C. This will make the reaction in Step 3 go faster, but it will have little effect on the speed of the overall reaction since Step 2 is the rate-determining step. If, however, the concentration of A is increased, then the overall reaction rate will increase because it will speed up Step 2. Knowing the reaction mechanism provides the basis for predicting the effect of a concentration change of a reactant on the overall rate of reaction. Another way of determining the effect of concentration changes is to do actual experimentation.

Chapter Summary

The following terms summarize all the concepts and ideas that were introduced in this chapter. You should be able to explain their meaning and how you would use them in chemistry. They appear in boldface type in this chapter to draw your attention to them. The boldface type also makes the terms easier for you to look up if you need to. You could also use a search engine on your computer to get a quick and expanded explanation of these terms, laws, and formulas.

TERMS YOU SHOULD KNOW

catalyst

collision theory

Law of Mass Action

reaction mechanism

Chapter 9 Review Exercises

1. List the five factors that affect the rate of a reaction:

    (1) _____

    (2) _____

    (3) _____

    (4) _____

    (5) _____

2. The addition of a catalyst to a reaction

    (A) changes the enthalpy
    (B) changes the entropy
    (C) changes the nature of the products
    (D) changes the activation energy

3. An increase in concentration

    (A) is related to the number of collisions directly
    (B) is related to the number of collisions inversely
    (C) has no effect on the number of collisions
    (D) makes the collisions that occur more effective

4. At the beginning of a reaction, the reaction rate for the reactants is

   (A) largest, then decreasing
   (B) largest and remains constant
   (C) smallest, then increasing
   (D) smallest and remains constant

5. The reaction rate law applied to the reaction $aA + bB \rightarrow AB$ gives the expression

   (A) $r \propto [A]^b[B]^a$
   (B) $r \propto [AB]^a[A]^b$
   (C) $r \propto [B]^a[AB]^b$
   (D) $r \propto [A]^a[B]^b$

Answers and Explanations

1. (1) Nature of the reactants
   (2) Surface area exposed
   (3) Concentrations
   (4) Temperature
   (5) Presence of a catalyst

2. **(D)** When a catalyst is added to a reaction, it decreases the activation energy needed for the reaction to occur. When the activation energy is decreased, the reaction occurs at a faster rate.

3. **(A)** When the concentration of reactants is increased, the number of collisions between reactants is directly affected since more reactant units (ions, atoms, or molecules) are available.

4. **(A)** At the beginning of a reaction, the reaction rate of the reactants is the highest because their concentration is the highest. As the reaction progresses, the concentration of the reactants decreases while the concentration of the products increases.

5. **(D)** For the reaction $aA + bB \rightarrow AB$, the reaction rate law shows r is proportional to $[A]^a[B]^b$.

CHEMICAL EQUILIBRIUM

CHAPTER OBJECTIVES

Upon completing this chapter, you will be able to:

- Explain the development of an equilibrium condition, explain how it is expressed as an equilibrium constant, and use it mathematically

- Describe Le Châtelier's Principle and how changes in temperature, pressure, and concentrations affect an equilibrium

- Solve problems dealing with ionization of water, finding the pH, solubility products, and the common ion effect

- Explain the relationship of enthalpy and entropy as driving forces in a reaction and how they are combined in the Gibbs Equation

Reversible Reactions and Equilibrium

In some reactions, no product is formed to allow the reaction to go to completion; that is, the reactants and products can still interact in both directions. This can be shown as

$$A + B \rightleftharpoons C + D$$

The double arrow indicates that C and D can react to form A and B, while A and B react to form C and D.

The reaction is said to have reached **equilibrium** when the forward reaction rate is equal to the reverse reaction rate. Notice that this is a dynamic condition, *not* a static one, although in appearance the reaction *seems* to have stopped. An example of an equilibrium is a crystal of copper sulfate in a saturated solution of copper sulfate. Although to the observer the crystal seems to remain unchanged, there is actually an equal exchange of crystal material with the copper sulfate in solution. As some solute comes out of solution, an equal amount goes into solution.

To express the rate of reaction in numerical terms, we can use the **Law of Mass Action**, discussed in Chapter 9, which states: The rate of a chemical reaction is proportional to the product of the concentrations of the reacting substances. The concentrations are expressed in moles of gas per liter of volume or moles of solute per liter of solution. Suppose, for example, that 1 mol/L of gas A_2 (diatomic molecule) is allowed to react with 1 mol/L of another diatomic gas, B_2, and they form gas AB; let R be the rate for the forward reaction forming AB. The bracketed symbols $[A_2]$ and $[B_2]$ represent the concentrations in moles per liter for these diatomic molecules. Then $A_2 + B_2 \rightarrow 2AB$ has the rate expression

$$R \propto [A_2] \times [B_2]$$

where $\propto$ is the symbol for "proportional to." When $[A_2]$ and $[B_2]$ are both 1 mol/L, the reaction rate is a certain constant value (k_1) at a fixed temperature.

$$R = k_1 \qquad (k_1 \text{ is called the rate constant})$$

For any concentrations of A and B, the reaction rate is

$$R = k_1 \times [A_2] \times [B_2]$$

If $[A_2]$ is 3 mol/L and $[B_2]$ is 2 mol/L, the equation becomes

$$R = k_1 \times 3 \times 2 = 6k_1$$

The reaction rate is six times the value for a 1 mol/L concentration of both reactants.

At the fixed temperature of the forward reaction, AB molecules are also decomposing. If we designate this reverse reaction as R', then, since

$$2AB \text{ (or } AB + AB) \rightarrow A_2 + B_2$$

two molecules of AB must decompose to form one molecule of A_2 and one of B_2. Thus the reverse reaction in this equation is proportional to the square of the molecular concentration of AB.

$$R' \propto [AB] \times [AB]$$
$$\text{or } R' \propto [AB]^2$$
$$\text{and } R' \propto k_2 \times [AB]^2$$

where k_2 represents the rate of decomposition of AB at the fixed temperature. Both reactions can be shown

$$A_2 + B_2 \rightleftharpoons 2AB \qquad \text{(note double arrows)}$$

When the first reaction begins to produce AB, some AB is available for the reverse reaction. If the initial condition is only the presence of A_2 and B_2 gases, then the forward reaction will occur rapidly to produce AB. As the concentration of AB increases, the reverse reaction will increase. At the same time, the concentrations of A_2 and B_2 will be decreasing and consequently the forward reaction rate will decrease. Eventually the two rates will be equal, that is, $R = R'$. At this point, equilibrium has been established, and

$$k_1[A_2] \times [B_2] = k_2[AB]^2$$

or

$$\frac{k_1}{k_2} = \frac{[AB]^2}{[A_2] \times [B_2]} = K_{eq}$$

The convention is that k_1 (forward reaction) is placed over k_2 (reverse reaction) to obtain this expression. Then k_1/k_2 can be replaced by K_{eq}, which is called the **equilibrium constant** for this reaction under the particular conditions.

In another general example, like

$$aA + bB \rightleftharpoons cC + dD$$

the reaction rates can be expressed as

$$R = k_1[A]^a \times [B]^b$$
$$R' = k_2[C]^c \times [D]^d$$

Note that the values of k_1 and k_2 are different, but that each is a constant for the conditions of the reaction. At the start of the reaction, [A] and [B] will be at their greatest values, and R will be large; [C], [D], and R' will be zero. Gradually R will decrease, and R' will become equal. At this point the reverse reaction is forming the original reactants just as rapidly as they are being used up by the forward reaction. Therefore, no further change in R, R', or any of the concentrations will occur.

If we set R' equal to R, we have

$$k_2 \times [C]^c \times [D]^d = k_1 \times [A]^a \times [B]^b$$

or

$$\frac{[C]^c \times [D]^d}{[A]^a \times [B]^b} = \frac{k_1}{k_2} = K_{eq}$$

This process of two substances A and B reacting to form products C and D, and the reverse, can be shown graphically to visualize what happens as equilibrium is established. This hypothetical equilibrium reaction is described by the following general equation:

$$aA + bB \rightleftharpoons cC + dD$$

At the beginning (at time t_0), the concentrations of C and D are zero and those of A and B are maximum. The following graph shows that over time the rate of the forward reaction decreases as A and D are used up. Meanwhile, the rate of the reverse reaction increases as

C and D are formed. When these two reaction rates become equal (at time t_1), equilibrium is established. The individual concentrations of A, B, C, and D no longer change if the conditions remain the same. To an observer, it appears that all reaction has stopped, when in fact both reactions are occurring at the same rate.

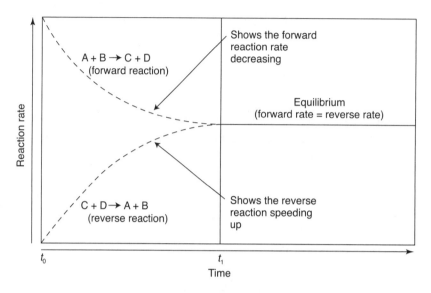

Rate Comparison for the Equilibrium Reaction System

We see that for the given reaction and the given conditions, K_{eq} is a constant, called the **equilibrium constant**. If K_{eq} is large, this means that equilibrium will not occur until the concentrations of the original reactants are small and those of the products large. A small value of K_{eq} means that equilibrium occurs almost at once and relatively little product is produced.

The equilibrium constant, K_{eq}, has been determined experimentally for many reactions, and the values are printed in chemical handbooks. (See Tables ① and ① in the Equations and Tables for Reference section.)

Suppose we find the K_{eq} for reacting H_2 and I_2 at 490°C to be equal to 45.9. Then the equilibrium constant for the reaction

$$H_2 + I_2 \rightleftharpoons 2HI \text{ at } 490°C$$

is

$$K_{eq} = \frac{[HI]^2}{[H_2][I_2]} = 45.9$$

Three moles of H_2 and 3 mol of I_2 are introduced into a 1 L box at a temperature of 490°C. Find the concentration of each substance in the box when equilibrium is established.

SOLUTION

Initial conditions:

$$[H_2] = 3 \text{ mol/L}$$
$$[I_2] = 3 \text{ mol/L}$$
$$[HI] = 0 \text{ mol/L}$$

The reaction proceeds to equilibrium and

$$K_{eq} = \frac{[HI]^2}{[H_2][I_2]} = 45.9$$

At equilibrium, then,

$[H_2]$ $= (3 - x)$ moles/liter (where x is the number of moles of H_2 that are in the form of HI at equilibrium)

$[I_2]$ $= (3 - x)$ moles/liter (the same x is used since 1 mol of H_2 requires 1 mol of I_2 to react to form 2 mol of HI)

$[HI]$ $= 2x$ moles/liter

and so

$$K_{eq} = \frac{(2x)^2}{(3 - x)(3 - x)} = 45.9$$

If

$$\frac{(2x)^2}{(3 - x)^2} = 45.9$$

then taking the square root of each side gives

$$\frac{2x}{3 - x} = 6.77$$

Solving for x,

$$x = 2.32$$

Substituting this x value into the concentration expressions at equilibrium, we have

$[H_2]$ $= (3 - x) = 0.68$ mol/L
$[I_2]$ $= (3 - x) = 0.68$ mol/L
$[HI]$ $= 2x$ $= 4.64$ mol/L

The crucial step in this type of problem is setting up the concentration expressions from your knowledge of the equation. Suppose that this problem had been as follows.

Find the concentrations at equilibrium for the same conditions as in the preceding example except that only 2 mol of HI are injected into the box.

$$[H_2] = 0 \text{ mol/L}$$
$$[I_2] = 0 \text{ mol/L}$$
$$[HI] = 2 \text{ mol/L}$$

At equilibrium

$$[HI] = (2 - x) \text{ moles/liter}$$

(For every mole of HI that decomposes, only 0.5 mol of H_2 and 0.5 mol of I_2 are formed.)

$$[H_2] = 0.5x$$
$$[I_2] = 0.5x$$

$$K_{eq} = \frac{(2 - x)^2}{(x/2)^2} = 45.9$$

Solving for x gives

$$x = 0.456$$

Then, substituting into the equilibrium conditions,

$$[HI] = 2 - x = 1.54 \text{ mol/L}$$

$$[I_2] = \tfrac{1}{2}x = 0.228 \text{ mol/L}$$

$$[H_2] = \tfrac{1}{2}x = 0.228 \text{ mol/L}$$

Le Châtelier's Principle

A general law, **Le Châtelier's Principle**, can be used to explain the results of applying any change of condition (stress) on a system in equilibrium. It states that if a stress is placed upon a system in equilibrium, the equilibrium is displaced in the direction that counteracts the effect of the stress. An increase in concentration of a substance favors the reaction that uses up that substance and lowers its concentration. A rise in temperature favors the reaction that absorbs heat and so tends to lower the temperature. These ideas are further developed in the following text.

Effects of Changing Conditions

EFFECT OF CHANGING CONCENTRATIONS

When a system at equilibrium is disturbed by adding or removing one of the substances (thus changing its concentration), all the concentrations will change until a new equilibrium point is reached with the same value of K_{eq}.

If the concentration of a reactant in the forward reaction is increased, the equilibrium is displaced to the right, favoring the forward reaction. If the concentration of a reactant in the reverse reaction is increased, the equilibrium is displaced to the left. Decreases in concentration will produce effects opposite to those produced by increases.

EFFECT OF TEMPERATURE ON EQUILIBRIUM

If the temperature of a given equilibrium reaction is changed, the reaction will shift to a new equilibrium point. If the temperature of a system in equilibrium is raised, the equilibrium is shifted in the direction that absorbs heat. Note that the shift in equilibrium as a result of temperature change is actually a change in the value of the equilibrium constant. This is different from the effect of changing the concentration of a reactant; when concentrations are changed, the equilibrium shifts to a condition that maintains the same equilibrium constant.

EFFECT OF PRESSURE ON EQUILIBRIUM

A change in pressure affects only equilibria in which a gas or gases are reactants or products. Le Châtelier's Principle can be used to predict the direction of displacement. If it is assumed that the total space in which the reaction occurs is constant, the pressure will depend on the total number of molecules in that space. An increase in the number of molecules will increase pressure; a decrease in the number of molecules will decrease pressure. If the pressure is increased, the reaction that will be favored is the one that will lower the pressure, that is, decrease the number of molecules.

An example of the application of these principles is the Haber process of making ammonia. The reaction is

$$N_2 + 3H_2 \rightleftharpoons 2NH_3 + \text{heat (at equilibrium)}$$

If the concentrations of the nitrogen and hydrogen are increased, the forward reaction is increased. At the same time, if the ammonia produced is removed by dissolving it in water, the forward reaction is again favored.

Since the reaction is exothermic, the addition of heat must be considered with care. Increasing the temperature causes an increase in molecular motion and collisions, thus allowing the product to form more readily. At the same time, the equilibrium equation shows that the reverse reaction is favored by the increased temperature, and so a compromise temperature of about 500°C is used to get the best yield.

An increase in pressure will cause the forward reaction to be favored since the equation shows that four molecules of reactants form two molecules of products. This effect tends to reduce the increase in pressure by the formation of more ammonia.

Equilibria in Heterogeneous Systems

The examples so far have been of systems made up of only gaseous substances. The expression of the K_{eq} of systems is changed with the presence of other phases.

EQUILIBRIUM CONSTANT FOR SYSTEMS INVOLVING SOLIDS

If the experimental data for this reaction are studied:

$$CaCO_3(s) \rightleftharpoons CaO(s) + CO_2(g)$$

it is found that at a given temperature an equilibrium is established in which the concentration of CO_2 is constant. It is also true that the concentrations of the solids have no effect on the CO_2 concentration as long as both solids are present. Therefore, the K_{eq}, which would conventionally be written as

$$K_{eq} = \frac{[CaO][CO_2]}{[CaCO_3]}$$

can be modified by incorporating the concentrations of the two solids. This can be done since the concentration of solids is fixed by the density. The K_{eq} becomes a new constant, K

$$K = [CO_2]$$

Any heterogeneous reaction involving gases does not include the concentrations of pure solids. As another example, K for the reaction

$$NH_4Cl(s) \rightleftharpoons NH_3(g) + HCl(g)$$

is

$$K = [NH_3][HCl]$$

ACID IONIZATION CONSTANTS

When a weak acid that does not ionize completely is placed in solution, an equilibrium is reached between the acid molecule and its ions. The mass action expression can be used to derive an equilibrium constant, called the **acid ionization constant**, for this condition. For example, an acetic acid solution ionizing is shown as

$$HC_2H_3O_2 + H_2O \rightleftharpoons H_3O^+ + C_2H_3O_2^-$$

$$K = \frac{[H_3O^+][C_2H_3O_2^-]}{[HC_2H_3O_2][H_2O]}$$

The concentration of water in mol/L is found by dividing the mass of 1 L of water (which is 1000 g at 4°C) by its gram-molecular mass, 18 g, giving H_2O a value of 55.6 mol/L. Since this number is so large compared to the other numbers involved in the equilibrium constant, it is practically constant and is incorporated into a new equilibrium constant designated K_a. Then the new expression is

$$K_a = \frac{[H_3O^+][C_2H_3O_2^-]}{[HC_2H_3O_2]}$$

Ionization constants have been found experimentally for many substances and are listed in chemical tables. The ionization constants of ammonia and acetic acid are about 1.8×10^{-5}. For boric acid $K_a = 5.8 \times 10^{-10}$, and for carbonic acid $K_a = 4.3 \times 10^{-7}$.

If the concentrations of the ions present in the solution of a weak electrolyte are known, the value of the ionization constant can be calculated. Also, if the value of K_i is known, the concentrations of the ions can be calculated.

A small value for K_a means that the concentration of the unionized molecule must be relatively large compared to the ion concentrations. Conversely, a large value for K_a means that the concentrations of ions are relatively high. Therefore, the smaller the ionization constant of an acid, the weaker the acid. Thus, for the three acids referred to above, the ionization constants show that the weakest of these is boric acid, and the strongest, acetic acid. It should be remembered that in all cases where ionization constants are used, the electrolytes must be weak in order to be involved in ionic equilibria.

IONIZATION CONSTANT OF WATER

Since water is a very weak electrolyte, its ionization constant can be expressed as

$$2H_2O \rightleftharpoons H_3O^+ + OH^-$$

(Equilibrium constant) $K = \dfrac{[H_3O^+][OH^-]}{[H_2O]^2}$

(Ionization constant) $K_w = [H_3O^+][OH^-] = 1 \times 10^{-14}$ at 25°C

From this expression, we see that for distilled water $[H_3O^+] = [OH^-] = 1 \times 10^{-7}$. Therefore, the pH, which is $-\log[H_3O^+]$, is

$$pH = -\log[1 \times 10^{-7}]$$
$$pH = -[-7] = 7 \text{ for a neutral solution}$$

A pH of less than 7 is acid, and a pH of greater than 7 is basic. (More information on pH is given on page 238.)

EXAMPLE (This sample incorporates the entire discussion of ionization constants, including finding the pH.)

Calculate (a) the $[H_3O^+]$, (b) the pH, and (c) the percentage dissociation for 0.100 M acetic acid at 25°C. The symbol K_a is used for the ionization of acids. K_a for $HC_2H_3O_2$ is 1.8×10^{-5}.

SOLUTION
(a) For this reaction

$$H_2O(\ell) + HC_2H_3O_2(\ell) \rightleftharpoons H_3O^+(aq) + C_2H_3O_2^-(aq)$$

and

$$K_a = \dfrac{[H_3O^+][C_2H_3O_2^-]}{[HC_2H_3O_2]} = 1.8 \times 10^{-5}$$

Let x = number of moles/liter of $HC_2H_3O_2$ that dissociate and reach equilibrium. Then

$$[H_3O^+] = x, [C_2H_3O_2^-] = x, [HC_2H_3O_2] = 0.1 - x$$

Substituting in the expression for K_a gives

$$K_a = 1.8 \times 10^{-5} = \dfrac{(x)(x)}{0.10 - x}$$

Since weak acids, like acetic acid, at concentrations of $0.01\ M$ or greater dissociate very little, the equilibrium concentration of the acid is very nearly equal to the original concentration; that is,

$$0.10 - x \cong 0.10$$

Therefore, the expression can be changed to

$$1.8 \times 10^{-5} = \frac{(x)(x)}{0.10}$$

$$x^2 = 1.8 \times 10^{-6}$$
$$x^2 = 1.3 \times 10^{-3} = [H_3O^+]$$

(b) Substituting this result into the pH expression gives

$$pH = -\log[H_3O^+] = -\log[1.3 \times 10^{-3}]$$
$$= 3 - \log 1.3$$
$$= 2.9$$

(c) The percentage of dissociation of the original acid may be expressed as

$$\%\ \text{dissociation} = \frac{\text{mol/L that dissociate}}{\text{original concentration}}(100)$$

$$\%\ \text{dissociation} \times \frac{1.3 \times 10^{-3}}{1.0 \times 10^{-1}}(100) = 1.3\%$$

SOLUBILITY PRODUCTS

A saturated solution of an ionic solid has been defined as an equilibrium condition between the solute and its ions. For example,

$$AgCl \rightleftharpoons Ag^+ + Cl^-$$

The equilibrium constant is

$$K = \frac{[Ag^+][Cl^-]}{[AgCl]}$$

Since the concentration of the solute remains constant for that temperature, the $[AgCl]$ is incorporated into K to give K_{sp}, called the **solubility product constant**:

$$K_{sp} = [Ag^+][Cl^-] = 1.2 \times 10^{-10}\ \text{at } 25°C$$

This setup can be used to solve problems in which the ionic concentrations are given and the K_{sp} is to be found or the K_{sp} is given and the ionic concentrations are to be determined.

EXAMPLE 1

Finding the K_{sp}

By experimentation it is found that a saturated solution of $BaSO_4$ at 25°C contains 3.9×10^{-5} mol/L of Ba^{2+} ions. Find the K_{sp} of this salt.

SOLUTION

Since $BaSO_4$ ionizes into equal numbers of Ba^{2+} and SO_4^{2-}, the barium ion concentration will equal the sulfate ion concentration. So the solution is

$$BaSO_4 \rightleftharpoons Ba^{2+} + SO_4^{2-}$$
$$K_{sp} = [Ba^{2+}][SO_4^{2-}]$$

Therefore,

$$K_{sp} = (3.9 \times 10^{-5})(3.9 \times 10^{-5}) = 1.5 \times 10^{-9}$$

EXAMPLE 2

Finding the solubility

If the K_{sp} of radium sulfate, $RaSO_4$, is 4×10^{-11}, calculate its solubility in pure water.

SOLUTION

Let $x =$ moles of $RaSO_4$ that dissolve per liter of water. Then, in the saturated solution,

$$[Ra^{2+}] = x \text{ mol/L}$$
$$[SO_4^{2-}] = x \text{ mol/L}$$
$$RaSO_4(s) \rightleftharpoons Ra^{2+} + SO_4^{2-}$$
$$[Ra^{2+}][SO_4^{2-}] = K_{sp} = 4 \times 10^{-11}$$

Let $x = [Ra^{2+}]$ and $[SO_4^{2-}]$. Then

$$(x)(x) = 4 \times 10^{-11}$$
$$x = 6 \times 10^{-6} \text{ mol/L}$$

Thus, the solubility of $RaSO_4$ is 6×10^{-6} mol/L of water, giving a solution 6×10^{-6} M in Ra^{2+} and 6×10^{-6} M in SO_4^{2-}.

EXAMPLE 3

Predicting the formation of a precipitate

In some cases the solubility products of solutions can be used to predict the formation of a precipitate.

Suppose we have two solutions. One contains 1.00×10^{-3} mol of silver nitrate, $AgNO_3$, per liter. The other solution contains 1.00×10^{-2} mol of sodium chloride, $NaCl$, per liter. If 1 L of each of these solutions is mixed to make a 2 L mixture, will a precipitate of $AgCl$ form?

SOLUTION

In the $AgNO_3$ solution, the concentrations are

$$[Ag^+] = 1.00 \times 10^{-3} \text{ mol/L} \quad \text{and} \quad [NO_3^-] = 1.00 \times 10^{-3} \text{ mol/L}$$

In the NaCl solution, the concentrations are

$$[Na^+] = 1 \times 10^{-2} \text{ mol/L} \quad \text{and} \quad [Cl^-] = 1.00 \times 10^{-2} \text{ mol/L}$$

When 1 L of each of these solutions is mixed to form a total volume of 2 L, the concentrations will be halved.

In the mixture then, the initial concentrations will be

$$[Ag^+] = 0.50 \times 10^{-3} \quad \text{or} \quad 5.0 \times 10^{-4} \text{ mol/L}$$
$$[Cl^-] = 0.50 \times 10^{-2} \quad \text{or} \quad 5.0 \times 10^{-3} \text{ mol/L}$$

In the K_{sp} of AgCl,

$$[Ag^+][Cl^-] = [5.0 \times 10^{-4}][5.0 \times 10^{-3}]$$
$$[Ag^+][Cl^-] = 25 \times 10^{-7} \quad \text{or} \quad 2.5 \times 10^{-6}$$

This is far greater than 1.7×10^{-10}, which is the K_{sp} of AgCl. These concentrations cannot exist, and Ag^+ and Cl^- will combine to form solid AgCl precipitate. Only enough Ag^+ ions and Cl^- ions will remain to make the product of the respective ions concentration equal 1.7×10^{-10}.

Common Ion Effect

When a reaction has reached equilibrium and an outside source adds more of one of the ions that is already in solution, the result is to cause the reverse reaction to occur at a faster rate and reestablish the equilibrium. This is called the **common ion effect**. For example, in the equilibrium reaction

$$NaCl(s) \rightleftharpoons Na^+ + Cl^-$$

the addition of concentrated HCl (12 M) adds H^+ and Cl^- both in a concentration of 12 M. This increases the concentration of the Cl^- and disturbs the equilibrium. The reaction will shift to the left and cause some solid NaCl to come out of solution.

The "common" ion is the one already present in an equilibrium before a substance is added that increases the concentration of that ion. The effect is to reverse the solution reaction and to decrease the solubility of the original substance, as shown in the above example.

EXAMPLE The ionization constant for $HC_2H_3O_2$ is 1.80×10^{-5} M. How many moles of hydrogen ions will be in a liter of 0.10 M $HC_2H_3O_2$ containing 0.20 mol of $NaC_2H_3O_2$?

Their ionization equations are

$$HC_2H_3O_2 \leftrightarrow H^+ + C_2H_3O_2^- \text{ and } NaC_2H_3O_2 \leftrightarrow Na^+ + C_2H_3O_2^-$$

Let x = concentration of $[H^+]$ at equilibrium

$0.20 + x$ = concentration of $[C_2H_3O_2^-]$ at equilibrium

$0.10 - x$ = concentration of $[HC_2H_3O_2]$ at equilibrium

$$K = \frac{[H^+] \times [C_2H_3O_2^-]}{[HC_2H_3O_2]} = 1.80 \times 10^{-5} \ M$$

Substituting the values of $[H^+]$, $[C_2H_3O_2^-]$, and $[HC_2H_3O_2]$ in the above equation gives:

$$K = \frac{[x] \times [0.20 + x]}{[0.10 - x]} = 1.80 \times 10^{-5} \ M$$

Since you know that acetic acid is weak, the value of x can be dropped from the terms $[0.20 + x]$ and $[0.10 - x]$ because it is so much smaller than the original concentrations.

Then the equation to solve is

$$\frac{[x] \times [0.20]}{[0.10]} = 1.80 \times 10^{-5} \ M$$

So $x = 9.00 \times 10^{-6}$ mol/L = $[H^+]$.

Factors Related to the Magnitude of *K*

RELATION OF MINIMUM ENERGY (ENTHALPY) TO MAXIMUM DISORDER (ENTROPY)

Some reactions are said to go to completion because the equilibrium condition is achieved when practically all the reactants have been converted to products. At the other extreme, some reactions reach equilibrium immediately, with very little product being formed. These two examples are representative of very large K values and very small K values, respectively. There are essentially two driving forces that control the extent of a reaction and determine when equilibrium will be established. These are the drive to the lowest heat content, or **enthalpy**, and the drive to the greatest randomness or disorder, which is called **entropy**. Reactions with negative ΔH values (enthalpy or heat content) are exothermic, and reactions with positive ΔS values (entropy or randomness) proceed to greater randomness.

The **Second Law of Thermodynamics** states that the entropy of the universe increases for any spontaneous process. This means that the entropy of a system may increase or decrease but that, if it decreases, then the entropy of the surroundings must increase to a greater extent so that the overall change in the universe is positive. In other words,

$$\Delta S_{universe} = \Delta S_{system} + \Delta S_{surroundings}$$

The following is a list of conditions in which ΔS is positive for the system:

1. When a gas is formed from a solid, for example,

$$CaCO_3(s) \rightarrow CaO(s) + CO_2(g)$$

2. When a gas is evolved from a solution, for example,

$$Zn(s) + 2H^+(aq) \rightarrow H_2(g) + Zn^{2+}(aq)$$

3. When the number of moles of gaseous product exceeds the number of moles of gaseous reactant, for example,

$$2C_2H_6(g) + 7O_2(g) \rightarrow 4CO_2(g) + 6H_2O(g)$$

4. When crystals dissolve in water, for example,

$$NaCl(s) \rightarrow Na^+(aq) + Cl^-(aq)$$

Looking at specific examples, we find that in some cases endothermic reactions occur when the products provide greater randomness or positive entropy. This reaction is an example:

$$CaCO_3(s) \rightleftharpoons CaO(s) + CO_2(g)$$

The production of the gas, and thus greater entropy, might be expected to take this reaction almost to completion. However, this is not true because another force is hampering this reaction. It is the absorption of energy, and thus the increase in enthalpy, as the $CaCO_3$ is heated.

The equilibrium condition, then, at a particular temperature, is a compromise between the increase in entropy and the increase in enthalpy of the system.

The Haber process of making ammonia is another example of this compromise of driving forces that affect the establishment of an equilibrium. In the reaction

$$N_2(g) + 3H_2(g) \rightleftharpoons 2NH_3(g) + heat$$

the forward reaction to reach the lowest heat content and thus release energy cannot go to completion because the force to maximum randomness is driving the reverse reaction.

CHANGE IN FREE ENERGY OF A SYSTEM—GIBBS EQUATION

Three factors can be combined in an equation that summarizes the change of **free energy** in a system. This is designated as ΔG. The relationship is

$$\Delta G = \Delta H - T\Delta S \qquad (T \text{ is temperature in kelvins})$$

and is called the **Gibbs Free Energy Equation**. The sign of ΔG can be used to predict the spontaneity of a reaction at constant temperature and pressure. If ΔG is negative, the reaction is (probably) spontaneous; if ΔG is positive, the reaction is improbable; and if ΔG is 0, the system is at equilibrium and there is no net reaction.

The ways in which the factors in the equation affect ΔG are shown in the table below.

ΔH	ΔS	Will It Happen?	Comment
Exothermic (−)	+	Yes	No exceptions
Exothermic (−)	−	Probably	At low temperature
Endothermic (+)	+	Probably	At high temperature
Endothermic (+)	−	No	No exceptions

This drive to achieve a minimum of free energy may be interpreted as the driving force of a chemical reaction.

Chapter Summary

The following terms summarize all the concepts and ideas that were introduced in this chapter. You should be able to explain their meaning and how you would use them in chemistry. They appear in boldface type in this chapter to draw your attention to them. The boldface type also makes the terms easier for you to look up if you need to. You could also use a search engine on your computer to get a quick and expanded explanation of these terms, laws, and formulas.

TERMS YOU SHOULD KNOW

acid ionization constant
common ion effect
enthalpy
entropy
equilibrium
equilibrium constant

Free Energy
Gibbs Free Energy Equation
Law of Mass Action
Le Châtelier's Principle
Second Law of Thermodynamics
solubility product constant

Chapter 10 Review Exercises

1. For the reaction $A + B \rightleftharpoons C + D$, the equilibrium constant can be expressed as

   (A) $K_{eq} = \dfrac{[A][B]}{[C][D]}$

   (B) $K_{eq} = \dfrac{[C][B]}{[A][D]}$

   (C) $K_{eq} = \dfrac{[C][D]}{[A][B]}$

   (D) $K_{eq} = \dfrac{C \cdot D}{A \cdot B}$

2. The concentrations in an expression of the equilibrium constant are given in

   (A) mol/mL
   (B) g/L
   (C) gram-equivalents/L
   (D) mol/L

3. In the equilibrium expression for the reaction $BaSO_4 \rightleftharpoons Ba^{2+} + SO_4^{2-}$, K_{sp} is equal to

   (A) $[Ba^{2+}][SO_4^{2-}]$

   (B) $\dfrac{[Ba^{2+}][SO_4^{2-}]}{BaSO_4}$

   (C) $\dfrac{[Ba^{2+}][SO_4^{2-}]}{[BaSO_4]}$

   (D) $\dfrac{[BaSO_4]}{[Ba^{2+}][SO_4^{2-}]}$

4. The K_w of water is equal to

    (A) 1×10^{-7}
    (B) 1×10^{-17}
    (C) 1×10^{-14}
    (D) 1×10^{-1}

5. The pH of a solution that has a hydrogen ion concentration of 1×10^{-4} mol/L is

    (A) 4
    (B) −4
    (C) 10
    (D) −10

6. The pH of a solution that has a hydroxide ion concentration of 1×10^{-4} mol/L is

    (A) 4
    (B) −4
    (C) 10
    (D) −10

7. A small value for K_{eq}, the equilibrium constant, indicates that

    (A) the concentration of the unionized molecules must be relatively small compared to the ion concentrations
    (B) the concentration of the ionized molecules must be larger than the ion concentrations
    (C) the substance ionizes to a large degree
    (D) the concentration of the unionized molecules must be relatively large compared to the ion concentrations

8. In the Haber process for making ammonia, an increase in pressure favors

    (A) the forward reaction
    (B) the reverse reaction
    (C) neither reaction
    (D) both reactions

9. A change in which of these conditions will change the K of an equilibrium?

    (A) temperature
    (B) pressure
    (C) concentration of reactants
    (D) concentration of products

10. If $Ca(OH)_2$ is dissolved in a solution of NaOH, its solubility, compared to that in pure water, is

    (A)   increased
    (B)   decreased
    (C)   unaffected

Questions 11–14, use one of the following choices to complete the sentences appropriately.

    (A)   the free energy is always negative.
    (B)   the free energy is negative at low temperatures.
    (C)   the free energy is negative at high temperatures.
    (D)   the free energy is never negative.
    (E)   the system is at equilibrium and there is no net reaction.

11. When enthalpy is negative and entropy is positive,

12. When enthalpy is positive and entropy is positive,

13. When enthalpy is negative and entropy is negative,

14. When enthalpy is positive and entropy is negative,

15. When the ΔG, free energy, is zero,

Answers and Explanations

1. **(C)** The correct setup of the K_{eq} is the product of the concentration of the products over the products of the reactants.

2. **(D)** The concentrations in an expression of the equilibrium constant are given in moles/liter (mol/L).

3. **(A)** This reaction is the equilibrium between a precipitate and its ions in solution. Since the $BaSO_4$ is a solid, it does not appear in the solubility product expression. Therefore, the $K_{sp} = [Ba^{2+}] [SO_4^{2-}]$.

4. **(C)** This is the K_w expression of water.

5. **(A)** Since the pH is defined as the negative of the log of the H^+ concentration, it is $-(\log$ of $10^{-4})$ which is $-(-4)$ or 4.

6. **(C)** $pH + pOH = 14$. In this problem the $pOH = -\log[OH] = -\log[10^{-4}] = -(-4) = 4$. By placing this value in the equation, you have $pH + 4 = 14$. Solving for pH, you get 10.

7. **(D)** For the K_{eq} to be a small value, the numerator of the expression must be small compared with the denominator. Since the numerator is the product of the concentrations of the ion concentrations and the denomination is the product of the unionized molecules, the concentration of the unionized molecules must be relatively large compared with the ion concentrations.

8. **(A)** In the reaction for the formation of ammonia,

$$N_2 + 3H_2 \leftrightarrow 2NH_3 + \text{heat (at equilibrium)}$$

An increase in pressure will cause the forward reaction to be favored since the equation shows that four molecules of reactants are forming two molecules of products. This effect tends to reduce the increase in pressure by the formation of more ammonia.

9. **(A)** Only a change in the temperature of the equilibrium reaction will change the K of the equilibrium.

10. **(B)** Because there already is a concentration of $(OH)^-$ ions from the NaOH in solution, this common ion effect will decrease the solubility of the $Ca(OH)_2$.

11. **(A)** When enthalpy is negative and entropy is positive, the free energy is always negative.

12. **(C)** When enthalpy is positive and entropy is positive, the free energy is negative at high temperatures.

13. **(B)** When enthalpy is negative and entropy is negative, the free energy is negative at low temperatures.

14. **(D)** When enthalpy is positive and entropy is negative, the free energy is never negative.

15. **(E)** When the ΔG, free energy, is zero, the system is at equilibrium and there is no net reaction.

ACIDS, BASES, AND SALTS

CHAPTER OBJECTIVES

Upon completing this chapter, you will be able to:

- Describe the properties of an Arrhenius acid and base, and know the name, formula, and degree of ionization of common acids and bases
- Explain the Brønsted-Lowry theory of acids and conjugate bases
- Draw examples and explain the Lewis theory of acids-bases
- Determine the pH and pOH of solutions
- Solve titration problems and the use of indicators in the process
- Describe how a buffer works
- Explain the formation and naming of salts
- Explain amphoteric substances in relation to acid-base theory

Definitions and Properties

ACIDS

There are some characteristic properties by which an acid may be defined. The most important are

1. *Water solutions of acids conduct electricity.* This conduction depends on their degree of ionization. A few acids ionize almost completely, while others ionize only to a slight degree. The following table indicates some common acids and their **degrees of ionization**.

DEGREE OF IONIZATION OF COMMON ACIDS

Completely or Nearly Completely Ionized	Moderately Ionized	Slightly Ionized
Nitric	Oxalic	Hydrofluoric
Hydrochloric	Phosphoric	Acetic
Sulfuric	Sulfurous	Carbonic
Hydriodic		Hydrosulfuric
Hydrobromic		(Most others)

2. *Acids will react with metals that are more active than hydrogen ions (see chart on page 265) to liberate hydrogen.* (Some acids are also strong oxidizing agents and will not release hydrogen. Somewhat concentrated nitric acid is such an acid.)

3. *Acids have the ability to change the color of indicators.* Some common indicators are **litmus** and phenolphthalein. Litmus is a dyestuff obtained from plants. When litmus is added to an acidic solution, or paper impregnated with litmus is dipped into an acid, the neutral purple color changes to pink-red. Phenolphthalein is red in a basic solution and becomes colorless in a neutral or acid solution.

4. *Acids react with bases so that the properties of both are lost to form water and a salt.* This is called *neutralization*. The general equation is

$$\text{Acid} + \text{base} \rightarrow \text{salt} + \text{water}$$

An example is

$$Mg(OH)_2 + H_2SO_4 \rightarrow MgSO_4 + 2H_2O$$

5. *If an acid is known to be a very weak solution, you might taste it and note the sour taste.* Vinegar is an example. Do not taste unknown acids.

6. *Acids react with carbonates to release carbon dioxide.* An example is

$$CaCO_3 + 2HCl \rightarrow CaCl_2 + H_2CO_3 \text{ (unstable and decomposes)}$$
$$H_2CO_3 \rightarrow H_2O + CO_2(g)$$

The most common theory used in first-year chemistry is the **Arrhenius Theory**, which states that an acid is a substance that yields hydrogen ions in an aqueous solution. Although we speak of hydrogen ions in solution, they are really not separate ions but become attached to the oxygen of the polar water molecule to form the H_3O^+ ion (the hydronium ion). So it is really this hydronium ion we are concerned with in an acid solution.

The general reaction for the dissociation of an acid, HX, is commonly written

$$HX \rightleftharpoons H^+ + X^-$$

To show the formation of the hydronium ion, H_3O^+, the complete equation is

$$HX + H_2O \rightleftharpoons H_3O^+ + X^-$$

A list of common acids and their formulas is given in Chapter 4, Table 7, with an explanation of the naming procedures for acids preceding the table.

BASES

Bases may also be defined by some operational definitions based on experimental observations. Some of the important ones are

1. ***Bases are conductors of electricity in an aqueous solution.*** Their degree of conduction depends on their degree of ionization. See the table of common bases below and their degree of ionization.

DEGREE OF IONIZATION OF COMMON BASES

Completely or Nearly Completely Ionized	Slightly Ionized
Potassium hydroxide	Ammonium hydroxide
Sodium hydroxide	(All others)
Barium hydroxide	
Strontium hydroxide	
Calcium hydroxide	

2. ***Bases cause a color change in indicators.*** Litmus changes from red to blue in a basic solution, and phenolphthalein turns pink from its colorless state.

3. ***Bases react with acids to neutralize each other and form a salt and water.***

4. ***Bases react with fats to form a class of compounds called soaps.*** Earlier generations used this method to make their own soap.

5. ***Aqueous solutions of bases feel slippery, and the stronger bases are very caustic to the skin.***

THE ARRHENIUS THEORY

Defines a base as a substance that yields hydroxide ions (OH^-) in an aqueous solution. Some of the common bases have common names. The list below gives some of these:

Sodium hydroxide—Lye, caustic soda

Potassium hydroxide—Caustic potash

Calcium hydroxide—Slaked lime, hydrated lime, limewater

Ammonium hydroxide—Ammonia water, household ammonia

Much of the sodium hydroxide produced today comes from the Hooker cell electrolysis apparatus. We have already discussed the electrolysis apparatus used for the decomposition of water. When an electric current is passed through a saltwater solution, hydrogen, chlorine, and sodium hydroxide are the products. The formula for this equation is

$$2NaCl + 2HOH \xrightarrow[\text{energy}]{\text{electrical}} H_2(g) + Cl_2(g) + 2NaOH$$

BROADER ACID-BASE THEORIES

Besides the common Arrhenius Theory of acids and bases discussed for aqueous solutions, two other theories, the Brønsted-Lowry Theory and the Lewis Theory, are widely used.

THE BRØNSTED-LOWRY THEORY (1923)

The **Brønsted-Lowry Theory** defines acids as proton donors, and bases as proton acceptors. This definition agrees with the aqueous solution definition of an acid giving up hydrogen ions in solution but goes beyond to other cases as well.

An example of this is dry HCl gas reacting with ammonia gas to form the white solid NH_4Cl.

$$HCl(g) + NH_3(g) \rightarrow NH_4^+(s) + Cl^-(s)$$

The HCl is the proton donor or acid, and the ammonia is a Brønsted-Lowry base that accepts the proton.

CONJUGATE ACIDS AND BASES

In an acid-base reaction, the original acid gives up its proton to become a **conjugate base**. In other words, after losing its proton, the remaining ion is capable of gaining a proton, thus qualifying as a base. The original base accepts a proton, and so it now is classified as a **conjugate acid** since it can release this newly acquired proton and thus behave like an acid.

Some examples of this are

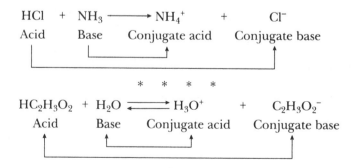

STRENGTH OF CONJUGATE ACIDS AND BASES

The extent of the reaction between a Brønsted-Lowry acid and base depends on the relative strengths of the acids and bases involved. Consider the following example. Hydrochloric acid is a strong acid. It gives up protons readily. Logically, the Cl^- ion has little tendency to attract and retain a proton. Consequently, the Cl^- ion is an extremely weak base.

$$HCl(g) + H_2O(\ell) \rightarrow H_3O^+(aq) + Cl^-(aq)$$

| strong acid | base | acid | weak base |

This observation leads to important conclusions. The stronger an acid is, the weaker its conjugate base is. The stronger a base is, the weaker its conjugate acid is. These concepts allow strengths of different acids and bases to be compared to predict the outcome of a reaction. As an example, consider the reaction of perchloric acid, $HClO_4$, and water.

$$HClO_4 (aq) + H_2O(\ell) \rightarrow H_3O^+(aq) + ClO_4^-(aq)$$

| stronger acid | stronger base | weaker acid | weaker base |

Another important general conclusion is that proton-transfer reactions favor the production of the weaker acid and the weaker base. For a reaction to approach completion, the reactants must be much stronger as an acid and as a base than the products.

Another example is boron trifluoride. It is an excellent Lewis acid. It forms a fourth covalent bond with many molecules and ions. Its reaction with a fluoride ion is shown below.

$$BF_3(aq) + F^-(aq) \rightarrow BF_4^-(aq)$$

The acid-base systems are summarized below.

Type	Acid	Base
Arrhenius	H^+ or H_3O^+ producer	OH^- producer
Brønsted-Lowry	proton (H^+) donor	proton (H^+) acceptor
Lewis	electron-pair acceptor	electron-pair donor

THE LEWIS THEORY (1916)

The **Lewis Theory** of acids and bases is defined in terms of the electron-pair concept. It is probably the most generally useful concept of acids and bases. According to the Lewis definition, an acid is an electron-pair acceptor, and a base is an electron-pair donor. An example of this is the formation of ammonium ions from ammonia gas and hydrogen ions follows.

x hydrogen electrons
o nitrogen electrons

Notice that the hydrogen ion is in fact accepting the electron pair of the ammonia, and so it is a Lewis acid. The ammonia is donating its electron pair, and so it is a Lewis base.

ACID CONCENTRATION EXPRESSED AS pH

Frequently, acid and base concentrations are expressed by means of the **pH system**. The pH can be defined as $-\log[H^+]$, where $[H^+]$ is the concentration of hydrogen ions expressed in mol/L. The logarithm is the exponent of 10 when the number is written in the base 10. For example,

$$100 = 10^2 \text{ and so logarithm of } 100, \text{ base } 10 = 2$$
$$10{,}000 = 10^4 \text{ and so logarithm of } 10{,}000, \text{ base } 10 = 4$$
$$0.01 = 10^{-2} \text{ and so logarithm of } 0.01, \text{ base } 10 = -2$$

The logarithms of more complex numbers can be found in a logarithm table. An example of a pH problem is the following:

Find the pH of a 0.1 M solution of HCl.

STEP 1:   Because HCl ionizes almost completely into H^+ and Cl^-, $[H^+] = 0.1$ mol/L.

STEP 2:   By definition,

$$pH = -\log[H^+]$$

so

$$pH = -\log[10^{-1}]$$

STEP 3:   The logarithm of 10^{-1} is -1,

so

$$pH = -(-1)$$

STEP 4:   The pH then is 1.

Because water has a normal H^+ concentration of 10^{-7} mol/L because of the slight ionization of water molecules, the water pH is 7 when it is neither acid nor base. The normal pH range is from 0 to 14.

	Acid	Neutral	Base	
0 ←		7 ←		→ 14

$$[H_3O]^+ = 10^0 \ \ldots\ldots\ldots\ldots\ldots\ 10^{-7} \ \ldots\ldots\ldots\ldots\ldots\ldots\ 10^{-14}$$

The pOH is the negative logarithm of the hydroxide ion concentration

$$pOH = -\log[OH^-]$$

If the concentration of the hydroxide ion is $10^{-9}\ M$, then the pOH of the solution is 9.

From the equation

$$[H^+][OH^-] = 1.0 \times 10^{-14} \text{ at } 25°C$$

the following relationship can be derived

$$pH + pOH = 14.00$$

In other words, the sum of the pH and pOH of an aqueous solution at 25°C must always equal 14.00. For example, if the pOH of a solution is 9.00, then its pH must be 5.00.

EXAMPLE   What is the pOH of a solution whose pH is 3.0?

SOLUTION

Substituting the 3.0 for pH in the expression

$$pH + pOH = 14.0$$
$$3.0 + pOH = 14.0$$
$$pOH = 14 - 3.0 = 11.0$$

Indicators

Some **indicators** can be used to determine pH because of their color changes somewhere along this pH scale. Some common indicators and their respective color changes are given below.

Indicator	pH Range of Color Change	Color below Lower pH	Color above Higher pH
Methyl orange	3.1–4.4	Red	Yellow
Bromthymol blue	6.0–7.6	Yellow	Blue
Litmus	4.5–8.3	Red	Blue
Phenolphthalein	8.3–10.0	Colorless	Red

The following is an example of how to read this chart. At pH values below 4.5, **litmus** is red; above 8.3, it is blue. Between these values, it is a mixture of the two colors.

Titration-Volumetric Analysis

Knowledge of the concentrations of solutions and the reactions they take part in can be used to determine the concentrations of "unknown" solutions or solids. The use of volume measurement in solving these problems is called **volumetric analysis-titration**.

A common example of a volumetric analysis uses acid-base reactions. If you are given a base of known concentration (standard solution), let us say 0.10 M NaOH, and you want to determine the concentration of an HCl solution, you could *titrate* the solutions in the following manner.

First, introduce a measured quantity, 25.0 mL, of the NaOH into a flask by using a pipet or buret. Next, introduce 2 drops of a suitable indicator. Since NaOH and HCl are considered a strong base and strong acid, respectively, an indicator that changes color in the middle pH range would be appropriate. Litmus solution would be one choice. It is blue in a basic solution but changes to red when the solution becomes acidic. Slowly introduce the HCl until the color change occurs. This is called the **endpoint**. This is the point at which enough acid solution is added to neutralize all the base in solution. It is also referred to as the **equivalence point**.

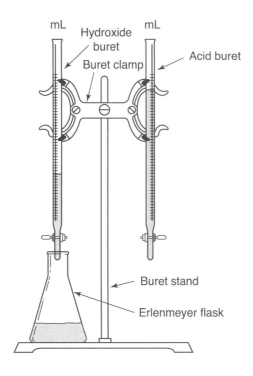

Buret Setup for Titration

Suppose 21.5 mL is needed to produce the color change; the reaction that occurs is

$$H^+(aq) + OH^-(aq) \rightarrow H_2O$$

until all the OH^- is neutralized, and then the excess H^+ causes the litmus paper to change color. To solve the question of the concentration of the NaOH, we use

$$M_{acid} \times V_{acid} = M_{base} \times V_{base}$$

Substituting the known amounts into this equation gives

$$xM_{acid} \times 21.5 \text{ mL} = 0.10 \text{ } M \times 25.0 \text{ mL}$$
$$x = 0.116 \text{ } M$$

In choosing an indicator for a titration, we need to consider whether the solution formed when the endpoint is reached has a pH of 7. Depending on the types of acids and bases used, the resulting hydrolysis of the salt formed may cause the solution to be slightly acidic, slightly basic, or neutral. If a strong acid and a strong base are titrated, the endpoint will be at pH 7, and practically any indicator can be used because adding a drop of either reagent will change the pH at the endpoint by about 6 units. For the titration of strong acids and weak bases, we need an indicator, such as methyl orange, that changes color between pH 3.1 and 4.4 in the acid region. When titrating a weak acid and a strong base, we should use an indicator that changes in the basic range. Phenolphthalein is a suitable choice for this type of titration since it changes color in the pH 8.3 to 10.0 range.

The process of the neutralization reaction can be represented by a titration curve like the one below, which shows the titration of a strong acid with a strong base.

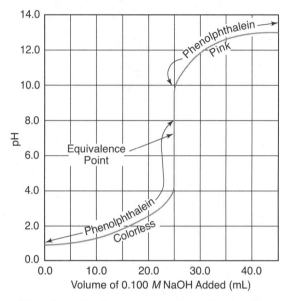

Titration of a Strong Acid with a Strong Base

EXAMPLE
1
Find the concentration of acetic acid in vinegar if 21.6 mL of 0.20 M NaOH is needed to titrate a 25 mL sample of the vinegar.

SOLUTION

Using the equation $M_{acid} \times V_{acid} = M_{base} \times V_{base}$, we have
$$xM \times 25 \text{ mL} = 0.2 \times 21.6 \text{ mL}$$
$$x = 0.17 \ M_{acid}$$

Another type of titration problem involves a solid and a titrated solution.

EXAMPLE
2
A solid mixture contains NaOH and NaCl. If a 0.100 g sample of this mixture required 10 mL of 0.100 M HCl to titrate the sample to its endpoint, what is the percent of NaOH in the sample?

SOLUTION

Since

$$\text{Molarity} = \frac{\text{No. of moles}}{\text{Liter of solution}}$$

then

$$M \times V = \frac{\text{No. of moles}}{\text{Liter of solution}} \times \text{liter of solution} = \text{No. of moles}$$

Substituting the HCl information into the equation, we have

$$\frac{0.100 \text{ mol}}{\text{Liter of solution}} \times 0.0100 \text{ L of solution} = 0.001 \text{ mol}$$

(*Note*: This is 10 mL expressed in liters)

Since 1 mol of HCl neutralizes 1 mol of NaOH, 0.001 mol of NaOH must be present in the mixture.

$$1 \text{ mol NaOH} = 40.0 \text{ g}$$
$$0.001 \text{ mol} \times 40.0 \text{ g/mol} = 0.04 \text{ g NaOH}$$

Therefore, 0.04 g of NaOH was in the 0.1 g sample of the solid mixture. The percent is 0.04 g/0.1 g $\times$ 100 = 40%.

In the explanations given to this point, the reactions that took place were between monoprotic acids (single hydrogen ions) and monobasic bases (one hydroxide ion per base). This means that each mole of acid had 1 mol of hydrogen ions available, and each mole of base had 1 mol of hydroxide ions available, to interact in the following reaction until the endpoint was reached:

$$H^+(aq) + OH^-(aq) \rightarrow 2H_2O$$

This is not always the case, however, and it is important to know how to deal with acids and bases that have more than one hydrogen ion and more than one hydroxide ion per formula. The following is an example of such a problem.

If 20.0 mL of an aqueous solution of calcium hydroxide, $Ca(OH)_2$, is used in a titration and an appropriate indicator is added to show the neutralization point (endpoint), the few drops of indicator that are added can be ignored in the volume considerations. Therefore, if 25.0 mL of standard 0.050 M HCl is required to reach the endpoint, what was the original concentration of the $Ca(OH)_2$ solution?

The balanced equation for the reaction gives the relationship of the moles of acid reacting to the moles of base:

$$\underbrace{2HCl}_{2 \text{ mol}} + \underbrace{Ca(OH)_2}_{1 \text{ mol}} \rightarrow CaCl_2 + 2H_2O$$

The mole relationship here is that the number of moles of acid is twice the number of moles of base:

Number of moles of acid = 2 times number of moles of base
↑ mole factor

Since the molar concentration of the acid times the volume of the acid gives the number of moles of acid:

$$M_a \times V_a = \text{moles of acid}$$

and the molar concentration of the base times the volume of the base gives the number of moles of base:

$$M_b \times V_b = \text{moles of base}$$

then, substituting these products into the mole relationship, we have

$$M_a V_a = 2M_b V_b$$

Solving for M_b gives

$$M_b = \frac{M_a V_a}{2V_b}$$

Substituting values, we obtain

$$M_b = \frac{0.050 \text{ mol/L} \times 0.0250 \text{ L}}{2 \times 0.0200 \text{ L}}$$
$$= 0.0312 \text{ mol/L or } 0.0312 \text{ } M$$

Buffer Solutions

Buffer solutions are equilibrium systems that resist changes in acidity and maintain a constant pH when acids or bases are added to them. A typical laboratory buffer can be prepared by mixing equal molar quantities of a weak acid such as $HC_2H_3O_2$ and its salt, $NaC_2H_3O_2$. When a small amount of a strong base such as NaOH is added to the buffer, the acetic acid reacts (and consumes) most of the excess OH^- ion. The OH^- ion reacts with the H^+ ion from the acetic acid, thus reducing the H^+ ion concentration in this equilibrium:

$$HC_2H_3O_2 \rightleftharpoons H^+ + C_2H_3O_2^-$$

This reduction of H^+ causes a shift to the right, forming additional $C_2H_3O_2^-$ ions and H^+ ions. For practical purposes, each mole of OH^- added consumes 1 mol of $HC_2H_3O_2$ and produces 1 mol of $C_2H_3O_2^-$ ions.

When a strong acid such as HCl is added to the buffer, the H^+ ions react with the $C_2H_3O_2^-$ ions of the salt and form more undissociated $HC_2H_3O_2$. This does not alter the H^+ ion concentration. Proportional increases and decreases in the concentrations of $C_2H_3O_2^-$ and $HC_2H_3O_2$ do not significantly affect the acidity of the solution.

Salts

A **salt** is an ionic compound containing positive ions other than hydrogen ions and negative ions other than hydroxide ions. The usual method of preparing a particular salt is by neutralizing the appropriate acid and base to form the salt and water.

The methods for preparing salts are

1. *Neutralization reaction.* An acid and base neutralize each other to form the appropriate salt and water. For example,

$$2HCl + Ca(OH)_2 \rightarrow CaCl_2 + 2H_2O$$
$$\text{Acid + base} \quad \rightarrow \text{Salt} \quad \text{+ water}$$

2. *Single replacement reaction.* An active metal replaces hydrogen in an acid. For example,

$$Mg(s) + H_2SO_4(aq) \rightarrow MgSO_4(aq) + H_2(g)$$

3. *Direct combination of elements.* An example of this method is the combination of iron and sulfur. In this reaction, small pieces of iron are heated with powdered sulfur.

$$Fe + S \rightarrow Fe\ S$$
$$\text{iron (II) sulfide}$$

4. *Double replacement reaction.* When solutions of two soluble salts are mixed to form an insoluble salt compound.

$$AgNO_3(aq) + NaCl(aq) \rightarrow NaNO_3(aq) + AgCl(s)$$

5. *Reaction of a metallic oxide with a nonmetallic oxide.*

$$MgO + SiO_2 \rightarrow MgSiO_3$$

The naming of salts is discussed on page 90.

Amphoteric Substances

Some substances, such as the HCO_3^- ion, the HSO_4^- ion, the H_2O molecule, and the NH_3 molecule can act as either a proton donor (acid) or a proton receiver (base) depending upon which substances they come in contact with. These substances are said to be **amphoteric**. Amphoteric substances donate protons in the presence of strong bases and accept protons in the presence of strong acids.

This can be illustrated by the reactions of the bisulfate ion, HSO_4^-:

With a strong acid, HSO_4^- accepts a proton:

$$HSO_4^- + H^+ \rightarrow H_2SO_4$$

With a strong base, the HSO_4^- donates a proton:

$$HSO_4^- + OH^- \rightarrow H_2O + SO_4^{2-}$$

Acid Rain—An Environmental Concern

Acid rain is currently a subject of great concern in many countries around the world because of the widespread environmental damage it reportedly causes. It forms when the oxides of sulfur and nitrogen combine with atmospheric moisture to yield sulfuric and nitric acids—both known to be highly corrosive, especially to metals. Once formed in the atmosphere, they can be carried long distances from their source before being deposited by rain. The pollution may also take the form of snow or fog or be precipitated in dry form. This dry form is just as damaging to the environment as the liquid form.

The problem of acid rain can be traced back to the beginning of the industrial revolution, and it has been growing ever since. The term "acid rain" has been in use for more than a century and is derived from atmospheric studies made in the region of Manchester, England. Although the severity of its effects has long been recognized in local settings, as indicated by spells of acid smog in heavily industrialized areas, its widespread destructiveness has been realized only in recent decades. One large area that has been studied extensively is northern Europe, where acid rain has eroded structures, injured crops and forests, and threatened or depleted life in freshwater lakes. In 1983, published reports claimed that 34% of the forested areas of West Germany had been damaged by acid rain. Many of the older monuments and buildings in Europe that have exposed metal have had to be refurbished. In 1989, the famous bronze horses on St. Mark's Basilica in Venice had to be removed for repair for this very reason.

Industrial emissions have been blamed as the major cause of acid rain. Industries have challenged this notion, because the chemical reactions involved in the production of acid rain in the atmosphere are complex and as yet little understood, and have stressed the need for further studies. Because of the high cost of pollution reduction, governments have tended to support this viewpoint. American studies released in the early 1980s, however, strongly implicated industries as the main source of acid rain in the eastern United States and Canada. In 1988, as part of the Long-Range Transboundary Air Pollution Agreement sponsored by the United Nations, the United States, along with 24 other countries, ratified a protocol freezing the rate of nitrogen oxide emissions at 1987 levels. The 1990 amendments to the Clean Air Act of 1967 put in place regulations to reduce the release of sulfur dioxide significantly from power plants. That achieved a 20 percent decrease in sulfur dioxide emissions. Attempts continue today to get all nations to join in this effort. Unfortunately, China and India are two countries where sulfur dioxide emissions are at levels similar to those of the United States and Western Europe in the 1980s.

Chapter Summary

The following terms summarize all the concepts and ideas that were introduced in this chapter. You should be able to explain their meaning and how you would use them in chemistry. They appear in boldface type in this chapter to draw your attention to them. The boldface type also makes the terms easier for you to look up if you need to. You could also use a search engine on your computer to get a quick and expanded explanation of these terms, laws, and formulas.

TERMS YOU SHOULD KNOW

acid
acid rain
amphoteric
Arrhenius Theory
base
Brønsted-Lowry Theory
buffer solution
conjugate acid
conjugate base
degree of ionization

endpoint
equivalence point
indicators
Lewis Theory
litmus
neutralization
pH system
salt
titration
volumetric analysis

Chapter 11 Review Exercises

1. The difference between HCl and $HC_2H_3O_2$ as acids is

   (A) the first has less hydrogen in solution
   (B) the second has more ionized hydrogen
   (C) the first is highly ionized
   (D) the second is highly ionized

2. The hydronium ion is represented as

   (A) H_2O^+
   (B) H_3O^+
   (C) HOH^+
   (D) H^-

3. H_2SO_4 is a strong acid because it is

   (A) slightly ionized
   (B) unstable
   (C) an organic compound
   (D) highly ionized

4. In the reaction of an acid with a base, the common ionic reaction involves ions of

   (A) hydrogen and hydroxide
   (B) sodium and chloride
   (C) hydrogen and hydronium
   (D) hydroxide and nitrate

5. The pH of an acid solution is

   (A) 3
   (B) 7
   (C) 9
   (D) 10

6. The pH of a solution with a hydrogen ion concentration of 1×10^{-3} is

   (A) +3
   (B) −3
   (C) ±3
   (D) +11

7. According to the Brønsted-Lowry Theory, an acid is

   (A) a proton donor
   (B) a proton acceptor
   (C) an electron donor
   (D) an electron acceptor

8. A buffer solution

   (A) changes pH rapidly with the addition of an acid
   (B) does not change pH at all
   (C) resists changes in pH
   (D) changes pH only with the addition of a strong base

9. The point at which a titration is complete is called the

   (A) endpoint
   (B) equilibrium point
   (C) calibrated point
   (D) chemical point

10. If 10 mL of 1 M HCl was required to titrate a 20 mL NaOH solution of unknown concentration to its endpoint, what was the concentration of the NaOH?

   (A) 0.5 M
   (B) 1.5 M
   (C) 2 M
   (D) 2.5 M

Answers and Explanations

1. **(C)** The strength of an Arrhenius acid is determined by the degree of ionization of the hydrogens in the formula. The HCl ionizes to a great degree and is considered to be a strong acid. In contrast, $HC_2H_3O_2$, acetic acid, which is found in vinegar, only ionizes to a small degree and is considered to be a weak acid.

2. **(B)** The hydronium ion is written as H_3O^+.

3. **(D)** H_2SO_4 is a strong acid because it is highly ionized.

4. **(A)** The basic reaction between an acid and a base is

$$H^+ + OH^- \rightarrow H_2O \text{ or}$$
$$H_3O^+ + OH^- \rightarrow 2H_2O$$

5. **(A)** The pH scale is 0 to 14, with the numbers below 7 indicating an acid solution and those above 7 indicating a basic solution.

6. **(A)** Because pH is defined as the negative of the log of the H_3O^+ concentration, it is $-(\log \text{ of } 10^{-3})$, which is $-(-3)$ or $+3$.

7. **(A)** By definition, a Brønsted-Lowry acid is a proton donor.

8. **(C)** A buffer solution resists the changes in pH.

9. **(A)** The point in a titration when the "unknown" solution has been neutralized in an acid-base titration by the "standard" solution of known concentration is called the endpoint or equivalence point.

10. **(A)** Use the equation

$$M_{acid} \times V_{acid} = M_{base} \times V_{base}$$

and change mL to L by dividing by 1000 mL/L. So 10 mL = 0.01 L and 20 mL = 0.02 L.

Then substitute to get

$$1\ M_{acid} \times 0.01\ L_{acid} = x\ M_{base} \times 0.02\ L_{base}$$

By solving for x, you get $x = 0.5\ M_{base}$.

OXIDATION-REDUCTION AND ELECTROCHEMISTRY

CHAPTER OBJECTIVES

Upon completing this chapter, you will be able to:

- Describe substances from an electrolyte perspective
- Identify a reaction as a redox process
- Assign oxidation states to substances in a reaction
- Understand combustion reactions from a redox perspective
- Describe galvanic cells and calculate their cell potential
- Explain the differences between galvanic and electrolytic cells
- Understand the operation of common batteries
- Describe and make calculations concerning electroplating
- Balance complex redox reactions

Ionization

In the early 1830s, Michael Faraday discovered that water solutions of certain substances conduct an electric current. He called these substances **electrolytes**. Our definition today of an electrolyte is much the same. It is a substance that dissolves in water to form a solution that will conduct an electric current.

The usual apparatus for testing this conductivity is a light bulb placed in series with two prongs immersed in the solution being tested.

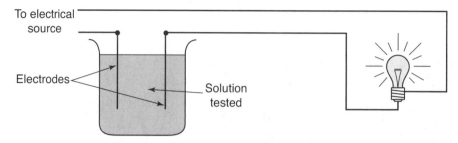

FIGURE 35. Conductivity of electrolytes.

Using this type of apparatus, we can classify substances that dissolve in water as strong electrolytes, weak electrolytes, or **nonelectrolytes**. Strong electrolytes added to water produce solutions that conduct electricity well. When placed in the conductivity device, this type of solution causes the bulb to light brightly. Solutes described as weak electrolytes produce water solutions that conduct electricity poorly. Solutions of weak electrolytes placed in the conductivity device only cause the bulb to light dimly. Nonelectrolytes dissolve in water, but the resulting solution fails to conduct electricity. The bulb in the conductivity apparatus doesn't light at all when containing a solution of a nonelectrolyte. Note that all types of electrolytes (strong, weak, or non-) dissolve in water due to their nature. This behavior is discussed in Chapter 7, where the rationale behind solution making is explained. The following table classifies some common substances as different types of electrolytes.

WATER SOLUTIONS TESTED FOR CONDUCTIVITY

Strong Electrolyte	Weak Electrolyte	Nonelectrolyte
Sodium chloride	Acetic acid	Sucrose
Hydrochloric acid	Ammonia	Ethanol

The reason strong and weak electrolytes produce solutions that conduct electricity was explained by the Swedish chemist Svante Arrhenius in the late 1800s. He claimed the conductivity was the result of the liberation of ions from these substances when dissolved in water. The varying degree of conductivity can be explained by the number of ions placed into solution by the dissolved substance.

In the process of making a solution between water and an ionically bonded solid, the ions that were previously held in fixed positions are now freed or **dissociated** from each other. These liberated ions are now immersed in the solvent and can move through the solution. Sodium chloride (NaCl) is an example of such a substance that dissociates completely when dissolved. This places large numbers of ions in solution and so sodium chloride is considered a strong electrolyte.

Molecular substances (i.e., those substances containing covalently bonded atoms) dissolved in water can also liberate ions in solution. Generally, for a molecule to dissolve in water to an appreciable degree, the molecule should be polar. Once dissolved, the atoms in the molecule may also dissociate from each other. In that process, bonds are broken, and electrons previously associated with one atom may now be associated with another. Consequently, these atoms with more or less electrons than normal (now ions) are free in the solution. In some cases, most of the molecules dissociate after dissolving and large numbers of ions are present in the solution. Hydrogen chloride (HCl) dissolved in water, also known as hydrochloric acid, is an example of a molecular substance that behaves this way and is considered a strong electrolyte. Hydrogen acetate ($HC_2H_3O_2$), also known as acetic acid, is a polar molecule that dissolves in water but does not dissociate well into ions. Although hydrogen chloride is seen to dissolve in water and dissociate completely (essentially 100%), hydrogen acetate typically dissociates after dissolving to a degree of only about 5%. That means that most of the particles dissolved in the water are molecules of hydrogen acetate. Since the number of ions present in the solution is relatively small, hydrogen acetate is considered a weak electrolyte. Nonelectrolytes, like sucrose or ethanol, are polar molecules that dissolve in water but do not dissociate into ions to an appreciable degree.

It should be easy now to envision how an electric current is carried through the solution in the device shown in Figure 35. The electricity causes one electrode to become positively charged, and one negatively charged. If the solution contains ions, they will be attracted to the electrode with the charge opposite their own. This means that the positive ions migrate to the cathode, the negative electrode in this situation, and so they are referred to as **cations**. The negative ions migrate toward the positive anode, and so they are called **anions**. When these ions arrive at the respective electrodes, the positive ions accept electrons, the negative ions give up electrons, and the movement of electrons can continue through the wire. This transfer of electrons constitutes a process that occurs often in other types of chemical reactions as well.

Oxidation-Reduction and Electrochemistry

ELECTROCHEMISTRY

The branch of chemistry that deals with electricity-related applications of reactions in which electrons are transferred is called **electrochemistry**. Many reaction categories discussed in Chapter 8 involve the transfer of electrons. The one in which this behavior is easiest to recognize is called **single replacement**. Recall that this type of reaction is characterized by a compound reacting with an element producing a new compound and a new element. It was shown that reactions such as these could be predicted to occur (i.e., be spontaneous) if the heat of formation of the compound in the products was negative and reasonably bigger than that of the compound in the reactants. These situations produce exothermic reactions, those in which the system under study becomes lower in energy, which nature tends to favor.

You can easily notice the exothermic nature of these reactions if the compound is dissolved in water to make a solution, and the element is placed directly into that solution. For example, take the reaction between a solution of copper (II) sulfate and zinc metal.

$$CuSO_4(aq) + Zn(s) \rightarrow ZnSO_4(aq) + Cu(s)$$

This reaction occurs by placing a strip of zinc metal into a solution of copper (II) sulfate as pictured below.

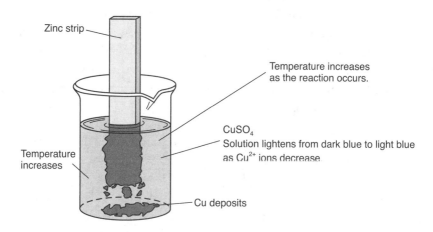

Notice that the temperature of the solution increases as the reaction occurs. When the substances involved in a single replacement reaction are placed in physical contact with one another, a transfer of heat accompanies the electron transfer. In this scenario, the exothermic nature of the reaction is a result of chemical energy being turned into thermal energy that is released.

Delving deeper into the single replacement reaction allows one to understand why the energy of the system goes down. The solution of copper (II) sulfate is actually composed of copper (II) ions (Cu^{2+}) and sulfate ions (SO_4^{2-}) surrounded by water molecules in solution. Likewise, the solution of zinc sulfate in the products contains ions of Zn^{2+} and SO_4^{2-}. The sulfate ions are found in both the reactants and products. As discussed in Chapter 4, they are referred to as **spectator ions**. Because spectator ions can be canceled out, the reaction can be simplified and written as the net ionic equation:

$$Cu^{2+}(aq) + Zn(s) \rightarrow Zn^{2+}(aq) + Cu(s)$$

Analysis of the reaction shows that, for the process to occur, the positively charged copper (II) ion in the reactants must turn into a copper atom in the products. Likewise, the zinc atom must become a positively charged zinc ion. How do these changes occur? The answer is simple and represents a major driving force for chemical change—the *transfer of electrons* from a substance that wants them less to a substance that wants them more.

The gain of 2 electrons

$$Cu^{2+}(aq) + Zn(s) \rightarrow Zn^{2+}(aq) + Cu(s)$$

The loss of 2 electrons

Whenever the transfer of electrons occurs during a chemical process, one substance must lose electrons in order for the other to gain them. The loss of electrons is called **oxidation**. Accordingly, the gaining of electrons is called **reduction**. Since one process can't happen without the other, these processes are often intertwined into one distinctly chemical term called **redox**. Redox reactions are those in which oxidation and reduction occur in complementary ways and is a major reaction category in chemistry. All single replacement reactions are redox reactions; however, all redox reactions are not single replacement reactions. Before we move on to other reaction types that can be seen as redox, let's take a further look at the single replacement category from a redox perspective.

The activity series of metals, whose use is described in Chapter 8 as another way to predict the spontaneity of single replacement reactions, can now be seen as a measure of the desire of certain metals to lose electrons compared to other metals. In the single replacement example reaction previously discussed, the process can be said to occur because the zinc atom has a greater desire to lose electrons than does the copper atom. In other words, the reaction will proceed in the forward direction as written (as opposed to going backward) because there is a natural push for zinc to lose electrons more so than the copper. Elements higher on the activity series, then, are simply those that have a stronger desire to lose electrons than the ones below them.

EXAMPLE   Using the activity series for metals found in Chapter 8, predict which of the following reactions is likely to occur as written. Also, for any reaction that does occur, describe the substances involved in the oxidation process, as well as those involved in the reduction process.

$$Au(s) + FeCl_3(aq) \rightarrow Fe(s) + AuCl_3(aq)$$

$$3\ Zn(s) + 2\ FeCl_3(aq) \rightarrow 2\ Fe(s) + 3\ ZnCl_2(aq)$$

SOLUTION

The first reaction will not occur (spontaneously). Because the activity of gold is lower than that of iron, it will not replace iron in a compound. In other words, gold has a lesser desire to lose electrons than does iron. The second reaction will occur. Zinc is more active than iron. In this reaction, the zinc will be oxidized, or lose electrons, as the reaction occurs. The recipient of electrons lost by the zinc is the iron. Therefore, the iron will be reduced.

OXIDATION STATES

Because the transfer of electrons in chemical reactions is so prevalent, chemists have developed a manner to keep track of their movement. It's important to note that for processes involving ionic species, of the type we have looked at so far, it is somewhat straightforward to distinguish redox from nonredox as the sign and magnitude of the charge on the ions involved allows for an easy view of what is taking place. Using the spontaneous reaction between zinc and iron (III) chloride referenced in the previous section, the zinc started as an atom with no net charge and turned into an ion with a 2+ charge. During chemical reactions, in order for the charge on particles to increase, negative electrons have to be lost and the process of oxidation occurs. A similar analysis of the iron in the reaction shows that the iron began as a 3+ ion (ferric) and ended as a neutral iron atom in the products. In other words, the iron went *down* in charge. Particles that go down in charge do so by gaining negative electrons. This description helps to explain why the process is referred to as reduction, even though electrons are being *gained*. Reduction refers to the *change in charge* involving ionic species, not in the change in the number of electrons possessed by a particle as a result of the process taking place. Since reduction involves a decrease in charge, oxidation, then, must involve an increase in charge (when ionic species are involved). Using the term *oxidation* may seem like a strange way to describe this process, but its use will be better understood from a historical context when **combustion** reactions are looked at as redox processes in an upcoming section of this chapter.

As discussed, reactions involving ions are relatively simple to recognize as redox (or not) because it is straightforward to identify changes in charges if they occur. It is not quite so simple when redox reactions involve molecular species. For these types of reactions, as well as those involving ionic species, a related but different term is used to describe the responsibility particles have for the electrons around them as a chemical reaction unfolds. To keep track of the transfer of electrons in all formulas (ionic or molecular), chemists have devised a system of electron bookkeeping called **oxidation states** (or **oxidation numbers**.) In this method, an oxidation state is assigned to each member of a formula using rules that recognize the degree to which electrons *practically belong* to a particular element in a given substance from an ionic bonding perspective. Basically, elements with higher electronegativities are assumed to have responsibility for the electrons in a bond, ionic or covalent, and the change in this responsibility will be noted by changes in their oxidation states. In this regard, the oxidation states method assumes an ionic perspective for *all* bonding, that is, electrons are not shared but *belong* to one element or the other in a chemical bond. Although we know that electrons *are* shared in a large number of chemical bonds, particularly those described as covalent (or molecular), assigning responsibility for the electrons in this way allows for easy recognition as to which atoms the electrons should be associated and how that changes during chemical processes.

Oxidation states are designated by a small number superscript *preceded* by a plus or minus sign. This is not to be confused with the ionic charges we have been using thus far that are shown as plus or minus signs *to the right* of the magnitude of ionic charge as a superscript.

THE RULES FOR ASSIGNING AN OXIDATION STATE

Below are the basic rules for assigning an oxidation state to an element in a substance's formula. By applying these simple rules, oxidation states can be assigned to elements in practically all substances you may encounter as a beginning chemistry student. To apply these rules, remember that the **sum of the oxidation states must be zero for an electrically neutral compound**. For a polyatomic ion, the **sum of the oxidation states must be equal to the charge on the ion**.

1. The oxidation state of an atom in an element is zero. Examples: 0 for Na (s), O_2 (g), and Hg (ℓ)

2. The oxidation state of a monatomic ion is the same as its charge. Examples: +1 for Na^+ and -1 for Cl^-.

3. The oxidation state for fluorine is -1 in its compounds. Examples: HF, hydrogen fluoride, has 1 H at +1 and 1 F at -1. PF_3 has 1 P at +3 and 3 F's at -1 each. (Note that in each compound the sum of the oxidation states is equal to zero.)

4. The oxidation state of oxygen is usually -2 in its compounds. Example: H_2O has 2 H's at +1 each and 1 O at -2. (Exceptions occur when the oxygen is bonded to fluorine and the oxidation state of fluorine takes precedence and in peroxide compounds where the oxidation state of oxygen is assigned the value of -1.)

5. The oxidation state of hydrogen in most compounds is +1. Examples: H_2O, HCl, and NH_3. (In hydrides, where hydrogen acts like an anion compounded with a metal, there is an exception, however. In this case, hydrogen is assigned the value of -1. Examples: LiH and KH.)

EXAMPLE 1 What is the oxidation state of each element in the compound Na_2SO_4?

SOLUTION

Sodium sulfate is an ionic substance containing the monatomic ion Na^+ and the polyatomic ion SO_4^{2-}. Therefore, the oxidation state of the sodium is +1 (by rule 2). In the polyatomic ion sulfate, the oxidation state of the oxygen is –2 (by rule 4). Since four oxygens are in the sulfate ion, the sum of their oxidation states is –8. Since the overall charge on the sulfate polyatomic ion is 2–, the oxidation state of the one sulfur must be +6 so that the overall charge matches the sum of the oxidation states numerically.

EXAMPLE 2 What is the oxidation state of carbon in the *molecule* CO_2?

SOLUTION

By rule 4, the oxidation state of oxygen is –2 and since there are two oxygen atoms in this formula, the total negative sum is –4. Since the positive and negative sums must add to zero, the oxidation state of carbon is +4 in this compound. Keep in mind that oxidation states are not "real" charges and carbon dioxide is not an ionic substance. In this case, the oxidation state of +4 for carbon indicates that for *electron bookkeeping purposes*, carbon will not be responsible for the electrons it is sharing in the bond between the carbon and oxygen atoms and that the responsibility will lie with the oxygen. When CO_2 is involved in a chemical process and carbon's responsibility for electrons changes (i.e., its oxidation state changes), a redox reaction will be recognized.

EXAMPLE 3 In the polyatomic ion dichromate ($Cr_2O_7^{2-}$), what is the oxidation state of chromium?

SOLUTION

In a polyatomic ion, the algebraic sum of the positive and negative oxidation states of all the elements must equal the charge on the ion.

$$Cr = 2 \times (x) = 2x$$

$$O = 7 \times (-2) = -14$$

Since the sum of these values must equal –2 (the charge on the polyatomic ion)

$$2x + (-14) = -2$$

$$x = +6$$

The oxidation state of chromium in dichromate is +6.

USING OXIDATION STATES TO RECOGNIZE REDOX REACTIONS

Once oxidation states can be assigned to elements in the substances involved in a chemical reaction, recognition of the process as being redox or not is straightforward. If the oxidation states change, a transfer of electrons is taking place. If the oxidation states remain the same, then a redox reaction is not occurring.

Consider these two reactions:

$$Na_2S(aq) + CuSO_4(aq) \rightarrow Na_2SO_4(aq) + CuS(s)$$

and

$$2\ Na(\ell) + Cl_2(g) \rightarrow 2\ NaCl(s)$$

The oxidation states of each of the elements in all the substances can be determined and is shown above each of them below:

$$\overset{+1\ -2}{Na_2S}(aq) + \overset{+2\ +6\ -2}{CuSO_4}(aq) \rightarrow \overset{+1\ +6\ -2}{Na_2SO_4}(aq) + \overset{+2\ -2}{CuS}(s)$$

and

$$\overset{0}{2\ Na}(\ell) + \overset{0}{Cl_2}(g) \rightarrow \overset{+1\ -1}{2\ NaCl}(s)$$

Because the oxidation states of the elements in the first reaction do not change, this reaction is not a redox process. You should recognize this as a precipitation reaction, as described in Chapter 8. The second reaction does exhibit a change in oxidation states and should be viewed as a redox reaction. In this reaction, the sodium is being oxidized and the chlorine is being reduced. The chlorine could not be reduced (i.e., gaining electrons) if the sodium was not present being oxidized (i.e., losing electrons.) In one regard, the sodium is acting as a facilitator for the reduction of the chlorine. As such, it is referred to as a **reducing agent**. Likewise, the sodium would not lose its electron if it had nowhere to go, and so the chlorine is referred to as the **oxidizing agent** in this reaction. Oxidizing agents, then, contain elements that are capable of being reduced by other substances (reducing agents) that contain elements that are capable of being oxidized. Sometimes the terms **oxidizer** and **reducer** are used as labels on the bottles of substances with the tendency to act as oxidizing and reducing agents, respectively.

EXAMPLE 1 Identify the elements that are being oxidized and reduced in the following reaction. Also, name the oxidizing and reducing agents.

$$2\ H_2O(\ell) + 2\ MnO_4^-(aq) + I^-(aq) \rightarrow 2\ MnO_2(aq) + IO_3^-(aq) + 2\ OH^-(aq)$$

SOLUTION

The oxidation states of the hydrogen and oxygen are not changing in the reaction. The oxidation states of the manganese and the iodine are changing as shown below.

$$\overset{+7}{\underset{\downarrow}{}} \qquad \overset{-1}{\underset{\downarrow}{}} \qquad \overset{+4}{\underset{\downarrow}{}} \qquad \overset{+5}{\underset{\downarrow}{}}$$

$$2\ H_2O(\ell) + 2\ MnO_4^-(aq) + I^-(aq) \rightarrow 2\ MnO_2(aq) + IO_3^-(aq) + 2\ OH^-(aq)$$

Since the manganese is changing from an oxidation state of +7 to that of +4, the manganese is being reduced. The iodine is being oxidized from a value of −1 to a value of +5. Manganese is gaining electron responsibility, while iodine is losing it. In that light then, the permanganate ion (MnO_4^-), which contains the manganese, is facilitating the process in which iodine is being oxidized and is referred to as the oxidizing agent. A typical source of the permanganate ion in chemical reactions like this one is from the compound potassium permanganate, $KMnO_4$, whose bottle is generally labeled with the term "oxidizer." Similarly, the iodide ion, I^-, can be called the reducing agent in this reaction.

EXAMPLE 2

Completely analyze the following reaction from a redox perspective.

$$3\ CO(g) + Fe_2O_3(s) \rightarrow 3\ CO_2(g) + 2\ Fe(s)$$

The change in the oxidation states of the elements is shown.

$$\overset{+2\ -2}{}\ \overset{+3\ -2}{}\ \overset{+4\ -2}{}\ \overset{0}{}$$
$$3\ CO(g) + Fe_2O_3(s) \rightarrow 3\ CO_2(g) + 2\ Fe(s)$$

Since the oxidation state of carbon is increasing from +2 to +4, it is being oxidized, its responsibility for electrons is decreasing, and the compound of which it is a part, carbon monoxide, is called the reducing agent. Carbon monoxide is a common reducer because of its chemical desire to turn into carbon dioxide when oxygen becomes available from other substances. On the other hand, the oxidation state of iron is decreasing from +3 to 0, it is being reduced, its responsibility for electrons is going up, and the compound of which it is a part, iron (III) oxide, is called the oxidizing agent.

COMBUSTION REACTIONS

Combustion reactions are those chemical processes in which substances (called fuels) are rapidly oxidized, accompanied by the release of heat and usually light. Combustion is also referred to as **burning**. Common and historically well-known combustion reactions involve using oxygen as the oxidizing agent—hence, the name for the process of losing electrons became known as *oxidation*. When a combustion reaction is complete, the elements in the burning fuel form compounds with the oxidizing agent and the elements with which the electrons are associated change.

The reaction between magnesium and oxygen is a common example of combustion. When the reaction occurs,

$$2\ Mg(s) + O_2(g) \rightarrow 2\ MgO(s)$$

the oxidation states of magnesium and oxygen change from 0 each to +2 and –2, respectively. A blinding light and large quantity of heat are released by the system as the reaction unfolds. The amount of heat released when 1 mole of a fuel burns is referred to as its *heat of combustion* and is symbolized ΔH_c. The ΔH_c for Mg is 602 kJ/mol. The *change in enthalpy* associated with the burning of 1 mole of carbon

$$C(s) + O_2(g) \rightarrow CO_2(g) + 393.5\ kJ$$

is –393.5 kJ, also written as $\Delta H_c = -393.5$ kJ and it represents the heat released in another combustion/redox reaction. The heats of combustion for many common substances can easily be found in reference tables. Some heats of combustion are found in Reference Table Ⓖ in the back of this book.

HYDROCARBON FUEL COMBUSTION REACTIONS

Another common combustion reaction involves the burning of hydrocarbon fuels. Methane, CH_4, is a typical hydrocarbon fuel and is the main component of **natural gas**. Analysis of the reaction

oxidation states:

$$CH_4(g) + 2\ O_2(g) \rightarrow CO_2(g) + 2\ H_2O(\ell)$$

shows that carbon is being oxidized and the oxygen is being reduced. This fuel is used to heat homes and cook food and should be familiar to all chemistry students as the gas used in their laboratory burners.

Describe the reaction that occurs when propane, C_3H_8, (the hydrocarbon fuel used in backyard barbeques) combusts, releasing 2221 kJ when one mole is burned.

SOLUTION

Note that when hydrocarbon fuels combust completely in the presence of oxygen, the products of the reaction are carbon dioxide and water. Therefore, the reaction is

oxidation states: −8/3*   0   +4 −2   −2

$$C_3H_8(g) + 5\ O_2(g) \rightarrow 3\ CO_2(g) + 4\ H_2O(\ell) + 2221\ kJ$$

Take note that oxidation states may be expressed in fractions.

Once again, the carbon is oxidized from the fraction −8/3 to +4 (*increasing from negative to positive*) and the oxygen is reduced from 0 to −2. Oxygen is the oxidizing agent and propane is the reducing agent. When this reaction occurs, light and heat is released and its heat of combustion, ΔH_c, is −2221 kJ.

GALVANIC (OR VOLTAIC) CELLS

In previous sections, we investigated reactions in which chemical energy, energy bound up in matter, is converted into heat via the transfer of electrons in redox processes. A **galvanic cell** is a device that enables the chemical energy to be converted into electrical energy instead—electrical energy that can be used to do work. Often galvanic cells, also known as **voltaic cells**, utilize spontaneous redox reactions in which the transfer of electrons is made to occur through a wire. Electrical devices attached to a galvanic cell can be made to operate, taking advantage of the electrical energy produced. A package of one or more galvanic cells is known as a battery.

In order to make the transfer of electrons take place through a wire, the substance being reduced must be separated from the one being oxidized. One way for this to occur is to place a porous barrier between the materials in a container that houses both. The barrier prevents the particles involved in the reduction process from mixing with the particles from which it will procure electrons. In contact, these substances liberate energy in the form of heat. Separated, the natural push to transfer the electrons from one substance to another is made through the conduit of the electrically conducting metal wire. The porous nature of the barrier allows for the charge that builds up in each compartment (due to the electron transfer through the wire) to dissipate as charged particles migrate through it. A specific example of the galvanic cell described above is shown below.

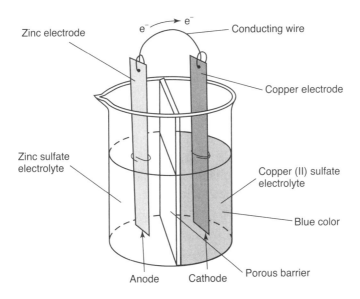

Zinc electrode

Conducting wire

Copper electrode

Zinc sulfate electrolyte

Copper (II) sulfate electrolyte

Blue color

Anode   Cathode   Porous barrier

Simple Galvanic Cell

The Zn strip is in an aqueous solution of $ZnSO_4$. The Cu strip is in an aqueous solution of $CuSO_4$. Both solutions conduct electricity as both $ZnSO_4$ and $CuSO_4$ are strong electrolytes. The zinc and copper strips are both electrodes. **Electrodes** are conductors used to establish electrical contact with a nonmetallic part of a circuit, such as an electrolytic solution. Each of the electrodes immersed in their respective solutions constitutes half of the cell where either oxidation or reduction will occur. In this case, the zinc electrode immersed in the zinc sulfate solution is where oxidation takes place and is called the **anode**. The copper strip immersed in the copper (II) sulfate solution is where reduction takes place and is called the **cathode**. The two halves of the cell together make the galvanic cell.

The half-reaction occurring at the anode is

$$Zn(s) \rightarrow Zn^{2+}(aq) + 2e^-$$

Zinc atoms, which make up the anode and turn into zinc ions, lose electrons that are transferred through the wire to the other compartment of the cell. As a result, the anode becomes less massive. The zinc ions produced enter into the solution of zinc sulfate. This causes the zinc ion concentration in the solution to increase, making the solution more positive. To mitigate the increase in positive charge, two processes occur to compensate for this. One is that the excess positive zinc ions migrate through the porous barrier to the other side of the cell. The other is that the negative sulfate ions from the other side of the cell migrate through the porous barrier into the zinc solution. If not for the migration of these ions, the cell would cease to operate in a very short time because the push to remove negative electrons from the more positive situation becomes less favored as the process occurs.

The half-reaction occurring at the cathode is

$$Cu^{2+}(aq) + 2e^- \rightarrow Cu(s)$$

Copper (II) ions in the solution gain the electrons transferred through the wire. These ions turn into copper atoms and deposit on the cathode. As a result, the cathode increases in mass. The solution of copper (II) sulfate contains less positive copper (II) ions and becomes more negative. Once again, the migration of ions through the porous barrier that eliminated the positive charge build-up on the oxidation side of the cell eliminates the charge build up that would occur here.

ELECTRODE POTENTIALS

The relative "desire" specific metals have to lose electrons cannot be measured independently or absolutely. This "push" to lose electrons for any particular metal can only be measured relative to the tendency other metals have to do the same. Since the process involves the movement of electrons from a situation where they exist at a higher potential energy (and therefore less wanted) to one where they exist at a lower potential energy (and therefore more wanted), the electron is experiencing a potential energy difference upon its transfer. Potential energy differences involving electrons are measured in **volts**. One volt is the potential (energy) difference when one *coulomb* of charge undergoes a change in energy of one *joule*. The "push" for metals to lose electrons (oxidize) relative to hydrogen doing so has been measured under *standard conditions* (i.e., a concentration of 1 M for solutions and a pressure of 1 atm for gases). Consequently, the "pull" to gain electrons relative to hydrogen is known as well and can be described as a **Standard Reduction Potential, $E°$**. These voltage values are shown in Table 10 and represent half of the redox reactions that may take place in a galvanic cell. You may notice that the order of metals in the list reflects that of the activity series for metals. Fundamentally, the more active a metal, the more negative the reduction potential associated with it.

To determine the *overall* potential of a galvanic cell, often referred to as the **cell potential** or the cell's **electromotive force** (EMF), the two "half-reaction" potentials must be added together. Since the reaction taking place in the cell is a redox process, one of the reactions found written in reduction in Table 10 must be reversed and shown as an oxidation reaction. Because thermodynamics mandates that galvanic cells have positive overall cell potentials, the reaction that must be run in oxidation is easily recognized.

TABLE 10

STANDARD REDUCTION POTENTIALS

Element			Standard Half-Cell Voltages		
			Reduction Electron Reaction		Electrode Potential, $E°$
Potassium			K^+	$+e^- \rightarrow K$	−2.93 V
Calcium			Ca^{2+}	$+2e^- \rightarrow Ca^-$	−2.87 V
Sodium	Increasing	Increasing	Na^+	$+e^- \rightarrow Na$	−2.71 V
Magnesium	tendency	tendency	Mg^{2+}	$+2e^- \rightarrow Mg^-$	−2.37 V
Aluminum	for	for	Al^{3+}	$+3e^- \rightarrow Al^-$	−1.67 V
Zinc	atoms	ions	Zn^{2+}	$+2e^- \rightarrow Zn$	−0.76 V
Iron	to	to gain	Fe^{2+}	$+2e^- \rightarrow Fe$	−0.44 V
Tin	lose	electrons	Sn^{2+}	$+2e^- \rightarrow Sn$	−0.14 V
Lead	electrons	and	Pb^{2+}	$+2e^- \rightarrow Pb$	−0.13 V
Hydrogen	and	form	$2H^+$	$+2e^- \rightarrow H_2$	0.00 V
Copper	form	atoms	Cu^{2+}	$+2e^- \rightarrow Cu$	+0.34 V
Mercury	positive	of the	Hg_2^{2+}	$+2e^- \rightarrow 2Hg$	+0.79 V
Silver	ions	metal	Ag^+	$+e^- \rightarrow Ag$	+0.80 V
Mercury			Hg^{2+}	$+2e^- \rightarrow Hg$	+0.85 V
Gold			Au^{3+}	$+3e^- \rightarrow Au$	+1.50 V

EXAMPLE Determine the standard cell potential for the galvanic cell described in the previous section composed of a zinc electrode immersed in a zinc sulfate solution and a copper electrode immersed in a copper (II) sulfate solution.

SOLUTION

Referring to Table 10, the reduction half-reaction involving zinc is more negative than that of copper. Therefore, the zinc reaction must be run in reverse to get an overall positive cell potential as reversing the reaction also switches the sign on the cell potential for the half-reaction as seen below.

Oxidation half-reaction:	$Zn \rightarrow Zn^{2+} + 2e^-$	$E° = +0.76$ V
Reduction half-reaction:	$Cu^{2+} + 2e^- \rightarrow Cu$	$E° = +0.34$ V
Net reaction:	$Zn + Cu^{2+} \rightarrow Zn^{2+} + Cu$	$E° = +1.10$ V

What this means is that if the concentrations of the Cu^{2+} and Zn^{2+} ions in solution were 1 molar and a voltmeter was attached to the galvanic cell, it would read 1.10 V. The positive cell potential means that electrical energy can be obtained from this galvanic cell to do work. As the cell runs, the concentrations of the ions change and so does the cell potential. The EMF eventually becomes zero, at which point no more work can be done and the cell is dead.

The type of device described in the example (pictured on page 263) is one manner in which a galvanic cell can be designed. Another common design is to use a **salt bridge** as opposed to the porous barrier as a means to eliminate the buildup of charge as the cell operates. This cell design is shown below involving a lead electrode immersed in a lead (II) nitrate solution and an aluminum electrode immersed in a solution of aluminum nitrate. The salt bridge, containing an electrolytic solution, is the tube immersed in each side of the cell. In this example, potassium chloride is the electrolyte. The salt bridge provides ions that can enter either solution to dissipate charge that develops as the cell operates.

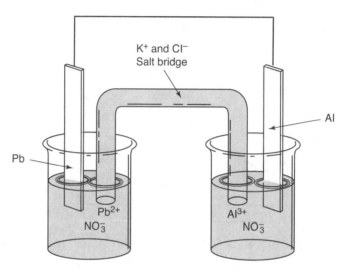

EXAMPLE   Determine the standard cell potential for the galvanic cell shown above. Also, describe the direction of the electron flow through the wire.

SOLUTION

Referring to Table 10, the reduction half-reaction involving aluminum is more negative than that of lead. Therefore, the aluminum reaction must be run in oxidation and the following values for the potentials of the half-reactions can be written.

Oxidation half-reaction:	$Al \rightarrow Al^{3+} + 3e^-$	$E° = +1.67$ V
Reduction half-reaction:	$Pb^{2+} + 2e^- \rightarrow Pb$	$E° = +0.13$ V

The individual half-reactions in this cell differ from the first example because they liberate and use different numbers of electrons with respect to each other. Consequently, the number of electrons must be equalized so that the number of electrons lost in the oxidation reaction equal the number of electrons gained in the reduction. This is easily done by realizing that if the oxidation reaction occurs twice, the reduction can occur three times. This can be represented by placing parentheses around each reaction and using coefficients to multiply each half-reaction to appropriately balance the charge transferred.

Oxidation half-reaction: $2 (Al \rightarrow Al^{3+} + 3e^-)$ $E° = +1.67$ V
Reduction half-reaction: $3 (Pb^{2+} + 2e^- \rightarrow Pb)$ $E° = +0.13$ V

When the oxidation half-reaction occurs twice, a total of 6 electrons are lost. When the reduction half-reaction occurs three times, a total of 6 electrons are gained in a complementary manner. Despite the half-reactions occurring more than once and for a different number of times, it is noteworthy that the half-reaction cell potentials are not modified. That's because cell potential simply describes a desire for the particles of the substances to lose or gain electrons and that desire is independent of how many particles of the substances are involved. Addition of the half-reactions above determines the overall reaction with the standard cell potential as shown.

Net reaction: $2Al + 3Pb^{2+} \rightarrow 2Al^{3+} + 3Pb$ $E° = +1.80$ V

In the net reaction, the 6 electrons being transferred are not shown as they would be found on both sides of the equation and in a sense cancel out. It should be recognized, though, that these electrons *are* involved in the process. The electrons flow through the wire from the anode (Al) to the cathode (Pb) as oxidation always occurs at the anode and reduction always occurs at the cathode. A representation commonly used to describe this in galvanic cells is called **line notation**. In line notation, the constituents of the anode system are written to the left of a double vertical line and that of the cathode written to the right. A single vertical line represents a phase boundary within the particular constituents of an electrode system. The double vertical line represents the salt bridge. For the cell on page 266, the line notation would be $Al \mid Al^{3+} \parallel Pb^{2+} \mid Pb$. This is read from left to right, as an aluminum atom in the anode is oxidized to an aluminum ion transferring electrons to the lead (II) ion, turning it into a lead atom at the cathode.

Nonmetal elements can also participate in galvanic cells. These substances have a natural tendency to gain electrons and so generally have positive reduction potentials. Table 11 describes the reduction reaction and voltage associated with members of the halogen family (Group 17 on the periodic table). As with metals, the reduction potential, $E°$, provides information concerning the relative activities of the elements. With nonmetals, a more positive reduction potential implies a more active element because nonmetals are more apt to gain electrons in chemical processes. It should not be surprising that fluorine has the most positive reduction potential as it is the most active of all nonmetals.

TABLE 11

ACTIVITY OF NONMETALS

Element			Reduction Electron Reaction	E^0_{red}
	↑	■		
Fluorine	Increasing tendency	Increasing tendency	$F_2 + 2e^- \rightarrow 2F^-$	+2.87 V
Chlorine	for atoms to gain	for ions to lose	$Cl_2 + 2e^- \rightarrow 2Cl^-$	+1.36 V
Bromine	electrons and form	electrons and form	$Br_2 + 2e^- \rightarrow 2Br^-$	+1.09 V
Iodine	negative ions	atoms	$I_2 + 2e^- \rightarrow 2I^-$	+0.54 V
	■	↓		

EXAMPLE Determine the standard cell potential for the galvanic cell with the line notation

$$Sn \mid Sn^{2+} \parallel Cl_2, Cl^- \mid Pt$$

SOLUTION

As described by the line notation, the anode electrode system in this cell is comprised of a strip of tin metal immersed in a 1 M solution containing tin (II) ions when under standard conditions. The cathode electrode system, in this case, though, is comprised of an inert electrode, Pt, immersed in a 1 M solution of chloride ions, over which a flow of Cl_2 gas is bubbled at a pressure of 1 atm. This more elaborate electrode system is needed; the cathode cannot be made of the materials involved in the reduction reaction because they are not in the solid state. The platinum does not participate in the reaction but aids in the transfer of the electrons. With information from Tables 10 and 11, the half-reactions are

Oxidation half-reaction:	$Sn \rightarrow Sn^{2+} + 2e^-$	$E° = +0.14$ V
Reduction half-reaction:	$Cl_2 + 2e^- \rightarrow 2Cl^-$	$E° = +1.36$ V

Since two electrons are lost by tin in the oxidation half-reaction and two electrons are gained by the chlorine in the reduction half-reaction, neither reaction needs to be multiplied by a coefficient. Therefore, the overall reaction is

$$Sn + Cl_2 \rightarrow Sn^{2+} + 2Cl^-$$

and the standard cell potential is +1.50 V.

ELECTROLYTIC CELLS

As described in previous sections of this chapter, galvanic (AKA voltaic) cells are devices that take advantage of the transfer of electrons that occurs during a redox reaction converting chemical energy to electrical energy. The electrical energy can be used to do work. Consequently, a galvanic cell is a *type* of **electrochemical cell** which requires a positive EMF that is associated with a spontaneous process. There is another type of electrochemical cell, called an **electrolytic cell**, in which the reaction is not spontaneous and must be forced to occur. This type of cell requires electrical energy be added to the system to promote chemical change that does not naturally occur. Generally, this is referred to as **electrolysis**. In this sense, an operating electrolytic cell is a galvanic cell made to run in reverse. Instead of electrical energy being released by the cell that could be used, a power supply is placed in the electrical circuit, forcing the electrons to flow in the direction opposite the spontaneous process. Therefore, these cells involve reactions with negative cell potentials.

An electrolytic cell is pictured in Figure 36. Without the power supply inserted into the circuit and turned on, electrons would naturally flow from the zinc electrode to the silver electrode. The cell potential for this spontaneous process is +1.56 V under standard conditions. This could be calculated as in the previous section, utilizing the half-reaction cell potentials in Table 10. The anode would be the zinc strip on the left and the cathode would be the silver strip on the right. When turned on, however, the power supply could overcome that natural flow of electrons and force them to travel in the opposite direction. The only requirement is that the voltage provided by the power supply pushing the electrons in the opposite direction be greater than 1.56 V. When this is the case, the zinc electrode becomes the cathode and the silver electrode the anode. This means the silver strip undergoes oxidation. Atoms of silver are converted into ions of silver (Ag^+) and placed into the solution as electrons are pulled toward the zinc electrode. As a result, the silver electrode becomes less massive. Zinc ions (Zn^{2+}) accept the electrons forced upon them, come out of solution, and deposit themselves on the zinc strip due to reduction. Consequently, that electrode, now the cathode, becomes more massive. The reaction that occurs is the reverse of the spontaneous process and is described here.

$$2Ag(s) + Zn^{2+}(aq) \rightarrow 2Ag^+(aq) + Zn(s) \quad E° = -1.56 \text{ V}$$

One use of an electrolytic cell is as a means of coating materials with metals. In this process, called **electroplating**, a layer of metal is deposited on the object, which itself is the cathode in the cell. Recall that reduction always occurs at the cathode; therefore, positive ions of a given metal in solution will adhere to the substance of which the cathode is made as they are reduced to atoms.

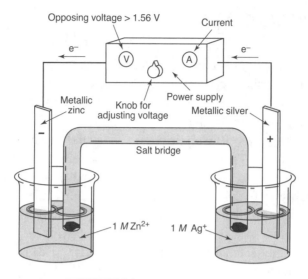

FIGURE 36. An electrolytic cell.

An example of an electroplating process is pictured in Figure 37. A solution of silver nitrate is placed in each compartment of the cell. One electrode is a bar of silver, while the other is a different metal, likely less expensive, in the form of a fork. The negative terminal on the power supply is hooked up to the fork making it the cathode in the electrolytic process. The positive terminal is attached to the silver bar, making it the anode. When the current is switched on, the positive silver ions in the solution are attracted to the fork. When the silver ions make contact, they are reduced and change from ions to atoms of silver. These atoms gradually form a metallic coating on the fork. At the anode, oxidation occurs, and the anode itself is oxidized. These two half-reactions are

$$\begin{aligned} \text{Cathode reaction:} \quad & Ag^+ + e^- \rightarrow Ag \\ \text{Anode reaction:} \quad & Ag \rightarrow Ag^+ + e^- \end{aligned}$$

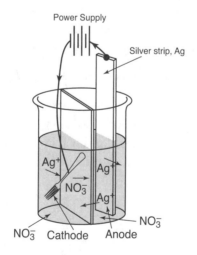

FIGURE 37. Electroplating a metal fork.

The silver ions formed at the anode replace those that are plated onto the fork during the cathode reaction.

QUANTITATIVE ASPECTS OF ELECTROLYTIC CELLS

The amount of metal plated out on the cathode in an electrolytic cell is directly related to the amount of electricity passed through the cell, as well as the particular reactions that occur at each electrode. The amount of electrical charge that passes through the cell in a given amount of time is referred to as the electric **current**. In an electrolytic cell, charge is carried by the electrons as they are transferred from the anode to the cathode. A single electron has a charge of 1.60×10^{-19} coulombs. Therefore, a mole of electrons carries a charge 6.02×10^{23} times larger, or 96,500 coulombs. This value is referred to as a **faraday**, F. The half-reactions associated with the electrolytic process in the cell describe the relationship between the moles of electrons involved in the reactions and the moles of the other substances. Therefore, the time needed to produce a certain amount of plating when a certain current is passed through the cell can be calculated.

EXAMPLE Considering the electrolytic cell previously described in which a fork was electro-plated with silver, determine the mass of silver that would be deposited on the fork if the cell ran for 4.0 seconds at a current of 15 A? (A is the symbol for an ampere, a unit of electric current.)

SOLUTION

Electric current describes the flow of charge. Current is measured by determining the amount of charge (in coulombs) passing through a wire in a given amount of time (in seconds). By definition, the ampere, a unit of electric current, is equal to a coulomb per second. Therefore, the total amount of charge associated with the electrolytic process can be determined by multiplying the current by the time.

$$15 \text{ A} \times 4.0 \text{ s} = 60. \text{ A} \cdot \text{s} = 60. \text{ C} \text{ (C is the symbol for a coulomb.)}$$

Using the definition of a faraday, the number of moles of electrons associated with the electrolysis can be found. Additionally, if you realize that one mole of electrons is associated with the oxidation of one mole of Ag atoms to one mole of Ag+ ions, and you know the molar mass of silver, you can find the mass of silver plated out on the fork.

$$\frac{60. \text{ C}}{1} \times \frac{1 \text{ mol e}^-}{96,500 \text{ C}} \times \frac{1 \text{ mol Ag}}{1 \text{ mol e}^-} \times \frac{108.9 \text{ g Ag}}{1 \text{ mol Ag}} = 0.068 \text{ g Ag}$$

Besides its use in electrolytic cells, the electrolysis process is commonly used to decompose compounds. A typical apparatus used to decompose water is shown in Figure 38.

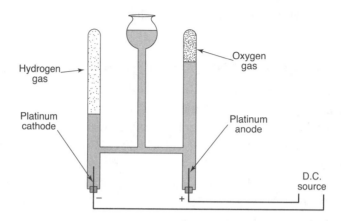

FIGURE 38. Electrolysis of water.

The solution in this apparatus contains distilled water and a small amount of H_2SO_4. The reason for adding H_2SO_4 is to make the solution conductive since distilled water alone will not conduct an electric current. The solution, therefore, contains ions of H^+, HSO_4^-, and SO_4^{2-}.

The hydrogen ions (H^+) migrate to the cathode where they are reduced to hydrogen atoms and form hydrogen molecules (H_2) in the form of a gas. The SO_4^{2-} and HSO_4^- migrate to the anode but are not oxidized since the oxidation of water occurs more readily. These ions then are merely spectator ions. The oxidized water reacts as shown in the following half-reactions.

Cathode reaction:
Reduction: $\qquad 4H_2O + 4e^- \rightarrow 2H_2(g) + 4OH^-$
Anode reaction:
Oxidation: $\qquad\qquad 2H_2O \rightarrow O_2(g) + 4H^+ + 4e^-$

Net reaction: $\qquad 2H_2O \rightarrow 2H_2(g) + O_2(g)$

Notice that the equation shows 2 vol of hydrogen gas are released while only 1 vol of oxygen gas is liberated. This agrees with the discussion of the composition of water in Chapter 7.

APPLICATIONS OF ELECTROCHEMICAL CELLS (COMMERCIAL GALVANIC CELLS)

One of the most common galvanic cells is the ordinary "dry cell" used in flashlights. Its general makeup is shown in the following drawing, along with oxidation and reduction reactions happening at the anode and cathode. The voltage associated with the overall redox reaction is 1.5 V.

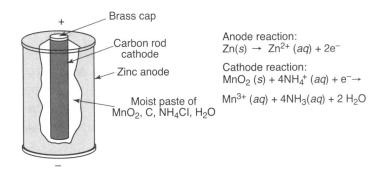

Brass cap

Carbon rod
cathode

Zinc anode

Moist paste of
MnO_2, C, NH_4Cl, H_2O

Anode reaction:
$Zn(s) \rightarrow Zn^{2+}(aq) + 2e^-$

Cathode reaction:
$MnO_2(s) + 4NH_4^+(aq) + e^- \rightarrow$

$Mn^{3+}(aq) + 4NH_3(aq) + 2 H_2O$

The standard dry cell consists of a cylindrical zinc container, which also serves as the anode. Next to the zinc, not pictured in the drawing but typically found, is a porous cardboard paper barrier that serves as a salt bridge to dissipate the buildup of charge. To the inside of the paper is a paste made of the solids manganese dioxide, carbon, ammonium chloride, and a little water. The small amount of water is the reason why it's called a "dry" cell. A graphite (carbon) rod is immersed in the paste at the center of the cell. Since the substances involved in the reduction reaction are not solid, the carbon rod acts as an inert electrode that provides a solid support where the reduction reaction can occur. This is called the **cathode**. In voltaic (galvanic) cells, the cathode is the positive electrode. The brass cap that sits atop the graphite rod is described, then, as the positive terminal of the dry cell. The bottom of the dry cell is an exposed section of the zinc casing (the anode) and is, therefore, the negative terminal.

The automobile lead storage battery is also a common galvanic cell. In reality, it is a series of galvanic cells. A series of galvanic cells is commonly referred to as a **battery**, but the term is also used to describe a single galvanic cell, as is the case of the dry cell previously discussed. In the common car battery, electrodes are made of Pb and $PbSO_4$, which are immersed in a sulfuric acid solution. A sulfuric acid solution mostly contains bisulfate (HSO_4^-) and hydrogen (H^+) ions. The oxidation and reduction reactions that happen at the anode and cathode are shown in the diagram below.

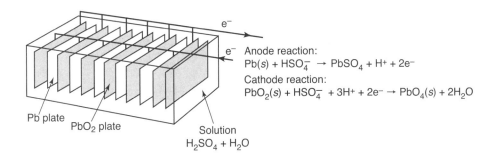

e^-

e^- Anode reaction:
$Pb(s) + HSO_4^- \rightarrow PbSO_4 + H^+ + 2e^-$
Cathode reaction:
$PbO_2(s) + HSO_4^- + 3H^+ + 2e^- \rightarrow PbO_4(s) + 2H_2O$

Pb plate
PbO_2 plate
Solution
$H_2SO_4 + H_2O$

The voltage associated with the overall redox reaction is about 2 V. Typically, each battery contains six cells, so the total voltage output on discharge is about 12 V. These types of batteries can also be *recharged* by forcing a current through the cell in the opposite direction of the spontaneous process. This is accomplished by the alternator in the car. This electrolytic process returns the battery to its conditions prior to the discharge so it can operate at its standard voltage as needed over and over again. The life of a car battery is limited, though, by physical stresses due to the conditions in which it operates, as well as side reactions that occur within the battery itself.

Balancing Redox Equations

In general, many chemical reactions are so simple that once the reactants and products are known, the equations can be readily written. Many redox reactions, however, are of such complexity that the process of writing the equations by trial and error is time-consuming. As a result, methods have been developed to aid in the balancing of these types of reactions.

THE HALF-REACTION METHOD

The half-reaction method presented here is appropriate for balancing complex redox reactions that occur in aqueous solution. This is the case for many of the reactions presented in this chapter that are used in galvanic cells. In addition to having information about the main reactants and products, you also need to know whether the reaction occurs in an acidic or basic environment. The general steps for balancing a redox reaction occurring in an *acid environment* are provided below, followed by examples detailing the use of the method.

1. Write the oxidation and reduction parts of the redox process as separate (half) reactions.
2. Balance the half-reactions individually, considering the following in order:
    a. Elements besides hydrogen and oxygen
    b. Oxygen by using water (H_2O)
    c. Hydrogen by using hydrogen ions (H^+)
    d. Charge by using electrons (e^-)
3. Equalize the charge on both half-reactions.
4. Add the balanced half-reactions, canceling out identical species found in both reactants and products

EXAMPLE
1

Balance the reaction that occurs in an acidic aqueous solution between dichromate and cobalt (II) ions producing chromium (III) and cobalt (III) ions.

SOLUTION

The basic reaction described is

$$Cr_2O_7^{2-}(aq) + Co^{2+}(aq) \rightarrow Cr^{3+}(aq) + Co^{3+}(aq)$$

Initial inspection indicates that the reaction cannot be balanced by normal means as there is no oxygen in the products to balance the oxygen in the reactants. Additionally, the amount of charge is not the same on both sides of the equation. The half-reaction method provides a means to balance both, using substances that are present to the reaction.

STEP 1: Identifying the half-reactions

Reduction: $Cr_2O_7^{2-}(aq) \rightarrow Cr^{3+}(aq)$

(The oxidation state of chromium is changed from +6 to +3.)

Oxidation: $Co^{2+}(aq) \rightarrow Co^{3+}(aq)$

(The oxidation state of cobalt is changed from +2 to +3.)

STEP 2: Balancing the half-reactions (using the order listed in the method on page 274)

Reduction: $6\,e^- + 14H^+(aq) + Cr_2O_7^{2-}(aq) \rightarrow 2\,Cr^{3+}(aq) + 7\,H_2O(\ell)$

Oxidation: $Co^{2+}(aq) \rightarrow Co^{3+}(aq) + e^-$

STEP 3: Equalizing the charge

Reduction: $6\,e^- + 14H^+(aq) + Cr_2O_7^{2-}(aq) \rightarrow 2\,Cr^{3+}(aq) + 7\,H_2O(\ell)$

Oxidation: $6\,(Co^{2+}(aq) \rightarrow Co^{3+}(aq) + e^-)$

(The oxidation reaction must occur six times for the reduction reaction to occur once, making the amount of charge transferred equal.)

STEP 4: Adding the half-reactions

$$14H^+(aq) + Cr_2O_7^{2-}(aq) + 6\,Co^{2+}(aq) \rightarrow$$
$$2\,Cr^{3+}(aq) + 7\,H_2O(\ell) + 6\,Co^{3+}(aq)$$

(The total charge on both sides of the reaction is balanced (at 24+), as is the number and type of particles.)

EXAMPLE
2

Balance the reaction that occurs in an automobile battery, described in a previous section, where lead reacts with lead (IV) oxide, in sulfuric acid, to make lead (II) sulfate.

SOLUTION

The basic reaction described is

$$Pb(s) + PbO_2(s) \rightarrow PbSO_4(s)$$

STEP 1: Identifying the half-reactions

Reduction: $PbO_2(s) \rightarrow PbSO_4(s)$

(The oxidation state of lead is changed from +4 to +2.)

Oxidation: $Pb(s) \rightarrow PbSO_4(s)$

(The oxidation state of lead is changed from 0 to +2.)

STEP 2: Balancing the half-reactions (using the order listed in the method on page 274)

Reduction:

$$2\ e^- + 3\ H^+(aq) + HSO_4^-(aq) + PbO_2(s) \rightarrow PbSO_4(s) + 2\ H_2O(\ell)$$

(HSO_4^- is added to the reactants to balance the sulfur. It can be added because it is present as a major species in the sulfuric acid solution. Adding water to the products balances the oxygen. Adding H^+ to the reactants balances the hydrogen. Adding electrons to the reactants balances the charge.)

Oxidation: $HSO_4^-(aq) + Pb(s) \rightarrow PbSO_4(s) + H^+(aq) + 2\ e^-$

STEP 3: Equalizing the charge

(Electrons transferred in the reactions is equalized as is.)

STEP 4: Adding the half-reactions

$$2H^+(aq) + 2\ HSO_4^-(aq) + PbO_2(s) + Pb(s) \rightarrow 2\ PbSO_4(s) + 2\ H_2O(\ell)$$

(There is a net of 2 H^+ ions in the reactants because there were three in the reactants of the reduction reaction but also one in the products of the oxidation reaction. There is a net of 2 HSO_4^- ions in the reactants because both half-reactions had one in the reactants. The same can be said for the $PbSO_4$ but in the products.)

Chapter Summary

The following terms summarize all the concepts and ideas that were introduced in this chapter. You should be able to explain their meaning and how you would use them in chemistry. They appear in boldface type in this chapter to draw your attention to them. The boldface type also makes the terms easier for you to look up if you need to. You could also use a search engine on your computer to get a quick and expanded explanation of these terms, laws, and formulas.

TERMS YOU SHOULD KNOW

anions
anode
battery
burning
cathode
cations
cell potential
combustion
combustion reactions
current
dissociation
electrochemical cell
electrochemistry
electrode
electrode potential
electrolysis
electrolyte
electrolytic cell
electrolytic reaction
electromotive force
electromotive series

electroplating
faraday
galvanic cell
half-cell
ionization
line notation
natural gas
nonelectrolyte
notation
oxidation
oxidation state
oxidizing agent
reducing agent
reduction
redox
salt bridge
single replacement
spectator ions
Standard Reduction Potential
standard voltage
voltaic cell

Chapter 12 Review Exercises

1. Which of the following when dissolved in water and placed in the conductivity apparatus will cause the light to glow?

   (A) table salt
   (B) ethyl alcohol
   (C) sugar
   (D) glycerine

2. The reason that a current can flow is because

   (A) ions combine to form molecules
   (B) molecules migrate to the charge plates
   (C) ions migrate to the charge plates
   (D) sparks cross the gap

3. The extent of ionization depends on the

   (A) nature of the solvent
   (B) nature of the solute
   (C) concentration of the solution
   (D) temperature of the solution
   (E) all the above

4. The oxidation state of manganese in the polyatomic ion MnO_4^- is

   (A) +1
   (B) +4
   (C) +7
   (D) +8

5. Which element is being reduced in the following reaction?

   $$6\,Fe^{2+}(aq) + Cr_2O_7^{2-}(aq) + 14\,H^+(aq) \rightarrow 6\,Fe^{3+}(aq) + 2\,Cr^{3+}(aq) + 7\,H_2O(\ell)$$

   (A) Fe
   (B) Cr
   (C) O
   (D) H

6. Which substance is the oxidizing agent in the following reaction?

$$3\ H_2O_2(aq) + 2\ Al(s) \rightarrow 2\ Al(OH)_3(s)$$

(A) H
(B) H_2O_2
(C) Al
(D) $Al(OH)_3$

7. Identify the combustion reaction(s) below.

 I. $Na(s) + O_2(g) \rightarrow Na_2O(s)$
 II. $CH_4(g) + O_2(g) \rightarrow CO_2(g) + H_2O(g)$
 III. $H_2(g) + Cl_2(g) \rightarrow HCl(g)$

(A) I
(B) II
(C) I and II
(D) II and III
(E) I, II, and III

8. In an electrochemical cell, reduction always occurs

(A) at the anode
(B) at the cathode
(C) in the salt bridge
(D) at both the anode and cathode

9. Using the reduction potentials given in Table 10, determine the standard cell potential for the reaction described by the following line notation.

$$Cu\ |\ Cu^{2+}\ ||\ Ag^+\ |\ Ag$$

(A) +0.34 V
(B) +0.46 V
(C) +0.80 V
(D) +1.14 V

10. How much time would it take for a current of 3.00 A to reduce Cu^{2+} ions in solution to produce 5.00 grams of copper metal?

(A) 31.7 minutes
(B) 36.1 minutes
(C) 42.3 minutes
(D) 50.6 minutes

11. An example of a galvanic cell is found when

(A) using a dry cell battery
(B) recharging a battery
(C) decomposing water with electricity
(D) forcing a nonspontaneous redox reaction to occur

12. Consider the reaction between ions of iron (II) and permanganate in an aqueous acid environment where ions of iron (III) and manganese (II) are produced. What would the coefficient in front of the H^+ ion be in the balanced redox reaction when considering the smallest whole-number coefficients possible.

(A) 5
(B) 6
(C) 7
(D) 8

Questions 13–15

The following elements are listed in order of decreasing reactivity as they appear in the electrochemical series.

Ca, Na, Mg, Zn, Fe, H, Cu, Hg, Ag, Au

13. Which element is the best reducing agent?

(A) Ca
(B) Au
(C) H
(D) Fe
(E) Cu

14. Of the following, which element does NOT react with hydrochloric acid to produce hydrogen gas?

(A) Zn
(B) Fe
(C) Hg
(D) Ca
(E) Mg

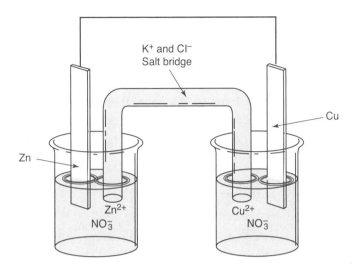

K⁺ and Cl⁻
Salt bridge

Cu

Zn

Zn²⁺
NO₃⁻

Cu²⁺
NO₃⁻

15. In the electrochemical cell shown above, which of the following half-reactions occurs at the anode?

   (A) $Cu^{2+} + e^- \rightarrow Cu+$
   (B) $Zn(s) \rightarrow Zn^{2+} + 2e^-$
   (C) $Zn^{2+} + 2e^- \rightarrow Zn(s)$
   (D) $Cu(s) \rightarrow Cu^{2+} + 2e^-$
   (E) $Cu^{2+} + 2e^- \rightarrow Cu(s)$

Answers and Explanations

1. **(A)** Table salt is the only substance listed that has ions when dissolved in water and will show a current is moving through the system by lighting the conductivity apparatus.

2. **(C)** The current is carried through the solution by ions migrating to the charged plates that act as the anode and cathode.

3. **(E)** The extent of ionization of a substance in a solvent depends on all four of the listed factors.

4. **(C)** The oxidation state of oxygen is –2. Since there are 4 oxygen atoms in the polyatomic ion MnO_4^- and the overall charge is 1–, the oxidation state of manganese must be +7.

5. **(B)** Chromium is being reduced in the reaction. The oxidation state of chromium is changing from +6 to +3.

6. **(B)** Hydrogen peroxide, H_2O_2, is the oxidizing agent as it is the substance associated with the element being reduced. The oxygen is being reduced from –1 in the peroxide to –2 in the $Al(OH)_3$.

7. **(C)** Reactions I and II are both combustion reactions. Combustion reactions occur when substances combine with oxygen liberating heat and/or light.

8. **(B)** By definition, the cathode is the location where reduction occurs in either type of electrochemical cell—galvanic or electrolytic.

9. **(B)** The overall cell potential is 0.46 V because the oxidation reaction $(Cu \rightarrow Cu^{2+} + 2e^-)$ has a cell potential of –0.34 V and the reduction reaction $(Ag^+ + e^- \rightarrow Ag)$ has a cell potential of +0.80 V. The addition of the half-reaction cell potentials results in the overall cell potential.

10. **(D)** The reaction for the reduction of copper (II) ions to copper atoms is

$$Cu^{2+} + 2e^- \rightarrow Cu$$

The amount of charge, in coulombs, to produce 5.00 grams of copper is

$$\frac{5.00 \text{ g Cu}}{1} \times \frac{1 \text{ mol Cu}}{63.5 \text{ g Cu}} \times \frac{2 \text{ mol e}^-}{1 \text{ mol Cu}} \times \frac{96,500 \text{ C}}{1 \text{ mol e}^-} = 15,200 \text{ C}$$

Since charge divided by time equals current, the time it takes to obtain that amount of charge with a current of 5.00 A is found by dividing the charge by the current.

$$\frac{15,200 \text{ C}}{5.00 \text{ A}} = 3040 \text{ s} = 50.6 \text{ minutes}$$

11. **(A)** A galvanic cell involves a spontaneous process with a positive cell potential. The only device or description with a positive cell potential is the dry cell.

12. **(D)** The coefficient in front of the H^+ is 8. This can be found when the redox reaction described is balanced using the half-reaction method. The balanced half-reactions are

Oxidation: $\quad 5 \text{ (Fe}^{2+} \rightarrow \text{Fe}^{3+} + e^-)$
Reduction: $\quad 5 e^- + 8 H^+ + MnO_4^- \rightarrow Mn^{2+} + 4 H_2O$

Addition of the half-reactions yields

$$5 \text{ Fe}^{2+} + 8H^+ + MnO_4^- \rightarrow 5 \text{ Fe}^{3+} + Mn^{2+} + 4 H_2O$$

13. **(A)** Ca is the most active. Therefore it loses electrons so that reduction can occur. Reducing agents are oxidized.

14. **(C)** Because Hg is below H in activity, Hg cannot displace H.

15. **(B)** Zn acts as the anode in this setup as it is shown as a galvanic cell. If a power source were in the wiring and the device were run as an electrolytic cell, the zinc would be the cathode.

PART VI

REPRESENTATIVE GROUPS AND FAMILIES

SOME REPRESENTATIVE GROUPS AND FAMILIES

CHAPTER OBJECTIVES

Upon completing this chapter, you will be able to:

- Describe the properties, both physical and chemical, of the major members of and compounds formed by the sulfur, halogen, and nitrogen families and by metals and their alloys

- Write equations for major reactions and processes involving these elements

In the following section, a brief description is given of some of the important and representative groups of elements usually referred to in most first-year chemistry courses. The presentation of each group will follow along this outline: the important element or elements, their occurrence, their preparation both in laboratory and in industrial processes, their properties, and some of their important uses.

Sulfur Family

Since we discussed oxygen in Chapter 5, the most important element left to discuss is sulfur.

Sulfur has three **allotropic forms**. Their differences are summarized in Table 12.

TABLE 12
ALLOTROPIC FORMS OF SULFUR

Characteristic	Allotropic Form		
	Rhombic	Monoclinic	Amorphous
Shape	Rhombic or octahedral crystals	Needle-shaped monoclinic crystals	Noncrystalline
Color	Pale-yellow, opaque, brittle	Yellow, waxy, translucent, brittle	Dark, tough, elastic
Preparation	Crystallizes from CS_2 solution of sulfur	Crystallizes from molten sulfur as it cools	Obtained by quick-cooling very hot liquid sulfur
Stability	Stable below 269 K	Stable between 269 K and 392 K	Unstable; slowly changes to rhombic

The solid sulfur is composed of S_8 molecules as shown below:

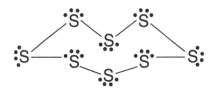

Since each covalent bond is sharing a pair of electrons, there is a potential for extending this structure: ($\cdot\ddot{S}\cdot$) which is found in the S_8 molecule. As heat is applied, the kinetic molecular motion increases and eventually breaks this ring structure, leading to a chain that becomes entangled with other chains and explains the increased viscosity of the liquid sulfur at high temperatures. Because the broken chain has unshared electrons on its ends, it is able to combine with other chains to form even longer chains and thus increase the viscosity further. The unpaired electrons are also associated with the change in color that occurs.

Sulfur is used in making sprays to control plant diseases and harmful insects. It is used in rubber to provide added hardness or to "vulcanize" it. Sulfur enters into the preparation of medicines, matches, gunpowder, and fireworks. Its most important use is making sulfuric acid, which is often referred to as the "king of chemicals" because of its widespread industrial use.

SULFURIC ACID

IMPORTANT PROPERTIES OF SULFURIC ACID

Sulfuric acid ionizes in two steps:

$$H_2SO_4 + H_2O \rightleftharpoons H_3O^+ + HSO_4^- \quad (Ka_1 \text{ is very large.})$$
$$HSO_4^- + H_2O \rightleftharpoons H_3O + SO_4^{2-} \quad (Ka_2 \text{ is very small.})$$

to form a strong acid solution. The ionization is more extensive in a dilute solution. Most hydronium ions are formed in the first step as indicated by the larger Ka. Salts formed with the HSO_4^- (bisulfate ion) are called **acids salts**; the SO_4^{2-} (sulfate ion) forms **normal salts**. The test for sulfate ion consists of adding a solution of HCl and $BaCl_2$ or $Ba(NO_3)_2$ to the solution to be tested. If sulfate is present, a white precipitate of $BaSO_4$ will form.

Sulfuric acid reacts like other acids, as shown below:

With active metals:	$Zn + H_2SO_4 \rightarrow ZnSO_4 + H_2(g)$ (for dilute H_2SO_4)
With bases:	$2NaOH + H_2SO_4 \rightarrow Na_2SO_4 + 2H_2O$
With metal oxides:	$MgO + H_2SO_4 \rightarrow MgSO_4 + H_2O$
With carbonates:	$CaCO_3 + H_2SO_4 \rightarrow CaSO_4 + H_2O + CO_2(g)$

Sulfuric acid has other particular characteristics as shown below:

As an oxidizing agent:

$$Cu + 2H_2SO_4 \text{ (concentrated)} \rightarrow CuSO_4 + SO_2(g) + 2H_2O$$

As a dehydrating agent with carbohydrates:

$$C_{12}H_{22}O_{11}(\text{sugar}) \xrightarrow[\text{H}_2\text{SO}_4]{\text{conc.}} 12C + 11H_2O$$

IMPORTANT USES OF H₂SO₄

1. Making other acids (because of its high boiling point, 611 K):

$$NaCl + H_2SO_4 \xrightarrow{\text{heat}} NaHSO_4 + HCl(g)$$

   (With more heat and an excess of salt, the second hydrogen ion can also be removed to form the salt Na_2SO_4.)

2. Washing objectionable colors from gasoline made by the "cracking" process.

3. Dehydrating agent in making explosives, dyes, and drugs.

4. Freeing metals of iron or steel of scale and rust. This is called "pickling."

5. Electrolyte in ordinary lead storage batteries.

6. Used in precipitating bath in making rayon.

This is a list of only the more important uses of sulfuric acid. It is by no means complete.

OTHER IMPORTANT COMPOUNDS OF SULFUR

Hydrogen sulfide is a colorless gas having an odor of rotten eggs. It is fairly soluble in water and poisonous in rather small concentrations. It can be prepared by reacting ferrous sulfide with an acid, such as dilute HCl:

$$FeS + 2HCl \rightarrow FeCl_2 + H_2S(g)$$

It does burn in excess oxygen to form compounds of water and sulfur dioxide. If insufficient oxygen is available, some free sulfur will form. It is only a weak acid in a water solution. Hydrogen sulfide is used widely in qualitative laboratory tests since many metallic sulfides precipitate with recognizable colors. These sulfides are sometimes used as paint pigments. Some common sulfides and their colors are

ZnS —White
CdS —Bright yellow
As_2S_3 —Lemon yellow
Sb_2S_3 —Orange
CuS —Black
HgS —Black
PbS —Brown-black

Another important compound of sulfur is *sulfur dioxide*. It is a colorless gas with a suffocating odor. It can be prepared by burning sulfur in air:

$$S + O_2 \rightarrow SO_2$$

or decomposing sulfites:

$$Na_2SO_3 + H_2SO_4 \rightarrow Na_2SO_4 + H_2O + SO_2(g)$$

Sulfur dioxide is the acid anhydride of sulfurous acid:

$$SO_2 + H_2O \rightarrow H_2SO_3$$

It is used as a bleach on moist straw, silk, wool, and paper because of its ability to reduce the coloring agent. This process is reversed by oxidation in sunlight as shown by the yellowing of straw and paper. Potassium permanganate can also be bleached by a water solution of SO_2 in the form of H_2SO_3:

$$5H_2SO_3 + 2KMnO_4 \rightarrow K_2SO_4 + 2MnSO_4 + 2H_2SO_4 + 3H_2O$$

The structure of sulfur dioxide is a good example of a **resonance structure**. Its molecule is depicted in Figure 39.

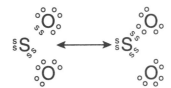

FIGURE 39. Sulfur dioxide molecule.

Notice that the covalent bonds between sulfur and oxygen are shown in one figure as single bonds and in the other as double bonds. This signifies that the bonds between the sulfur and oxygens have been shown by experimentation to be neither single nor double bonds but "hybrids" of the two. Sulfur trioxide also is a resonance structure.

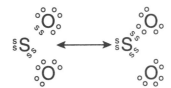

"——" indicates a covalent bond

Halogen Family

The common members of the halogen family are shown in Table 13 with some important facts concerning them.

TABLE 13
HALOGEN FAMILY

Item	Fluorine	Chlorine	Bromine	Iodine
Molecular formula	F_2	Cl_2	Br_2	I_2
Atomic number	9	17	35	53
Activity	Most active	◄——— to ———►		Least active
Outer energy level structure	7 electrons	7 electrons	7 electrons	7 electrons
State and color at room temperature	Gas—pale yellow	Gas—green	Liquid— dark red	Solid— purplish black crystals

Because all the halogens lack one electron in their outer principal energy level, they usually are acceptors of electrons (oxidizing agents). Fluorine is the most active nonmetal in the periodic chart.

TESTING FOR HALIDES

This ability of chlorine to replace I^- and Br^- ions is also used in testing for iodides and bromides. The test involves putting the salt into a water solution and then adding fresh chlorine water and some carbon tetrachloride. The resulting mixture is shaken vigorously and observed for a purple or red color in the carbon tetrachloride, which will sink to the bottom of the container. Purple indicates a positive test for iodine; deep red shows the presence of bromine. The reactions that release these halogens are as follows:

$$2I^- + Cl_2 \rightarrow 2Cl^- + I_2 \text{ (purple in CCl}_4)$$
$$2Br^- + Cl_2 \rightarrow 2Cl^- + Br_2 \text{ (deep red in CCl}_4)$$

Test for fluoride Add sulfuric acid and test if the gas released, which will be hydrogen fluoride (HF), will etch glass. (Use **caution** not to get on hands.)

Test for chloride Add a solution of silver salt (usually silver nitrate) and look for a white precipitate that is soluble in ammonium hydroxide but insoluble in nitric acid. The white precipitate darkens to gray when exposed to light.

SOME IMPORTANT HALIDES AND THEIR USES

Hydrochloric acid Common acid prepared in the laboratory by reacting sodium chloride with concentrated sulfuric acid. It is used in many important industrial processes.

Silver bromide and silver iodide Halides used on photographic film. Light intensity is recorded by developing as black metallic silver those portions of the film upon which the light fell during exposure.

Hydrofluoric acid Acid used to etch glass by reacting with the SiO_2 to release silicon fluoride gas. Also used to frost light bulbs.

USES OF HALOGENS

Fluorides are added to drinking water and to toothpaste to reduce tooth decay. The halogen chlorine is used to purify water supplies, to bleach, and to prepare pure hydrochloric acid, along with many other industrial applications. Its reaction as a bleach is

$$Cl_2 + H_2O \rightarrow HCl + HClO \text{ (hypochlorous acid)}$$

It is the hypochlorous acid that does the bleaching by oxidation.

$$\text{Colored material} + HClO \rightarrow \text{decolorized material} + HCl$$

Chlorine also has the ability to withdraw hydrogen from hydrocarbons. An example is:

$$C_{10}H_{16} + 8Cl_2 \rightarrow 10C + 16HCl$$

$$\uparrow \qquad\qquad\qquad \uparrow$$

Warm	Carbon released
turpentine	as dense, black
	smoke

The use of bromine containing compounds has been curtailed in recent years, however, due to environmental concerns with such materials.

Iodine is most widely known for its use as an antiseptic in an alcohol solution of iodine called tincture of iodine. Iodine is also used in small amounts to treat goiter conditions.

Nitrogen Family

The most common member of this family is nitrogen itself. It is a colorless, odorless, tasteless, and rather inactive gas that makes up about four-fifths of the air in our atmosphere. The inactivity of N_2 gas can be explained by the fact that the two atoms of nitrogen are bonded by three covalent bonds ($:N\!::\!N:$) that require a great deal of energy to break. Since nitrogen must be "pushed" into combining with other elements, many of its compounds tend to decompose violently with a release of the energy that went into forming them.

Nitrogen can be prepared in the laboratory by the equation

$$NaNO_2 + NH_4Cl \rightarrow NaCl + 2H_2O + N_2(g)$$

It is most often prepared industrially by the fractional distillation of liquid air. This product, of course, contains some of the inert gases but is pure enough for most uses.

Nature "fixes" nitrogen, or makes nitrogen combine, by a nitrogen-fixing bacteria found in the roots of beans, peas, clover, and other leguminous plants. Discharges of lightning also cause some nitrogen fixation with oxygen to form nitrogen oxides.

NITRIC ACID

An important compound of nitrogen is nitric acid. This acid is useful in making dyes, celluloid film, and many of the lacquers used on cars.

Its physical properties are as follows: It is a colorless liquid (when pure), it is $1\frac{1}{2}$ times as dense as water, it has a boiling point of 86°C, the commercial form is about 68% pure, and it is miscible with water in all proportions.

Its outstanding chemical properties are: the dilute acid shows the usual properties of an acid except it rarely produces hydrogen when it reacts with metals, and it is quite unstable and decomposes as follows:

$$4HNO_3 \rightarrow 2H_2O + 4NO_2(g) + O_2(g)$$

Because of this ease of decomposition, nitric acid is a good oxidizing agent. When it reacts with metals, the nitrogen product formed will depend on the conditions of the reaction, especially the concentration of the acid, the activity of the metal, and the temperature. If the nitric acid is concentrated and the metal is copper, the principal reduction product will be nitrogen dioxide (NO_2), a heavy, red-brown gas with a pungent odor.

$$Cu + 4HNO_3 \rightarrow Cu(NO_3)_2 + 2NO_2(g) + 2H_2O$$

With dilute nitric acid, this reaction is

$$3Cu + 8HNO_3 \rightarrow 3Cu(NO_3)_2 + 4H_2O + 2NO(g)$$

The product, called nitric oxide (NO), is colorless and is immediately oxidized in air to NO_2 gas.

With still more dilute nitric acid, considerable quantities of nitrous oxide (N_2O) are formed; with an active metal like zinc, the product may be the ammonium ion (NH_4^+).

When nitric acid is mixed with hydrochloric acid, the mixture is called *aqua regia* because of its ability to dissolve gold. The great activity of aqua regia is due to the formation of the nitrosyl chloride (NOCl) and chlorine (Cl_2). The Cl_2 reacts with gold to form the soluble compound of gold.

Nitric acid reacts with protein to give a yellow color and also reacts with other organic compounds to form unstable compounds, such as nitrobenzene, trinitrotoluene (TNT), and picric acid. The last two are powerful explosives. Another explosive is made by reacting glycerine with nitric acid to form glyceryl trinitrate ("nitroglycerine").

The preparation of nitric acid in the laboratory involves reacting a nitrate salt with sulfuric acid in a retort.

The industrial preparation, called the Ostwald process, is the greatest source of nitric acid. Ammonia and air are passed through a heated platinum gauze, which acts as a catalyst, to form nitric oxide (NO).

$$4NH_3 + 5O_2 \rightarrow 4NO(g) + 6H_2O$$

The nitric oxide is further oxidized to NO_2.

$$2NO + O_2 \rightarrow 2NO_2(g)$$

Then $3NO_2 + H_2O \rightarrow 2HNO_3 + NO(g)$ forms the nitric acid.

OTHER IMPORTANT COMPOUNDS OF NITROGEN

Ammonia is one of the oldest known compounds of nitrogen. In times past, it was prepared by distilling leather scraps, hoofs, and horns. It can be prepared in the laboratory by heating an ammonium salt with a strong base. An example is

$$Ca(OH_2) + (NH_4)_2SO_4 \rightarrow CaSO_4 + 2H_2O + 2NH_3(g)$$

Industrial methods include the **Haber process**, which combines nitrogen and hydrogen (discussed further on page 219)

$$3H_2 + N_2 \rightarrow 2NH_3$$

and the destructive distillation of coal, which gives off NH_3 as a by-product.

The physical properties of ammonia are: it is a colorless, pungent gas that is extremely soluble in water; it is lighter than air and has a high critical temperature (the critical temperature of a gas is the temperature above which it cannot be liquefied by pressure alone).

Because ammonia can be liquefied easily and has a high heat of vaporization, one of its uses is as a refrigerant. Also, a water solution of ammonia is used as a household cleaner to cut grease. Another use of ammonia is as a fertilizer.

Some other nitrogen compounds are summarized in Table 14.

TABLE 14
SOME NITROGEN COMPOUNDS

Name	Formula	Notes of Importance
Nitrate ion	NO_3^-	1. Nitrates of metals can be decomposed with heat to release oxygen. 2. Nitrate test—add freshly prepared ferrous sulfate solution; hold test tube at an angle and slowly add concentrated sulfuric acid down the side; if a brown ring forms just above the sulfuric acid level, it indicates a positive test for nitrate ion. (Test can be repeated using acetic instead of sulfuric acid; if brown ring again forms, it indicates a nitrite.)
Nitrous oxide	N_2O	1. Colorless; 1.5 times denser than air; slight sweet smell and odor; supports combustion much like oxygen. 2. Sometimes used as an anesthetic called "laughing gas."
Nitric oxide	NO	1. Colorless; slightly denser than air; does not support combustion; is easily oxidized to NO_2 when exposed to air.
Nitrogen dioxide	NO_2	1. Reddish-brown gas; 1.6 times heavier than air; very soluble in water; extremely poisonous; has unpleasant odor.
Nitrogen tetroxide	N_2O_4	1. When NO_2 is cooled, its brown color fades to a pale yellow (N_2O_4). Two molecules of NO_2 combine to form this new compound.

OTHER MEMBERS OF THE NITROGEN FAMILY

As you proceed down this family (see Table 15), the elements change from gases to volatile solids and then to less volatile solids. At the same time, acid-forming properties are decreasing while base-forming properties are increasing. These changes can be attributed to the increased atomic radii and the decreased ability to attract and hold the outer energy level electrons.

TABLE 15
THE NITROGEN FAMILY

Name	Symbol	Atomic Number	Properties and Uses
Nitrogen	N	7	(Discussed previously)
Phosphorus	P	15	Two allotropic forms: white (kindling temperature near room temperature, thus kept under water) and red. When burned in oxygen, forms P_2O_5, the acid anhydride of phosphoric acid. When P_2O_5 reacts with water, it first forms metaphosphoric acid, HPO_3; then more water reacts to form the orthophosphoric acid. Phosphorus is used in making matches.
Arsenic	As	33	Slightly volatile, gray solid. Most abundant compound is As_2O_3, which is used as a preservative, in medicines, and in glassmaking (highly poisonous).
Antimony	Sb	51	Lustrous, gray solid. Forms very weak acid, mostly basic solutions.
Bismuth	Bi	83	Almost wholly metallic properties. Shiny, dense metal with slight reddish tinge. Low melting point; therefore used in alloys for electric fuses and fire sprinkler systems. "Woods metal" is the most outstanding of these alloys (melting point 70°C).

Metals

PROPERTIES OF METALS

Some physical **properties of metals** are: they have metallic luster, they can conduct heat and electricity, they can be pounded into sheets (malleable), they can be drawn into wires (ductile), most have a silvery color, they have densities usually between 7 and 14 g/cm^3, and none is soluble in any ordinary solvent without a chemical change.

The general chemical properties of metals are: they are electropositive, and the more active metallic oxides form bases, although some metals form amphoteric hydroxides, which can react as both acid and base.

Metals can be removed from their ores by various means depending on their degree of activity (see Table 16).

TABLE 16
MEANS OF EXTRACTING METALS

Occurrence	Activity Decreasing	Metal	Ease of Reduction	Reduced by
In combined form only		Potassium Calcium Sodium Magnesium Aluminum	Very difficult	Electrolysis
		Manganese Zinc Chromium	Difficult	Carbon with diminishing difficulty
Usually combined		Iron Cadmium Cobalt Nickel		
		Tin Lead		
		HYDROGEN	Easy	Carbon
Free and combined		Antimony Bismuth Arsenic Copper		
		Mercury Silver	Very easy	By heating oxides
Free		Platinum Gold	Easiest	All compounds decomposed by heat

In some cases the ore is concentrated before it is reduced. This might be done by a merely physical means, such as flotation, in which the ore is crushed, mixed with a water solution containing a suitable oil (e.g., pine oil), and then vigorously agitated so that a frothy mass of oil rises to the top with the adhering metal-bearing portion of the ore. The froth is removed from the top of the machine and filtered to recover the metal ore.

A chemical process is used to convert sulfide or carbonate ores to oxides. It is called **roasting** and consists of heating the compounds in air. For example,

$$2ZnS + 3O_2 \rightarrow 2ZnO + 2SO_2(g)$$

and

$$PbCO_3 \rightarrow PbO + CO_2(g)$$

These oxides can then be treated with appropriate reducing agents to free the metal.

The following table summarizes the properties of the first two families of metals found in Group IA and Group IIA or 2 of the periodic chart. The table shows the distinct similarities in the properties and bonding of these metals. These similarities are directly related to the similarities that exist in the outer orbital electron configuration.

Element	Outer Orbital	Bonding and Charge	Physical Properties
Alkali Metals			
Lithium (Li)	$2s^1$	Ionic, +1	Silvery white
Sodium (Na)	$3s^1$	Ionic, +1	Silvery white
Potassium (K)	$4s^1$	Ionic, +1	Silvery white
Rubidium (Rb)	$5s^1$	Ionic, +1	Silvery white
Cesium (Cs)	$6s^1$	Ionic, +1	Silvery white
Francium (Fr)	$7s^1$	Ionic, +1	Silvery white
Alkaline-Earth Metals			
Beryllium (Be)	$2s^2$	Ionic, +2	Gray-white
Magnesium (Mg)	$3s^2$	Ionic, +2	Silvery white
Calcium (Ca)	$4s^2$	Ionic, +2	Silvery white
Strontium (Sr)	$5s^2$	Ionic, +2	Silvery white
Barium (Ba)	$6s^2$	Ionic, +2	Yellow white; lumpy
Radium (Ra)	$7s^2$	Ionic, +2	White; radioactive

SOME IMPORTANT REDUCTION METHODS

The **Hall process** is the electrolytic method of preparing **aluminum**. It uses molten cryolite (Na_3AlF_6) as the solvent for bauxite ore ($Al_2O_3 \cdot 2H_2O$). This mixture is electrolyzed in a rectangular iron tank lined with carbon that acts as the cathode. The anodes are heavy carbon rods suspended in the tank. The aluminum released at the cathode is tapped periodically and allowed to flow out of the tank so that the process can continue.

The outstanding properties of aluminum are its light weight with high strength, its ability to form an aluminum oxide coating that protects this metal from further oxidation, and its ability to conduct an electric current.

The **Dow process** is an electrolytic method of preparing **magnesium** from seawater. The magnesium salts are precipitated with $Ca(OH)_2$ solution to form $Mg(OH)_2$. The hydroxide is then converted to a chloride by adding hydrochloric acid to the hydroxide. This chloride is mixed with potassium chloride, melted, and electrolyzed to release the magnesium at the cathode and chlorine at the anode.

Magnesium has the important property of being a light, rigid, and inexpensive metal. It, too, can form a protective adherent oxide coating.

Another major metal prepared by electrolysis is **copper**. After the ore is concentrated by flotation and roasted to change sulfides to oxides, it is heated in a converter to obtain impure copper called "blister" copper. This copper is then used as anodes in the electrolysis apparatus, in which pure copper is used to make the cathode plates. As the process progresses, the copper in the anode becomes ions in the solution, which are reduced at the cathode as pure copper.

Copper has wide usage in electrical appliances and wires because of its high conductivity. It is widely employed in **alloys** of **bronze** (copper and tin) and **brass** (copper and zinc).

Iron ore is refined by reduction in a blast furnace, that is, a large, cylinder-shaped furnace charged with iron ore (usually hematite, Fe_2O_3), limestone, and coke. A hot-air blast, often enriched with oxygen, is blown into the lower part of the furnace through a series of pipes called tuyeres. The chemical reactions that occur can be summarized as follows:

Burning coke:	$2C + O_2 \rightarrow 2CO(g)$
	$C + O_2 \rightarrow CO_2(g)$
Reduction of CO_2:	$CO_2 + C \rightarrow 2CO(g)$
Reduction of ore:	$Fe_2O_3 + 3CO \rightarrow 2Fe + 3CO_2(g)$
	$Fe_2O_3 + 3C \rightarrow 2Fe + 3CO(g)$
Formation of slag:	$CaCO_3 \rightarrow CaO + CO_2(g)$
	$CaO + SiO_2 \rightarrow CaSiO_3$

The molten iron from the blast furnace is called "**pig iron**."

From pig iron, the molten metal might undergo one of three steel-making processes that burn out impurities and set the contents of carbon, manganese, sulfur, phosphorus, and silicon. Often nickel and chromium are alloyed in steel to give particular properties of hardness needed for tool parts. The three most important means of making steel involve the basic oxygen, the open-hearth, and the electric furnace. The first two methods are the most common.

The basic oxygen furnace uses a lined "pot" into which the molten pig iron is poured. Then a high-speed jet of oxygen is blown from a watercooled lance into the top of the pot. This "burns out" impurities to make a batch of steel rapidly and cheaply.

The open-hearth furnace is a large oven containing a dish-shaped area to hold the molten iron, scrap steel, and other additives with which it is charged. Alternating blasts of flame are directed across the surface of the melted metal until the proper proportions of additives are established for that "heat" so that the steel will have the particular properties needed by the customer. The tapping of one of these furnaces holding 50 to 400 tons of steel is a truly beautiful sight.

The final method of making steel involves the electric furnace. This method uses enormous amounts of electricity through graphite cathodes that are lowered into the molten iron to purify it and produce a high grade of steel.

ALLOYS

An **alloy** is a mixture of two or more metals. In a mixture, certain properties of the metals involved are affected. Some of these are

Melting point The melting point of an alloy is lower than that of its components.

Hardness An alloy is usually harder than the metals that compose it.

Crystal structure The size of the crystalline particles in the alloy determine many of the physical properties. The size of these particles can be controlled by heat treatment. If the alloy cools slowly, the crystalline particles tend to be larger. Thus, by heating and cooling an alloy, its properties can be altered considerably.

Common alloys are

Brass—which is made of copper and zinc.

Bronze—which is made of copper and tin.

Steel—which has controlled amounts of carbon, manganese, sulfur, phosphorus, and silicon that is alloyed with nickel and chromium.

Sterling silver—which is silver alloyed with copper.

METALLOIDS

In the preceding sections, representative metals and nonmetals have been reviewed, along with the properties of each. Some elements, however, are difficult to classify as one or the other. One example is carbon. The diamond form of carbon is a poor conductor, yet the graphite form conducts fairly well. Neither form looks metallic, and so carbon is classified as a nonmetal.

Silicon looks like a metal. However, its conductivity properties are closer to those of carbon.

Since some elements are neither distinctly metallic nor clearly nonmetallic, a third class, called the **metalloids**, is recognized.

The properties of metalloids are intermediate between those of metals and those of nonmetals. Although most metals form ionic compounds, metalloids as a group may form ionic or covalent bonds. Under certain conditions pure metalloids conduct electricity but do so poorly and are thus termed **semiconductors**. This property makes them important in microcircuitry.

The metalloids are located in the periodic table along the heavy dark line that starts alongside boron and drops down in steplike fashion between the elements found lower in the table (see the chart below).

	13 IIIA	14 IVA	15 VA	16 VIA	17 VIIA
	5 B	6 C	7 N	8 O	9 F
12 IIB	13 Al	14 Si	15 P	16 S	17 Cl
30 Zn	31 Ga	32 Ge	33 As	34 Se	35 Br
48 Cd	49 In	50 Sn	51 Sb	52 Te	53 I
80 Hg	81 Ti	82 Pb	83 Bi	84 Po	85 At

Location of Metalloids

Chapter Summary

The following terms summarize all the concepts and ideas that were introduced in this chapter. You should be able to explain their meaning and how you would use them in chemistry. They appear in boldface type in this chapter to draw your attention to them. The boldface type also makes the terms easier for you to look up if you need to. You could also use a search engine on your computer to get a quick and expanded explanation of these terms, laws, and formulas.

TERMS YOU SHOULD KNOW

acid salt
allotropic forms
alloys
aluminum
amorphous
brass
bronze
copper
Haber process
iron
magnesium

metalloids
monoclinic
normal salt
pig iron
properties of metals
resonance structure
roasting
semiconductor
steel
sterling silver

Chapter 13 Review Exercises

1. The most active nonmetallic element is

    (A) chlorine
    (B) fluorine
    (C) oxygen
    (D) sulfur

2. The order of decreasing activity of the halogens is

    (A) Fl, Cl, I, Br
    (B) F, Cl, Br, I
    (C) Cl, F, Br, I
    (D) Cl, Br, I, F

3. A light-sensitive substance used on photographic films has the formula

    (A) $AgBr$
    (B) CaF_2
    (C) $CuCl$
    (D) $MgBr_2$

4. Sulfur dioxide is the anhydride of

    (A) hydrosulfuric acid
    (B) sulfurous acid
    (C) sulfuric acid
    (D) hyposulfurous acid

5. The charring action of sulfuric acid is due to its being

   (A) a strong acid
   (B) an oxidizing agent
   (C) a reducing agent
   (D) a dehydrating agent

6. Ammonia is prepared commercially by the

   (A) Oswald process
   (B) arc process
   (C) Haber process
   (D) contact process

7. A nitrogen compound that has a color is

   (A) nitric oxide
   (B) nitrous oxide
   (C) nitrogen dioxide
   (D) ammonia

8. If students heat a mixture of ammonium chloride and calcium hydroxide in a test tube, they will detect

   (A) no reaction
   (B) the odor of ammonia
   (C) the odor of rotten eggs
   (D) a sweet smell like bananas

9. The difference between ammonia and ammonium is

   (A) an electron
   (B) a neutron
   (C) a proton
   (D) an –OH

10. An important ore of iron is

   (A) bauxite
   (B) galena
   (C) hematite
   (D) ironite

11. The material added to the charge of a blast furnace to react with the sand is

   (A) calcium carbonate
   (B) coke
   (C) silicon dioxide
   (D) slag

12. The process of extracting aluminum is called

   (A) the Dow process
   (B) the Haber process
   (C) the Hall process
   (D) the Frasch process

13. The order that halogen family elements are placed in the periodic table explains which of the following statements?

   I.   The most active nonmetallic element in the periodic table is fluorine.
   II. The normal physical state of the halogens goes from solid to gaseous as you go down the family.
   III. The halogen elements become ions by filling the outermost d orbital.

   (A) I only
   (B) II only
   (C) I and II
   (D) II and III
   (E) I and III

14. Which of the following statements concerning metals is/are true?

   I.   Aluminum is produced by the electrolytic Hall process from bauxite ore, $Al_2O_3 \cdot 2\, H_2O$.
   II. Brass is an alloy of copper and zinc.
   III. Silicon looks like a metal, but its conductivity is closer to that of carbon.

   (A) I only
   (B) II only
   (C) I and II
   (D) II and III
   (E) I, II, and III

15. Nitrogen for commercial use is generally obtained from

   (A) ammonia
   (B) liquid air
   (C) sodium nitrate
   (D) nitrogen-fixing bacteria
   (E) liquefaction of nitrogen oxide

Answers and Explanations

1. **(B)** Fluorine is the most active nonmetallic element because of its atomic structure. It needs only one electron to complete its outer shell and has the highest electronegativity.

2. **(B)** The order of activity of the halogens is from the smallest atomic radius to the largest straight down the family on the periodic table.

3. **(A)** Silver bromide, like many of the halogen salts, is light sensitive and used in photographic films.

4. **(B)** The reaction of SO_2 with water forms sulfurous acid. The equation is

$$SO_2 + H_2O \rightarrow H_2SO_3$$

5. **(D)** Sulfuric acid is a strong dehydrating agent. It draws water to itself so strongly that it can char sucrose by withdrawing the hydrogen and oxygen from $C_{12}H_{22}O_{11}$.

6. **(C)** The Haber process is used to prepare ammonia commercially.

7. **(C)** The only nitrogen compound in the list that has a reddish brown color is nitrogen dioxide.

8. **(B)** The reaction of these two chemicals results in the production of ammonium hydroxide, which is unstable and forms ammonia gas and H_2O. The reaction is

$$2NH_4Cl + Ca(OH)_2 \rightarrow 2NH_4OH + CaCl_2$$

Because the ammonium hydroxide is unstable at room temperatures, this reaction then occurs:

$$NH_4OH \rightarrow NH_3\uparrow + H_2O$$

9. **(C)** The ammonia molecule is trigonal pyramidal with an unshared pair of electrons in one corner of the pyramid. This negative charge attracts a H^+ (which is a proton) to form the ammonium ion NH_4^+.

10. **(C)** An important ore of iron is hematite.

11. **(A)** Calcium carbonate, also known as limestone, decomposes with heat and produces CaO, which then reacts with sand to produce slag.

12. **(C)** The famous Hall process is used to produce aluminum electrolytically.

13. **(A)** Since the halogen family's physical state goes from a gas to a solid down the table and halogens form ions by completing the outer p orbital, the only true statement is I, that fluorine is the most active.

14. **(E)** All three statements are true.

15. **(B)** Since nitrogen makes up about four-fifths of the air in our atmosphere, it is prepared industrially by the fractional distillation of liquid air.

PART VII

ORGANIC AND NUCLEAR CHEMISTRY

CARBON AND ORGANIC CHEMISTRY

CHAPTER OBJECTIVES

Upon completing this chapter, you will be able to:

- Describe the bonding patterns of carbon and its allotropic forms
- Explain the structural pattern of, draw, and name the alkanes, alkenes, and alkynes and their isomers
- Show graphically how hydrocarbons can be changed and the development of these functional groups, their structures, and names: alcohols, aldehydes, ketones, ethers, esters, amines, and carbohydrates
- Explain the development of polymers and their primary types

Carbon

FORMS OF CARBON

The element carbon occurs in three allotropic forms: diamond, graphite, and amorphous (although some evidence shows the amorphous forms have some crystalline structure). In the mid-1980s, **fullerenes** were identified as a new allotropic form of carbon. They are found in soot that forms from burning carbon-containing materials with limited oxygen. Their structure consists of near-spherical cages of carbon atoms resembling geodesic domes.

The **diamond** form has a close-packed crystal structure that gives it its property of extreme hardness. In it each carbon is bonded to four other carbons in a tetrahedron arrangement as shown on page 308. These covalent solids form crystals that can be viewed as a single giant molecule made up of an almost endless number of covalent bonds. Because all of the bonds in this structure are equally strong, covalent solids are often very hard and they are difficult to melt. Diamond is the hardest natural substance. It melts at 3550°C.

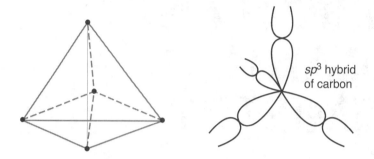

sp³ hybrid of carbon

It has been possible to make synthetic diamonds in machines that subject carbon to extremely high pressures and temperatures. Most of these diamonds are used for industrial purposes, such as dies.

The **graphite** form is made up of planes of hexagonal structures that are weakly bonded to the planes above and below. This explains graphite's slippery feeling and makes it useful as a dry lubricant. Its structure can be seen below. Graphite also has the property of being an electrical conductor.

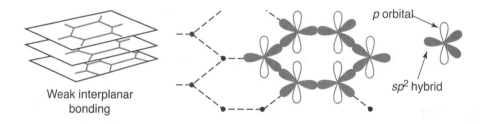

Weak interplanar bonding

p orbital

sp² hybrid

Some of the common amorphous forms of carbon are shown in Table 17 along with their properties and uses.

The process of **destructive distillation** is used in the preparation of both charcoal and coke. This process involves the heating of a substance in an enclosed container so that volatile materials, which would cause wood and coal to burn with a flame, can be forced out. After these substances are removed, the charcoal and coke merely glow when burned. Many of these volatile substances have proved to be useful by-products of destructive distillation. From the destructive distillation of coal, we get coal gas, coal tar, and ammonia. Coal gas can be used as a fuel; coal tars are the primary source of most synthetic dyes and many other useful organic compounds; and ammonia is also a useful chemical.

TABLE 17
ALLOTROPES OF CARBON

Forms	Occurrence or Preparation	Properties	Uses
1. Crystalline			
a. Diamond	South Africa; machines using very high pressures and temperatures	Hardest substance Brilliant: reflects and refracts light	Abrasive: drills, saws, polish bearings, gems, dies
b. Graphite	From hard coal in an electric furnace	Soft, gray, greasy Electrical conductor Refractory: high melting and kindling points	Lubricant, crucibles, electrodes, lead pencils, atomic pile
2. Amorphous (noncrystalline)—no definite shape			
a. Charcoal	Destructive distillation of wood	Burns with a glow and no flame; porous Adsorbs gases Reducing agent	Fuel Gas masks (activated) Reducing agent
b. Coke	Destructive distillation of soft coal	Burns with little smoke or flame; porous gray Reducing agent	Fuel and reducing agent in metallurgy Manufacture of water, gas, SiC, CaC_2, CS_2
c. Bone black	Destructive distillation of bones	Adsorbs coloring matter	Decolorizing sugar
d. Lampblack (carbon black)	Incomplete combustion of natural gas (or oil)	Soft, black, minute particles	Rubber tires Pigment: India and printing ink, shoe polish
e. Anthracite coal	Pennsylvania	Almost pure carbon, burns with little smoke	Fuel, production of water gas

CARBON 309

CARBON DIOXIDE

Carbon dioxide is a widely distributed gas that makes up 0.04% of the air. There is a cycle that keeps this figure relatively stable. It is shown in Figure 40.

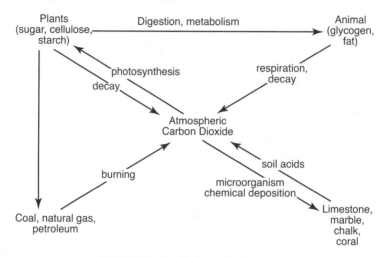

FIGURE 40. Carbon dioxide cycle.

LABORATORY PREPARATION OF CO$_2$

The usual laboratory preparation consists of reacting calcium carbonate (marble chips) with hydrochloric acid, although any carbonate or bicarbonate and any common acid can be used. The gas is collected by water displacement or air displacement. (See Figure 41.)

The test for carbon dioxide consists of passing it through limewater (Ca(OH)$_2$). If CO$_2$ is present, the limewater turns cloudy because of the formation of a white precipitate of finely divided CaCO$_3$:

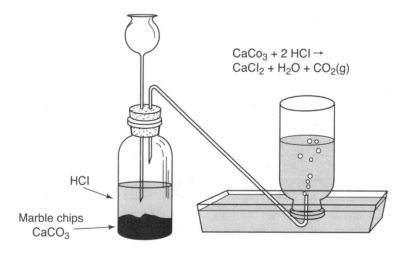

FIGURE 41. Laboratory preparation of carbon dioxide.

Continued passing of CO_2 into the solution will clear the cloudy condition because the insoluble $CaCO_3$ becomes soluble calcium bicarbonate $[Ca(HCO_3)_2]$.

$$CaCO_3(s) + H_2O + CO_2 \rightarrow Ca^{2+}(HCO_3)_2^{-}$$

This reaction can easily be reversed with increased temperature or decreased pressure. This is the way stalagmites and stalactites form on the floors and roofs of caves, respectively. The groundwater containing calcium bicarbonate is deposited on the roof and floor of the cave and decomposes into solid calcium carbonate formations.

IMPORTANT USES OF CO_2

1. Since CO_2 is the acid anhydride of carbonic acid, it forms the acid when reacted with soft drinks, thus marking them "carbonated" beverages.

$$CO_2 + H_2O \rightarrow H_2CO_3$$

2. Solid carbon dioxide ($-78°C$), or "dry ice," is used as a refrigerant because it has the advantage of not melting into a liquid but sublimes and in the process absorbs 3 times as much heat per gram as ice.

3. Fire extinguishers make use of CO_2 because it is $1\frac{1}{2}$ times heavier than air and does not support ordinary combustion. It is used in extinguishers that release CO_2 from a steel cylinder in the form of a gas to smother the fire.

4. Leavening agents are usually made up of a dry acid-forming salt, dry baking soda ($NaHCO_3$), and a substance to keep these ingredients dry, such as starch or flour. Some of the acid-forming salts are alum $[NaAl(SO_4)_2]$ in alum baking powders, potassium acid tartrate ($KHC_4H_4O_6$) in cream of tartar baking powders, and sodium hydrogen phosphate (NaH_2PO_4) or calcium hydrogen phosphate $[Ca(H_2PO_4)_2]$ in phosphate baking powders. In the moist batter, the acid reacts with the baking soda to release CO_2 as the leavening agent. This gas is trapped and raises the batter, which hardens in this raised condition.

   Baking soda, when used independently, must have some source of acid, such as buttermilk, molasses, or sour milk, to cause the release of CO_2. The simple ionic equation is

$$H^+ + HCO_3^{-} \rightarrow H_2O + CO_2(g)$$

5. Plants use up CO_2 in the photosynthesis process. In this process, chlorophyll (the catalyst) and sunlight (the energy source) must be present. The reactants and products of this reaction are

$$6CO_2 + 6H_2O \rightarrow C_6H_{12}O_6 + 6O_2(g)$$
$$\text{simple}$$
$$\text{sugar}$$

or

$$6CO_2 + 5H_2O \rightarrow C_6H_{10}O_5 + 6O_2(g)$$
$$\text{cellulose}$$

6. The Solvay process for making $NaHCO_3$ and Na_2CO_3 uses CO_2. The CO_2 is combined NH_3 and water, and the product is reacted with NaCl. Sodium bicarbonate ($NaHCO_3$) precipitates out of solution since its solubility in the NaCl solution is exceeded. The overall equation can be shown as

$$NaCl + H_2O + NH_3 + CO_3 \rightarrow NaHCO_3(s) + NH_4Cl$$

Sodium carbonate is made by heating the $NaHCO_3$:

$$2Na^+HCO_3^- \rightarrow Na_2^+CO_3^{2-} + H_2O + CO_2(g)$$

Organic Chemistry

Organic chemistry may be defined simply as the chemistry of the compounds of carbon. Since Friedrich Wöhler synthesized urea in 1828, chemists have synthesized thousands of carbon compounds in the areas of dyes, plastics, textile fibers, medicines, and drugs. The number of organic compounds has been estimated to be in the neighborhood of a million and is constantly increasing.

The carbon atom (atomic number 6) has four electrons in its outermost energy level, which show a tendency to be shared (electronegativity of 2.5) in covalent bonds. By this means, carbon bonds to other carbons, hydrogens, halogens, oxygen, and other elements to form the many compounds of organic chemistry.

HYDROCARBONS

Hydrocarbons, as the name implies, are compounds containing only carbon and hydrogen in their structure. The simplest hydrocarbon is methane, CH_4. As previously mentioned, this type of formula, which shows the kinds of atoms and their respective numbers, is called an *empirical* formula. In organic chemistry this is not sufficient to identify the compound it is used to represent. For example, the **empirical formula** C_2H_6O can denote either an ether or an ethyl alcohol. For this reason, a **structural formula** is used to indicate how the atoms are arranged in the molecule. The ether of C_2H_6O looks like

<pre>
        H   H
        | |
   H — C — O — C — H
        | |
        H   H
</pre>

whereas the ethyl alcohol is represented by the structural formula

$$
\begin{array}{c}
\quad\;\; H \quad\; H \\
\quad\;\; | \quad\;\; | \\
H - C - C - OH \\
\quad\;\; | \quad\;\; | \\
\quad\;\; H \quad\; H
\end{array}
$$

To avoid ambiguity, structural formulas are more often used than empirical formulas in organic chemistry. The structural formula of methane is

$$
\begin{array}{c}
\quad\;\; H \\
\quad\;\; | \\
H - C - H \\
\quad\;\; | \\
\quad\;\; H
\end{array}
$$

ALKANE SERIES (SATURATED)

Methane is the first member of a **hydrocarbon series** called the **alkanes** (or paraffin series). The general formula for this series is C_nH_{2n+2}, where n is the number of carbons in the molecule. Table 18 provides some essential information about this series. Since many other organic structures use the stem of the alkane names, you should learn these names and structures well. Notice that as the number of carbons in the chain increases, the boiling point also increases. The first four alkanes are gases at room temperature; the subsequent compounds are liquid and then become more viscous with increasing length of the chain.

Since the chain is increased by a carbon and two hydrogens in each subsequent molecule, the alkanes are referred to as a **homologous series**. They are **saturated** because they are linked exclusively by single bonds.

The alkanes are found in petroleum and natural gas. They are usually extracted by **fractional distillation**, which separates the compounds by varying the temperature so that each vaporizes at its respective boiling point. Methane, which forms 90% of natural gas, can also be prepared in the laboratory by heating soda lime (containing NaOH) with sodium acetate:

$$NaC_2H_3O_2 + NaOH \rightarrow CH_4(g) + Na_2CO_3$$

TABLE 18

THE ALKANES

Name	Formula	Number of Structural Isomers	Structure	Boiling Point (°C)
Methane	CH_4	1	$$\begin{array}{c} H \\ | \\ H-C-H \\ | \\ H \end{array}$$	−162
Ethane	C_2H_6	1	$$\begin{array}{cc} H & H \\ | & | \\ H-C-C-H \\ | & | \\ H & H \end{array}$$	−89
Propane	C_3H_8	1	$$\begin{array}{ccc} H & H & H \\ | & | & | \\ H-C-C-C-H \\ | & | & | \\ H & H & H \end{array}$$	−42
n-Butane	C_4H_{10}	2	$$\begin{array}{cccc} H & H & H & H \\ | & | & | & | \\ H-C-C-C-C-H \\ | & | & | & | \\ H & H & H & H \end{array}$$	0
n-Pentane	C_5H_{12}	3	$$\begin{array}{ccccc} H & H & H & H & H \\ | & | & | & | & | \\ H-C-C-C-C-C-H \\ | & | & | & | & | \\ H & H & H & H & H \end{array}$$	36
n-Hexane	C_6H_{14}	5	$CH_3-CH_2-CH_2-CH_2-CH_2-CH_2$	69
n-Heptane	C_7H_{16}	7	$CH_3-CH_2-CH_2-CH_2-CH_2-CH_2-CH_3$	98
n-Octane	C_8H_{18}	18	$CH_3-CH_2-CH_2-CH_2-CH_2-CH_2-CH_2-CH_3$	126
n-Nonane	C_9H_{20}	35	$CH_3-CH_2-CH_2-CH_2-CH_2-CH_2-CH_2-CH_2-CH_3$	151
n-Decane	$C_{10}H_{22}$	75	$CH_3-CH_2-CH_2-CH_2-CH_2-CH_2-CH_2-CH_2-CH_2-CH_3$	174

Gas at room temperatures (arrow spanning Methane through n-Butane)

Liquid state (bracket spanning n-Pentane through n-Decane)

When the alkanes are burned with sufficient air, the compounds formed are CO_2 and H_2O. An example is

$$2C_2H_6 + 7O_2 \rightarrow 4CO_2(g) + 6H_2O(g)$$

The alkanes can be reacted with halogens so that hydrogens are replaced by a halogen atom:

$$H-\underset{\underset{\displaystyle H}{|}}{\overset{\overset{\displaystyle H}{|}}{C}}-H + Br_2 \rightarrow H-\underset{\underset{\displaystyle H}{|}}{\overset{\overset{\displaystyle H}{|}}{C}}-Br + HBr$$

Some common substitution compounds of methane are

$$H-\underset{\underset{\displaystyle H}{|}}{\overset{\overset{\displaystyle H}{|}}{C}}-Cl \qquad Cl-\underset{\underset{\displaystyle Cl}{|}}{\overset{\overset{\displaystyle H}{|}}{C}}-Cl \qquad Cl-\underset{\underset{\displaystyle Cl}{|}}{\overset{\overset{\displaystyle Cl}{|}}{C}}-Cl$$

methyl chloride	chloroform	carbon tetrachloride
chloromethane	trichloromethane	tetrachloromethane

Naming alkane substitutions When an alkane hydrocarbon has an end hydrogen removed, it is referred to as an alkyl substituent or group. The respective name of each is the alkane name with **-ane** replaced by **-yl**.

Alkane	Alkyl Group	Compound
methane	methyl	methyl bromide
butane	butyl	butyl chloride

One method of naming a substitution product is to use the alkyl substituent or group name for the respective chain and the halide as shown above.

The IUPAC system uses the name of the longest carbon chain as the parent chain. The carbon atoms are numbered in the parent chain to indicate where branching or substitution takes place. The direction of numbering is chosen so that the lowest numbers possible are given to the side chains. The complete name of the compound is arrived at by first naming the attached group, each one being prefixed by the number of the carbon to which it is attached, and then the parent alkane. If a particular group appears more than once, the appropriate prefix (*di-*, *tri-*, and so on) is used to indicate how many times it appears. A carbon atom number must be used to indicate the position of each such group. If two or more of the same group are attached to the same carbon atom, the number of the carbon atom is repeated. If two or more different substituted groups are in a name, they are arranged alphabetically.

EXAMPLE 1   2,2-Dimethylbutane

SOLUTION

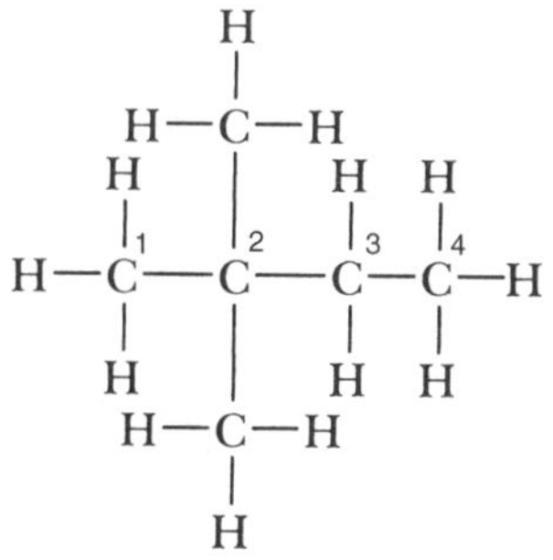

EXAMPLE 2   1,1-Dichloro-3-ethyl-2,4-dimethylpentane

SOLUTION

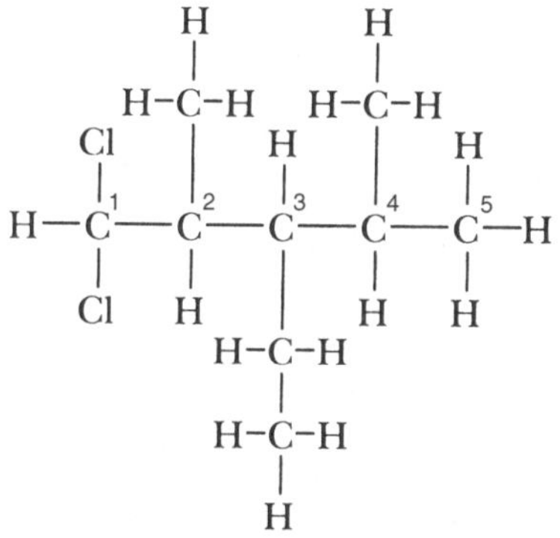

EXAMPLE 3   2-Iodo-2-methylpropane

SOLUTION

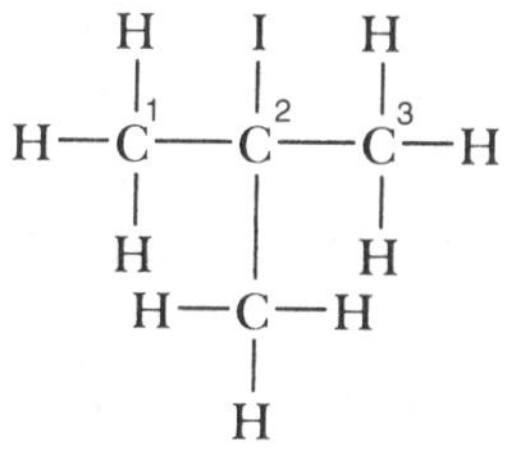

Cycloalkanes Starting with propane in the alkane series, it is possible to obtain a ring form by attaching the two chain ends. This reduces the number of hydrogens by two.

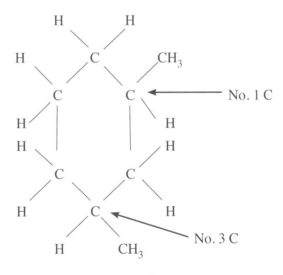

This hydrocarbon is called cyclopropane.

Cycloalkanes are named by adding the prefix *cyclo-* to the name of the straight–chain alkane.

An alkyl group can attach to this ring structure by taking the place of one of the hydrogens. When this occurs in only one position, there is no need to identify the carbon. If two or more alkyl groups are attached to the ring, number the carbon atoms in the ring by assigning position 1 to the alkyl group that comes first in alphabetical order. Then number in the direction that gives the rest of the alkyl groups the lowest numbers possible. This means that one of the alkyl groups will always be in position 1. The general formula is C_nH_{2n}. Here is an example:

1,3-dimethylcyclohexane

Since all the members of the cycloalkanes have single covalent bonds, this series and all such structures are also said to be **saturated**.

If the hydrocarbon molecule contains double or triple covalent bonds, it is referred to as **unsaturated**.

PROPERTIES AND USES OF ALKANES

Properties for some straight-chain alkanes are listed in Table 17. The trends in these properties can be explained by examining the structure of alkanes. The carbon-hydrogen bonds of alkanes are nonpolar. The only forces of attraction between nonpolar molecules are weak intermolecular forces, or London dispersion forces. These forces increase as the mass of a molecule increases.

The table also shows the physical states of alkanes. Smaller alkanes exist as gases at room temperature, whereas larger ones exist as liquids. Gasoline and kerosene consist mostly of liquid alkanes. It is not until there are 17 carbons in the chain that the solid form occurs. Paraffin wax contains solid alkanes.

Making use of the difference in the boiling point of mixtures of the liquid alkanes found in petroleum, it is possible to separate the various components through fractional distillation. This is the major industrial process used in refining petroleum into gasoline, kerosene, lubricating oils, and several other minor components.

ALKENE SERIES (UNSATURATED)

The **alkene** series has a double covalent bond between two adjacent carbon atoms. The general formula of this series is C_nH_{2n}. In naming these compounds, the suffix of the alkane is replaced by *ene*. Two examples are

$$
\begin{array}{c}
\overset{\displaystyle H \quad H}{\underset{\displaystyle |\quad\;|}{}} \\
H-C=C-H \qquad \text{ethene (common name: ethylene)}
\end{array}
$$

$$
\begin{array}{c}
H \quad H \quad H \\
|\quad\;|\quad\;| \\
H-C=C-C-H \qquad \text{propene (common name: propylene)} \\
| \\
H
\end{array}
$$

If the double bond occurs on an interior carbon, the chain is numbered so that the position of the double bond is designated by the lowest possible number assigned to the first doubly bonded carbon. For example,

$$
\begin{array}{c}
H \quad H \qquad\quad H \\
|\quad\;| \qquad\quad | \\
H-C-C-C=C-C-H \qquad \text{2-pentene} \\
|\quad\;|\quad\;|\quad\;|\quad\;| \\
H \quad H \quad H \quad H \quad H
\end{array}
$$

A more complex example is

$$CH_2-CH_3$$
$$|$$
$$CH_2=C-CH_2-CH_2-CH_3$$

The position number and name of the alkyl group are in front of the double-bond position number. The alkyl group above is an ethyl group. It is on the second carbon atom of the parent hydrocarbon.

The name of this compound is 2-ethyl-1-pentene.

The bonding is more complex in the double covalent bond than in the single bonds in the molecule. Using the orbital pictures of the atom, we can show this as the orbital depiction of ethane.

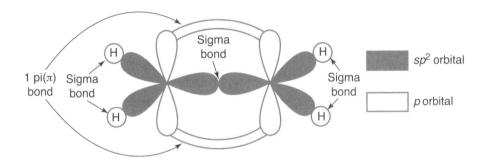

The two p lobes attached above and below constitute *one* bond called a pi (π) bond.

The sp^2 orbital bonds between the carbons and with each hydrogen are referred to as sigma (σ) bonds.

ALKYNE SERIES (UNSATURATED)

The **alkyne** series has a triple covalent bond between two adjacent carbons. The general formula of this series is C_nH_{2n-2}. In naming these compounds, the alkane suffix is replaced by *-yne*. A few examples are

$$H-C\equiv C-H \qquad \text{ethyne (common name: acetylene)}$$

$$\begin{array}{c} H \\ | \\ H-C\equiv C-C-H \\ | \\ H \end{array} \quad \text{propyne}$$

The orbital structure can be shown as

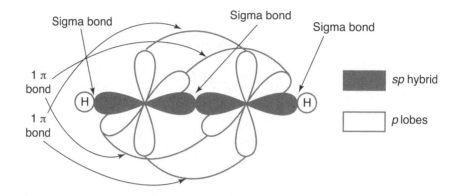

The bonds formed by the p orbitals and the one bond between the sp orbitals make up the triple bond.

The above examples show only one triple bond. If there is more than one triple bond, modify the suffix to indicate the number of triple bonds. For example, if there are 2 triple bonds, it would be called a diyne, if there were 3, it would be a triyne, and so on. Next, add the names of the alkyl groups if they are attached. Number the carbon atoms in the chain so that the first carbon atom in the triple bond nearest the end of the chain has the lowest number. If numbering from both ends gives the same positions for two triple bonds, then number from the end nearest the first alkyl group. Then place the position numbers of the triple bonds immediately before the name of the parent hydrocarbon alkyne, and place the alkyl group position numbers immediately before the name of the corresponding alkyl group.

Two more examples of alkynes are

$$CH_3-CH_2-CH_2-C\equiv CH \qquad CH\equiv C\text{--}CH\text{--}CH_3$$
$$\vert$$
$$CH_3$$

1-pentyne   3-methyl-1-butyne

A more complex example is

$$CH_2-CH_3$$
$$\vert$$
$$CH\equiv C-CH_2-CH_2-CH_3$$

The position number and name of the alkyl group are placed in front of the triple-bond position number. The previous alkyl group is an ethyl group. It is on the second carbon atom of the parent hydrocarbon. The name is

$$2\text{-ethyl-1-pentyne}$$

ADDITIONS TO ALKENES AND ALKYNES

Unsaturates of the alkenes and alkynes can add to their structures by breaking the double or triple bond present. For example,

$$
\begin{array}{ccc}
& \text{H} \ \ \text{H} & \text{Br} \ \ \text{Br} \\
& | \ \ \ | & | \ \ \ \ | \\
\text{H} - \text{C} = \text{C} - \text{H} + \text{Br}_2 \longrightarrow & \text{H} - \text{C} - \text{C} - \text{H} \\
& & | \ \ \ \ | \\
& & \text{H} \ \ \text{H}
\end{array}
$$

1,2-dibromoethane

The 1,2- means that on the first and second carbons in the chain a bromine atom is bonded. 1,1-dibromoethane is

$$
\begin{array}{c}
\text{Br} \ \ \text{H} \\
| \ \ \ | \\
\text{H} - \text{C} - \text{C} - \text{H} \\
| \ \ \ | \\
\text{Br} \ \ \text{H}
\end{array}
$$

An addition to ethyne could be

$$
\begin{array}{c}
\text{Br} \ \ \text{Br} \\
| \ \ \ \ | \\
\text{H} - \text{C} \equiv \text{C} - \text{H} + 2\text{Br}_2 \longrightarrow \text{H} - \text{C} - \text{C} - \text{H} \\
| \ \ \ \ | \\
\text{Br} \ \ \text{Br}
\end{array}
$$

Alkadienes have two double covalent bonds in each molecule. The *-ene* indicates a double bond, and the *di-* indicates two such bonds. Names of this type are also derived from the alkanes, and so butadiene has four carbons and two double bonds.

$$
\begin{array}{c}
\text{H} \quad \ \text{H} \ \ \text{H} \quad \ \text{H} \\
\diagdown \quad | \ \ \ | \quad \diagup \\
\text{C} = \text{C} - \text{C} = \text{C} \\
\diagup \quad \quad \quad \diagdown \\
\text{H} \quad \quad \quad \quad \text{H}
\end{array}
$$

(1,3-butadiene is a more precise name and indicates that the double bonds follow the first and third carbons.)

This compound is used in making synthetic rubber.

ALICYCLIC HYDROCARBONS

All the hydrocarbons discussed to this point have been **aliphatic**, that is, open-chain hydrocarbons with either straight or branched structures. Hydrocarbons in which the carbon atoms are arranged in a closed ring structure are cyclic hydrocarbons. The two main groups of cyclic hydrocarbons are the **alicyclic** hydrocarbons and the **aromatic** hydrocarbons. Aromatic hydrocarbons are discussed in the next section.

In the alicyclic hydrocarbons, the carbon atoms are linked by single or double bonds and are arranged in a closed structure. As previously shown with alkanes, the cycloalkanes begin with cyclopropane.

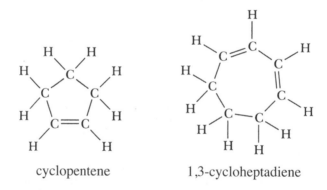

cyclopropane

Similarly, cycloalkenes contain one double bond between a pair of carbon atoms in the ring structure. Cycloalkadienes contain two double bonds between two different pairs of carbon atoms. Some examples are

cyclopentene   1,3-cycloheptadiene

AROMATIC HYDROCARBONS

The aromatic compounds are unsaturated ring structures. The basic formula of this series is C_nH_{2n-6}, and the simplest compound is benzene (C_6H_6). The benzene structure is a resonance structure represented as

The orbital structure can be represented as

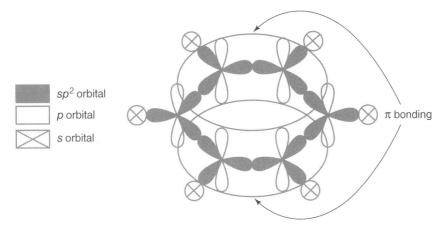

Most of the aromatics have an aroma, thus the name **aromatic**.

The C_6H_5 group is a substituent called phenyl. This is the benzene structure with one hydrogen missing. If the phenyl substituent adds a methyl group, the compound is called toluene or methylbenzene:

Two other members of the benzene series and their structures are

$C_{10}H_8$ naphthalene

$C_{14}H_{10}$ anthracene

The IUPAC system of naming benzene derivatives, like that for chain compounds, involves numbering the carbon atoms in the ring in order to pinpoint the locations of the side chains. However, if only two groups are substituted in the benzene ring, the compound formed will be a benzene derivative having three possible isomeric forms. In such cases, the prefixes **orth-**, **meta-**, and **para-**, abbreviated as **o-**, **m-**, and **p-**, may be used to name the isomers. In the ortho structure, the two substituted groups are located on adjacent carbon atoms. In the meta structure, they are separated by one carbon atom. In the para structure, they are separated by two carbon atoms.

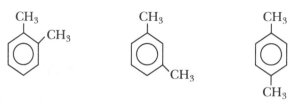

1,2-dimethylbenzene   1,3-dimethylbenzene   1,4-dimethylbenzene
o-xylene   *m*-xylene   *p*-xylene

ISOMERS

Many of the chain hydrocarbons can have the same formula, but their structures may differ. For example, butane is the first compound that can have two different structures or **isomers** for the same formula.

n-butane   isobutane

This isomerization can be shown by the equation

$$CH_3 - CH_2 - CH_2 - CH_3 \xrightarrow[70-100°C]{AlCl_2} CH_3 - \underset{\underset{CH_3}{|}}{CH} - CH_3$$

butane   isobutane

The isomers have different properties, both physical and chemical, from the normal structure.

CHANGING HYDROCARBONS

CRACKING

Under proper conditions of temperature and pressure and often in the presence of a catalyst, long chains of hydrocarbons may be made into more useful smaller molecules. This process is called **cracking**. An example is

$$C_{16}H_{34} \rightarrow C_8H_{18} + C_8H_{16}$$
hexadecane   octane   octene

Notice that the products formed by cracking a saturated hydrocarbon of the alkanes are a saturated hydrocarbon and an unsaturated alkene.

ALKYLATION

Alkylation is the combining of a saturated alkane with an unsaturated alkene. An example is

$$C_4H_{10} \quad + \quad C_4H_8 \quad \xrightarrow{H_2SO_4} \quad C_8H_{18}$$

isobutane	isobutylene	isooctane
2-methylpropane	2-methylpropene	2,2,4-trimethylpentane

Notice that the product formed by alkylation is a useful chain hydrocarbon without the previous double bond.

POLYMERIZATION

The combination of two or more unsaturated molecules to form a larger chain molecule is called **polymerization**. An example is

In the product name, 2,4,4-trimethylpentene-1, the first three numbers indicate the carbons that have the trimethyl (three methyl substituents) attached—one each. The -1 at the end indicates that the first carbon has the double bond attaching it to the next carbon. In numbering compounds of the alkenes and alkynes, always number the carbon atoms from the end that has the double or triple bond.

HYDROGENATION

Hydrogenation is the process of adding hydrogens to an unsaturated hydrocarbon in the presence of a suitable catalyst. It is often used to make liquid unsaturated fats into solid saturated fats. An example for hydrocarbons is

$$\underset{\text{2-methylpropene-1}}{CH_3 - \overset{\overset{\displaystyle CH_3}{|}}{C} = CH_2 + H_2} \quad \rightarrow \quad \underset{\text{2-methylpropane}}{CH_3 - \overset{\overset{\displaystyle CH_3}{|}}{\underset{\underset{\displaystyle H}{|}}{C}} - CH_3}$$

DEHYDROGENATION

Dehydrogenation is the removal of hydrogens from a chain molecule in the presence of a catalyst to form an unsaturated molecule. An example is

$$\underset{\text{butane}}{CH_3 - CH_2 - CH_2 - CH_3} \rightarrow \underset{\text{1-butene}}{CH_2 = CH - CH_2 - CH_3} + H_2$$

AROMATIZATION

Hydrocarbons of the alkane series having six or more carbons in the chain can be made to form an aromatic hydrocarbon with the loss of hydrogen. This must be done at high temperatures and with a suitable catalyst. An example is

$$\underset{\text{heptane}}{CH_3 - CH_2 - CH_2 - CH_2 - CH_2 - CH_2 - CH_3} \rightarrow$$

toluene $+ 4H_2$

FISCHER-TROPSCH PROCESS

This process was developed to make gasoline from natural gas, coal, or lignite by converting the original carbon compound to water gas. It has gained interest (because of the high price of gasoline) as an emerging technology to make fuels from gas and coal. For example,

$$\underset{\text{water gas}}{2CH_4 + O_2 \rightarrow 2CO + 4H_2}$$

This water gas is then enriched with hydrogen and reacted at high temperatures with a catalyst to form a mixture of hydrocarbons, one of which is gasoline.

HYDROCARBON DERIVATIVES

ALCOHOLS

The simplest **alcohols** are alkanes that have one or more hydrogen atoms replaced by the hydroxyl group, —OH. This is called its functional group.

Methanol

Methanol is the simplest alcohol. Its structure is

$$
\begin{array}{c}
\text{H} \\
| \\
\text{H}-\text{C}-\text{OH} \\
| \\
\text{H}
\end{array}
\quad \text{methanol or wood alcohol}
$$

Properties and uses Methanol is a colorless, flammable liquid with a boiling point of 65°C. It is miscible with water, is exceedingly poisonous, and can cause blindness if taken internally. It can be used as a fuel, as a solvent, and as a denaturant to make denatured ethyl alcohol, unsuitable for drinking.

Ethanol

Ethanol is the best known and most used alcohol. Its structure is

$$
\begin{array}{c}
\text{H} \quad \text{H} \\
| \quad\; | \\
\text{H}-\text{C}-\text{C}-\text{OH} \\
| \quad\; | \\
\text{H} \quad \text{H}
\end{array}
\quad \text{ethanol}
$$

(Notice that the alcohol names are derived from the alkane names by replacing the *e* with *-ol*.)

Its common names are ethyl alcohol and grain alcohol.

Simple sugars can be converted to ethanol by the action of an enzyme (zymase) found in yeast.

$$
\underset{\text{sugar}}{C_6H_{12}O_6} \rightarrow \underset{\text{ethanol}}{2C_2H_5OH} + 2CO_2(g)
$$

After fermentation, the alcohol is distilled off. Grains, potatoes, and other starch plants, which can be treated with an acid-water solution to form sugars, can be converted into ethyl alcohol.

Properties and uses Ethanol is a colorless, flammable liquid with a boiling point of 351 K. It is miscible with water and is a good solvent for a wide variety of substances (these solutions are often referred to as "tinctures"). It is used as an antifreeze because of its low freezing point, 158 K (or −115°C), and for making acetaldehyde and ether. One of its growing uses today is as an additive to gasoline, and it is projected to be an alternative to gasoline. Ethanol is one of the rational sources of fuel because it is renewable. Studies have shown this fuel source could take the lead one day as the top source of liquid fuel because ethanol could be derived from virtually limitless feedstock, which could be replaced after each harvest. On the other hand, petroleum and gas sources will be depleted.

Other Alcohols

Isomeric alcohols have similar formulas but different properties because of their differences in structure. If the —OH is attached to an end carbon, the alcohol is called a primary alcohol. If attached to a "middle" carbon, it is called a secondary alcohol. Some examples are

$$\underset{\substack{\text{1-propanol}\\\text{(propyl alcohol)}}}{\underset{\substack{|\quad|\quad|\\H\ \ H\ \ H}}{\overset{\substack{H\ \ H\ \ H\\|\quad|\quad|}}{H-C-C-C-OH}}} \xleftrightarrow{\text{isomers}} \underset{\substack{\text{2-propanol}\\\text{(isopropyl alcohol)}}}{\underset{\substack{|\quad|\quad|\\H\ \ OH\ H}}{\overset{\substack{H\ \ H\ \ H\\|\quad|\quad|}}{H-C-C-C-H}}}$$

$$\underset{\substack{\text{1-butanol}\\\text{(sometimes called}\\n\text{-butanol or normal butanol)}\\\text{butyl alcohol}}}{CH_3-CH_2-CH_2-\overset{\overset{\displaystyle H}{|}}{\underset{\underset{\displaystyle H}{|}}{C}}-OH} \xleftrightarrow{\text{isomers}} \underset{\substack{\text{2-butanol}\\\ \\\text{isobutyl alcohol}}}{CH_3-CH_2-\overset{\overset{\displaystyle H}{|}}{\underset{\underset{\displaystyle OH}{|}}{C}}-CH_3}$$

Phenols are alcohols derived from the aromatic hydrocarbons. An important example is

Structure	Name	Properties or Uses
⬡—OH	Phenol Carbolic acid Hydroxybenzene	Slightly acidic, extremely corrosive, poisonous. Used to make synthetic resins, plastics, drugs, dyes, and photographic developers. Good disinfectant.

Other alcohols with more than one —OH group are

H H
| |
H—C—C—H   ethylene glycol
| | 1,2-ethanediol
OH OH

A colorless liquid, high boiling point, low freezing point. Used as permanent antifreeze in automobiles.

H H H
| | |
H—C—C—C—H   glycerine
| | | glycerol
OH OH OH   1,2,3-propanetriol

Colorless liquid, odorless, viscous, sweet taste. Used to make nitrogylcerine, resins for paint, and cellophane.

ALDEHYDES

The functional group of an **aldehyde** is the $-C\overset{O}{\underset{H}{\big\langle}}$, formyl group. The general formula

is RCHO, where R represents a hydrocarbon substituent.

Preparation from an alcohol Aldehydes can be prepared by the oxidation of an alcohol. This can be done by inserting a hot copper wire into the alcohol. A typical reaction is

$$\begin{array}{c} H \\ | \\ H-C-OH \\ | \\ H \end{array} \overset{\text{mild oxidizing}}{\underset{\text{agent}}{+ [O] \longrightarrow}} \left[\begin{array}{c} H \\ | \\ H-C-OH \\ | \\ OH \end{array}\right] \longrightarrow H-C\overset{O}{\underset{H}{\big\langle}} + H_2O$$

methanol
(methyl alcohol)

methanal
(formaldehyde)

The middle structure is an intermediate structure, but because two hydroxyl groups do not stay attached to the same carbon, it changes to the aldehyde by a water molecule "breaking away."

The aldehyde name is derived from the alcohol name by dropping the *-ol* and adding *-al*. Ethanol forms ethanal (acetaldehyde) in the same manner.

ORGANIC ACIDS OR CARBOXYLIC ACIDS

The functional group of an organic acid is $-C\overset{O}{\underset{OH}{\big\langle}}$, the **carboxyl** group. The general

formula is R—COOH.

Preparation from an aldehyde Organic acids can be prepared by the mild oxidation of an aldehyde. The simplest acid is methanoic acid, which is present in ants, bees, and other insects. A typical reaction is

$$H-C\overset{O}{\underset{H}{\diagup}} + [O] \rightarrow H-C\overset{O}{\underset{OH}{\diagup}}$$

methanal   methanoic acid
(formaldehyde)   (formic acid)

Notice that in the IUPAC system the name is derived from the alkane stem by adding *-oic*.

Ethanal can be oxidized to ethanoic acid:

$$H-\overset{H}{\underset{H}{C}}-C\overset{O}{\underset{H}{\diagup}} + [O] \rightarrow H-\overset{H}{\underset{H}{C}}-C\overset{O}{\underset{OH}{\diagup}}$$

ethanal   ethanoic acid
(acetic acid)

Acetic acid, as ethanoic acid is commonly called, is a mild acid which in the concentrated form is called glacial acetic acid. Glacial acetic is used in many industrial processes, such as making cellulose acetate. Vinegar is a 4% to 8% solution of acetic acid, which can be made by fermenting alcohol.

$$C_2H_5OH + O_2 \rightarrow CH_3COOH + H_2O$$
ethanol   ethanoic acid
(acetic acid)

One of the aromatic acids is benzoic acid with a carboxyl group replacing one of the hydrogens in the benzene ring:

$$\overset{O}{\underset{OH}{C\diagup}}$$

Benzoic acid

COMPLEX ORGANIC ACIDS

There are some more complex organic acids that are important to mention. These are listed in Table 19.

TABLE 19
SOME COMPLEX ORGANIC ACIDS

Acid	Source	Structure	Notes
Palmitic	Fats	$C_{15}H_{31}-C\begin{smallmatrix}O\\\\OH\end{smallmatrix}$	Saturated
Stearic	Fats	$C_{17}H_{35}-C\begin{smallmatrix}O\\\\OH\end{smallmatrix}$	Saturated
Oleic	Fats	$C_{17}H_{33}-C\begin{smallmatrix}O\\\\OH\end{smallmatrix}$	Unsaturated—has a double bond in the chain
Citric	Citrus fruits	$COOH-CH_2-COHCOOH-CH_2-COOH$	
Lactic	Sour milk	$CH_3-CHOH-COOH$	
Oxalic	Rhubarb	$COOH-COOH$	
Salicylic	Oil of wintergreen	(benzene ring with $C\begin{smallmatrix}O\\\\OH\end{smallmatrix}$ and OH)	

SUMMARY OF OXYGEN DERIVATIVES

Functional Group

$R-H \rightarrow R-Cl \rightarrow R-OH \rightarrow R^1CHO \rightarrow R^1-COOH$

hydrocarbon   chlorine   alcohol   aldehyde   acid
         substitution
         product   (ending -*ol*) (ending -*al*) (ending -*oic*)

Note: R^1 indicates a hydrocarbon chain different from R by one less carbon in the chain.

An actual example using ethane is

$$C_2H_6 \xrightarrow{Cl_2} C_2H_5Cl \xrightarrow{NaOH} C_2H_5OH \xrightarrow{[O]} CH_3CHO \xrightarrow{[O]} CH_3COOH$$

ethane   chloro-ethane   ethanol   ethanal   ethanoic acid

KETONES

When a secondary alcohol is slightly oxidized, it forms a compound having the

functional group $\begin{smallmatrix} R-C-R^1 \\ \| \\ O \end{smallmatrix}$, called a **ketone**. The R^1 indicates that this group need not

be the same as R. An example is

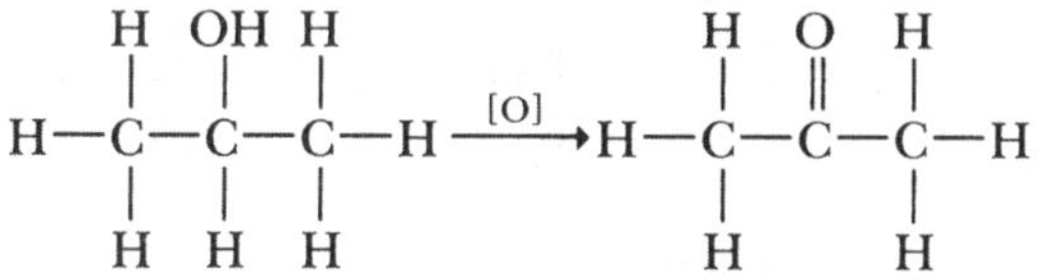

2-propanol   2-propanone
(isopropyl alcohol)   (acetone)

In the IUPAC method, the name of the ketone has the ending *-one* with a digit indicating the carbon that has the double-bonded oxygen. Another method of designating a ketone is to name the radicals on either side of the ketone structure and use the word *ketone*. In the preceding reaction, the product would be dimethyl ketone.

ETHERS

When a primary alcohol, such as ethanol, is dehydrated with sulfuric acid, an **ether** forms. The functional group is $R-O-R^1$, in which R^1 may be the same hydrocarbon group, as shown below in the first example, or a different hydrocarbon group, as shown in the second example.

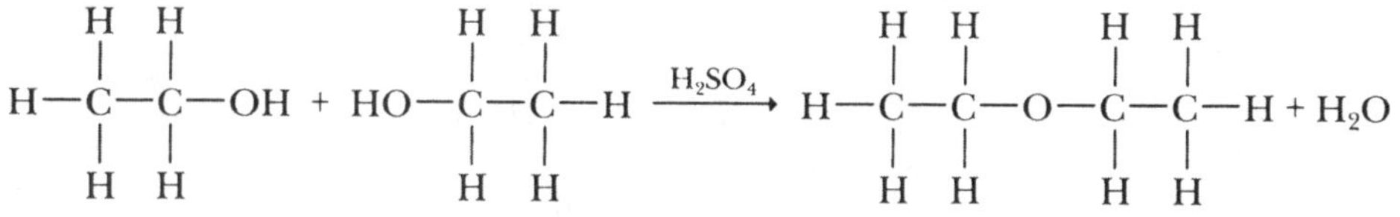

ethanol   ethanol   ethyl ether or diethyl ether

Another ether with unlike groups, $R-O-R^1$:

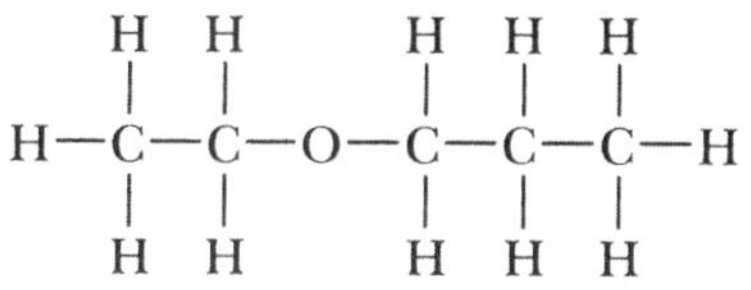

ethyl propyl ether

In the IUPAC method of naming compounds, the alkyl stem is named first and then the word *ether* is added. If the alkyls are not the same, then they are named in alphabetical order followed by the word *ether* as shown in the example.

AMINES AND AMINO ACIDS

The group NH_2^- is found in the amide ion or the amino group. Under the proper conditions, this ion can replace a hydrogen in a hydrocarbon compound. The resulting compound is called an **amine**. Some examples are

$$H-\underset{\underset{H}{|}}{\overset{\overset{H}{|}}{C}}-NH_2 \qquad \text{methylamine}$$

$$\text{(benzene ring)}-NH_2 \qquad \text{aniline}$$

In *amides*, the NH_2^- group replaces a hydrogen in the carboxyl group. When naming amides, the *-ic* of the common name or the *-oic* of the IUPAC name of the parent acid is replaced by *-amide*. See below.

Amino acids are organic acids that contain one or more amino groups. The simplest uncombined amino acid is glycine, or amino acetic acid, NH_2CH_2-COOH. More than 20 amino acids are known, about half of which are necessary in the human diet because they are needed to make up the body proteins.

$$H-\underset{\underset{H}{|}}{\overset{\overset{H}{|}}{C}}-\underset{\underset{N}{\diagdown}}{\overset{\overset{O}{\diagup}}{C}}\diagup^H \qquad \begin{array}{l}\text{acetamide}\\ \text{(ethanamide)}\end{array}$$

ESTERS

Esters are often compared to inorganic salts because their preparations are similar. To make a salt, you react the appropriate acid and base. To make an ester, you react the appropriate organic acid and alcohol. For example,

$$\underset{\begin{array}{c}\text{ethanoic}\\\text{acid}\end{array}}{CH_3-\overset{\overset{O}{\|}}{C}+OH} + \underset{\text{ethanol}}{H+O-C_2H_5} \xrightarrow[\begin{array}{c}\text{dehydrating}\\H_2SO_4\end{array}]{\text{(forms } H_2O)} \underset{\begin{array}{c}\text{ethyl ethanoate}\\\text{(ethyl acetate)}\end{array}}{CH_3-\overset{\overset{O}{\|}}{C}-O-C_2H_5} + H_2O$$

The name is made up of the alkyl substituent of the alcohol and the acid name, in which -ic is replaced with -ate.

The general equation is

$$RO-H + R^1CO-OH \rightarrow R^1COO-R + HOH$$

alcohol   acid   ester

The asterisk identifies the oxygen atom.

Esters usually have a sweet smell and are used in perfumes and flavor extracts. Esters can be compared to inorganic salts because the formation equation of an ester is similar to the neutralization equation of an inorganic acid and a base to form a salt and water.

SOME COMMON ESTERS

Name	Formula	Characteristic Odor
Ethyl butyrate	$C_3H_7COO \cdot C_2H_5$	Pineapple
Amyl acetate	$CH_3COO \cdot C_5H_{11}$	Banana, pear
Methyl salicylate	$C_6H_4(OH)COO \cdot CH_3$	Wintergreen
Amyl valerate	$C_4H_9COO \cdot C_5H_{11}$	Apple
Octyl acetate	$CH_3COO \cdot C_8H_{17}$	Orange
Methyl anthranilate	$C_3H_4(NH_2)COO \cdot CH_3$	Grape

Some esters found in the seeds of plants and the bodies of animals are fats. Stearin is such an ester.

$$
\begin{array}{ll}
C_{17}H_{35}-COOH & HO-CH_2 \\
C_{17}H_{35}-COOH + HO-CH \longrightarrow 3\,H_2O + \\
C_{17}H_{35}-COOH & HO-CH_2
\end{array}
\quad
\begin{array}{l}
C_{17}H_{35}COO-CH_2 \\
C_{17}H_{35}COO-CH \\
C_{17}H_{35}COO-CH_2
\end{array}
$$

stearic acid   glycerol   glycerol stearate (stearin)

Some other examples are

Olein or glyceryl oleate $= (C_{17}H_{35}COO)_3C_3H_5$

Butyrin or glyceryl butyrate $= (C_3H_7COO)_3C_3H_5$

Soap was once made by housewives using a fat (like stearin) and lye (sodium hydroxide). This saponification reaction is

$$(C_{17}H_{35}COO)_3C_3H_5 + 3NaOH \rightarrow 3C_{17}H_{35}COONa + C_3H_5(OH)_3$$

$$\text{stearin} \qquad \text{lye} \qquad \underset{\text{(sodium stearate)}}{\text{soap}} \qquad \text{glycerine}$$

The following chart summarizes the functional groups and general formulas of the classes of organic compounds discussed in the section "Hydrocarbon Derivatives."

CLASSES OF ORGANIC COMPOUNDS

Class	Functional Group	General Formula
alcohol	$-$ OH	$R - OH$
alkyl halides	$-$ X, X = F, Cl, Br, I	$R - X$
ether	$-$ O $-$	$R - O - R^1$
aldehyde	$\overset{\displaystyle O}{\overset{\displaystyle \|}{-C-H}}$	$\overset{\displaystyle O}{\overset{\displaystyle \|}{R-C-H}}$
ketone	$\overset{\displaystyle O}{\overset{\displaystyle \|}{-C-}}$	$\overset{\displaystyle O}{\overset{\displaystyle \|}{R-C-R^1}}$
carboxylic acid	$\overset{\displaystyle O}{\overset{\displaystyle \|}{-C-OH}}$	$\overset{\displaystyle O}{\overset{\displaystyle \|}{R-C-OH}}$
ester	$\overset{\displaystyle O}{\overset{\displaystyle \|}{-C-O-}}$	$\overset{\displaystyle O}{\overset{\displaystyle \|}{R-C-O-R^1}}$
amine	$\underset{\displaystyle \|}{-N-}$	$\underset{\underset{\displaystyle R^1}{\displaystyle \|}}{R-N-R^2}$

CARBOHYDRATES

Carbohydrates are made up of carbon, hydrogen, and oxygen. Usually the hydrogen-to-oxygen ratio is 2 : 1. The simple carbohydrates are more-or-less sweet and are referred to as saccharides (meaning "sweet").

Monosaccharides and Disaccharides

The **monosaccharides** are simple compounds having six carbons in the formula, but their properties differ because of structural differences. This is shown in dextrose and levulose, which have the same formula (are isomers) but different properties.

dextrose (glucose)

$$
\begin{array}{c}
H \\
| \\
C=O \\
| \\
H-C-OH \\
| \\
HO-C-H \\
| \\
H-C-OH \\
| \\
H-C-OH \\
| \\
H-C-OH \\
| \\
H
\end{array}
$$

Not as sweet as ordinary sugar (sucrose); does not need to be digested.

levulose (fructose)

$$
\begin{array}{c}
H \\
| \\
H-C-OH \\
| \\
C=O \\
| \\
HO-C-H \\
| \\
H-C-OH \\
| \\
H-C-OH \\
| \\
H-C-OH \\
| \\
H
\end{array}
$$

Twice as sweet as dextrose; must be changed to dextrose in the body before available for body use.

The sugars that contain an aldehyde structure ($-C\!\!\overset{H}{=}\!\!O$), such as dextrose, or a group that readily changes to this structure, act as mild reducing agents and are called reducing sugars. They can be identified by their ability to reduce copper (II) hydroxide, in Fehling's solution or Benedict's solution, to copper (I) oxide, which is recognized by its brick-red color. This test is often used to detect the disease known as diabetes.

The "double" sugars, those having twelve carbons in their structures, are called disaccharides. Some of these are sucrose, lactose (found in milk), and maltose.

Sucrose can be converted to simple sugars by treating a solution of it with a little acid and boiling. This process is called inversion, and the products, dextrose and levulose, are called *invert* sugars. The general name for this type of reaction with disaccharides is hydrolysis.

$$C_{12}H_{22}O_{11} + H_2O \rightarrow C_6H_{12}O_6 + C_6H_{12}O_6$$

sucrose   dextrose   levulose

Polysaccharides

These complex carbohydrates are made up of some multiple of $C_6H_{10}O_5$. This is shown in the general formula $(C_6H_{10}O_5)_n$, in which n represents the variable multiple.

This group includes starch, dextrin, and cellulose.

POLYMERS

Polymers are very large organic compounds made of repeating units. The term *polymer* comes from two Greek roots: *poly* meaning "many," and *mer* meaning "part." The repeating units in a polymer are called *monomers*. (*Mono* means "one" in Greek.) You can compare a polymer to a long string of beads, and a monomer to an individual bead.

Almost all living organisms make and use different polymers. Plants use glucose as a monomer to form the polymers starch, an important food source, and cellulose, an important structural compound in plants and the principal component of paper. Different amino acids link together to form proteins, which are also polymers. Depending on the sequence of amino acids, the protein might be the hair on your head, a muscle in your arm, or an enzyme that helps you to digest food.

One of the first completely synthetic polymers was nylon. Other synthetic polymers have a wide variety of different properties and uses. Polyethylene is a lightweight, inexpensive polymer used to make such items as trash bags and plastic containers. Polyvinyl chloride is used as plastic wrap because it can be made into a thin film that adheres well to itself. Polymethylmethacrylate is a polymer valued for its transparency and resistance to shattering. It is used as a substitute for glass. Another well-known polymer is Teflon, which is used as a nonstick finish on metal cookware.

Polymers can be classified by the way they behave when heated. A **thermoplastic** *polymer* melts when heated and can be reshaped many times. A **thermosetting** *polymer* does not melt when heated but keeps its original shape. The molecules of a linear polymer are free to move. They slide back and forth against each other easily when heated, and so they are called thermoplastics. The molecules of a branched polymer contain side chains that prevent the molecules from sliding across each other easily. More heat is required to melt a branched polymer than a linear polymer, but they are still likely to be thermoplastic. In cross-linked polymers, adjacent molecules have formed bonds with each other. Individual molecules are not able to slide past each other when heated, and so they retain their shape when heated and are thermosetting polymers.

The two principal methods of synthesizing polymers are addition **polymerization** and condensation polymerization.

An addition polymer is a polymer formed by chain addition reactions between monomers containing a double bond. Molecules of ethene can polymerize with each other to form polyethene, commonly called polyethylene.

$$n\,CH_2{=}CH_2 \xrightarrow[\text{(catalyst)}]{} (-CH_2{-}CH_2{-})_n$$

ethene polyethylene

The letter n shows that the addition reaction can be repeated multiple times to form a polymer n monomers long. This can be repeated hundreds of times.

A condensation polymer is a polymer formed by condensation reactions. Monomers of condensation polymers must contain two functional groups. This allows each monomer to link with two other monomers by condensation reactions. Condensation polymers are usually copolymers with two monomers in alternating order.

One example of a condensation polymer is shown below. The hexanediamines with two amine groups react with adipic acids with two carboxyl groups to form nylon 66 and water.

$$n\,\overset{\overset{\displaystyle H}{|}}{H-N}-CH_2-\ CH_2-\ CH_2-\ CH_2-\ CH_2-\ CH_2-\overset{\overset{\displaystyle H}{|}}{N}-H\ +$$

hexanediamine

$$n\,OH-\overset{\overset{\displaystyle O}{\|}}{C}-CH_2-CH_2-CH_2-CH_2-\overset{\overset{\displaystyle O}{\|}}{C}-OH \rightarrow$$

adipic acid

$$(-\overset{\overset{\displaystyle H}{|}}{N}-CH_2-CH_2-CH_2-CH_2-CH_2-CH_2-\overset{\overset{\displaystyle H}{|}}{N}-\overset{\overset{\displaystyle O}{\|}}{C}-CH_2-CH_2-CH_2-CH_2-\overset{\overset{\displaystyle O}{\|}}{C}-)_n + n\,H_2O$$

nylon 66 water

The product contains two kinds of monomers, an adipic acid monomer and a hexanediamine monomer. This copolymer is known as nylon 66 because each of the monomers contains six carbon atoms. It is the most widely used of all synthetic polymers.

Chapter Summary

The following terms summarize all the concepts and ideas that were introduced in this chapter. You should be able to explain their meaning and how you would use them in chemistry. They appear in boldface type in this chapter to draw your attention to them. The boldface type also makes the terms easier for you to look up if you need to. You could also use a search engine on your computer to get a quick and expanded explanation of these terms, laws, and formulas.

TERMS YOU SHOULD KNOW

alcohol
aldehyde
alicyclic
aliphatic
alkane
alkylation
alkyne
amine
amino acid
aromatics
carbohydrate
carboxyl
cracking
dehydrogenation
destructive distillation
diamond
empirical formula
ester
ether

fractional distillation
fullerenes
graphite
homologous series
hydrocarbon
hydrocarbon series
hydrogenation
isomer
ketone
monosaccharide
polymerization
polymers
polysaccharide
saturated
structural formulas
thermoplastic
thermosetting
unsaturated

Chapter 14 Review Exercises

1. Carbon atoms usually

   (A) lose 4 electrons
   (B) gain 4 electrons
   (C) form 4 covalent bonds
   (D) share the 2 electrons in the first principal energy level
   (E) lose 4 electrons to form ionic bonds

2. Coke is produced from bituminous coal by

   (A) cracking
   (B) synthesis
   (C) substitution
   (D) destructive distillation
   (E) replacement reactions

3. The usual method for preparing carbon dioxide in the laboratory is

   (A) heating a carbonate
   (B) fermentation
   (C) reacting an acid and a carbonate
   (D) burning carbonaceous materials
   (E) fractional distillation of air

4. The precipitate formed when carbon dioxide is bubbled into limewater is

   (A) $CaCl_2$
   (B) H_2CO_3
   (C) CaO
   (D) $CaCO_3$
   (E) $Ca(OH)_2$

5. The "lead" in a lead pencil is

   (A) bone black
   (B) graphite and clay
   (C) lead oxide
   (D) lead peroxide
   (E) bone black and clay

6. The newest allotropic form of carbon discovered is

    (A) diamond
    (B) graphite
    (C) fullerenes
    (D) monoclinic
    (E) carborundum

7. A common ingredient of baking powder is

    (A) NaCl
    (B) $NaHCO_3$
    (C) Na_2CO_3
    (D) NaOH
    (E) Na_2O

8. The first and simplest alkane is

    (A) ethane
    (B) methane
    (C) C_2H_2
    (D) methene
    (E) CCl_4

9. Slight oxidation of a primary alcohol gives

    (A) a ketone
    (B) an organic acid
    (C) an ether
    (D) an aldehyde
    (E) an ester

10. The characteristic group of the organic ester is

    (A) $-CO-$
    (B) $-COOH$
    (C) $-CHO$
    (D) $-O-$
    (E) $-COO-$

11. Fermentation of glucose gives

    (A) CO_2 and H_2O
    (B) CO and alcohol
    (C) CO_2 and CH_3OH
    (D) CO and C_2H_5OH
    (E) CO_2 and C_2H_5OH

12. The organic acid that can be made from ethanol is

    (A) acetic acid
    (B) formic acid
    (C) C_3H_7OH
    (D) found in bees and ants
    (E) butanoic acid

13. An ester can be prepared by the reaction of

    (A) two alcohols
    (B) an alcohol and an aldehyde
    (C) an alcohol and an organic acid
    (D) an organic acid and an aldehyde
    (E) an acid and a ketone

14. Phenol is a derivative of an

    (A) alkane
    (B) alkene
    (C) aliphatic (chain) hydrocarbon
    (D) aromatic hydrocarbon
    (E) alkyne

15. Sucrose is a

    (A) reducing sugar
    (B) monosaccharide
    (C) sugar substitute with a ketone group
    (D) disaccharide
    (E) sugar with an aldehyde group

16. Compounds that have the same composition but differ in structural formulas

    (A) are used for substitution products
    (B) are called isomers
    (C) are called polymers
    (D) have the same properties
    (E) are usually alkanes

17. Ethene is the first member of the

    (A) alkane series
    (B) saturated hydrocarbons
    (C) alkyne series
    (D) unsaturated hydrocarbons
    (E) aromatic hydrocarbons

Each of the following sets of lettered choices refers to the numbered questions immediately below it. For each numbered item, choose the one lettered choice that fits it best. Each choice in a set may be used once, more than once, or not at all.

Questions 18–22

(A) $CH_3 - CH_2 - CH_3$

(B) $CH_3 - C \overset{\displaystyle O}{\underset{\displaystyle OH}{\big<}}$

(C) $CH_3 - O - C_3H_7$

(D) $CH_3 - \overset{\displaystyle O}{\overset{\|}{C}} - CH_3$

(E) $CH_3 - CH_2 - N \overset{\displaystyle H}{\underset{\displaystyle H}{\big<}}$

18. Which organic structure is ethylamine?

19. Which organic structure is methyl propyl ether?

20. Which organic structure is propane?

21. Which organic structure is ethanoic acid?

22. Which organic structure is 2-propanone?

Questions 23–25

Using the same choices, match the functional groups named to the structure that contains it.

23. Which structure contains an organic acid functional group?

24. Which structure contains a ketone grouping?

25. Which structure contains an amine group?

Answers and Explanations

1. **(C)** Because carbon has 4 electrons in its outer energy level, it usually forms 4 covalent bonds to fill each of the four sp^3 orbitals.

2. **(D)** Bituminous coal has too many gaseous impurities to burn at the high temperature needed to refine iron ore. Instead, it is heated in coke ovens to form the hotter and cleaner burning coke.

3. **(C)** The reaction of an acid and a carbonate is the usual way to prepare CO_2.

4. **(D)** The reaction is
$$CO_2 + Ca(OH)_2 \rightarrow CaCO_3\downarrow + H_2O$$

5. **(B)** The lead in a lead pencil is a mixture of graphite and clay.

6. **(C)** Fullerenes were the most recently discovered form of carbon.

7. **(B)** The $NaHCO_3$ decomposes when heated in an oven to give off CO_2, which causes the batter to rise.

8. **(B)** The first and simplest of the alkanes is methane.

9. **(D)** The oxidation of a primary alcohol gives an aldehyde.

10. **(E)** The –COO– is the functional group for an ester.

11. **(E)** This is the fermentation that occurs in the making of wine. The carbon dioxide is given off and leaves the ethyl alcohol behind.

12. **(A)** Ethanol can be oxidized into the organic acid ethanoic acid, which has the common name of acetic acid.

13. **(C)** The formation of an ester is from the reaction of an organic acid and an alcohol. The general equation is
$$RO-H + R^1CO-OH \rightarrow R^1COO-R + HOH$$
$$\text{alcohol} \qquad \text{acid} \qquad\qquad \text{ester}$$

14. **(D)** Phenol is an aromatic (attached to a benzene ring) with an alcohol (–OH) attached.

15. **(D)** Sucrose is a disaccharide

16. **(B)** Isomers are compounds that have the same composition but differ in their structural formulas.

17. **(D)** Ethene is the first member of the alkene series that has one double bond. Because it has a double bond, it is said to be unsaturated. The alkane series, which has all single bonds between the carbon atoms in the chain, is called a saturated series.

18. **(E)** The $-NH_2$ group, called an amine group, makes the structure ethylamine. This type of organic structure is called an amide.

19. **(C)** The methyl (CH_3-) group and the propyl ($-C_3H_7$) group attached to a center oxygen (–O–) makes this methyl propyl ether.

20. **(A)** Propane is the third member of the alkane series, which is made up of a chain of single-bonded carbons and hydrogens with the general formula of C_nH_{2n+2}.

21. **(B)** Ethanoic acid is composed of a methyl group attached to the carboxyl group (–COOH). The latter is the functional group for an organic acid.

22. **(D)** The *propan-* part of the name that tells you its basic structure is from propane, which is a three carbon alkane. The *-one* part tells you it is a ketone that has a double-bonded oxygen attached to the (2- in the name) second carbon in the chain.

23. **(B)** The ethanoic acid contains the carboxyl group (–COOH). This is the identifying functional group for organic acids.

24. **(D)** The functional group for ketones is a carbon in the chain double-bonded to an oxygen atom (–CO–). The number in the front tells you which carbon has the double-bonded oxygen attached to it.

25. **(E)** The amine group is the nitrogen with two hydrogens ($-NH_2$) attached to a chain carbon. These are basic to the amino acid structures in the body.

NUCLEONICS

CHAPTER OBJECTIVES

Upon completing this chapter, you will be able to:

- Explain the nature of radioactivity, including types, the characteristics of each, and the inherent dangers
- Describe methods of detection of alpha, beta, and gamma rays
- Explain the changes that occur in a decay series
- Calculate the age of a substance using half-life
- Explain nuclear fission and fusion
- Describe how a nuclear reactor operates

Radioactivity

The discovery of **radioactivity** came in 1896 (less than two months after Wilhelm Röntgen announced the discovery of X-rays). Röntgen had pointed out that the X-rays came from a spot on a glass tube where a beam of electrons, in his experiments, was hitting, and that this spot simultaneously showed strong **fluorescence**. It occurred to Henri Becquerel and others that X-rays might in some way be related to fluorescence (emission of light when exposed to some exciting agency) and to phosphorescence (emission of light *after* the exciting agency is removed).

Becquerel accordingly tested a number of phosphorescent substances to determine whether they emitted X-rays while phosphorescing. He had no success until he tried a compound of uranium. Then he found that the uranium compound, whether or not it was allowed to phosphoresce by exposure to light, continuously emitted something that could penetrate lightproof paper and even thicker materials.

Becquerel determined that the compounds of uranium and the element itself produced **ionization** in the surrounding air. Thus, either the ionizing effect, as indicated by the rate of discharge of a charged electroscope, or the degree of darkening of a photographic plate, could be used to measure the intensity of the invisible emission. Moreover, the emission from the uranium was continuous, perhaps even permanent, and required no energy from any external source. Yet, probably because of the current interest and excitement over X-rays, Becquerel's work received little attention until early in 1898 when Marie and Pierre Curie entered the picture. (Pierre Curie was one of Becquerel's colleagues in Paris.)

Searching for the source of the intense radiation in uranium ore, they used tons of it to isolate very small quantities of two new elements, radium and polonium, both radioactive. Along with Becquerel, the Curies shared the Nobel Prize in Physics in 1903. Unfortunately, Pierre died soon afterward in a tragic accident involving a horse-drawn street cart. Marie Curie carried on their work and was chosen to succeed him at the Sorbonne as the first woman professor at that university. In 1911, she received a second Nobel Prize for her work in the chemistry of the new elements, radium and polonium. She founded the Radium Institute in Paris and devoted much of her time to the application of radioactivity to different mediums.

THE NATURE OF RADIOACTIVE EMISSIONS

While the early separation experiments were in progress, an understanding was slowly being gained about the nature of the spontaneous emission from the various radioactive elements. Becquerel thought at first that there were simply X-rays, but *three* different kinds of radioactive emission, now called **alpha particles**, **beta particles**, and **gamma rays**, were soon found. We now know that alpha particles are positively charged particles of helium nuclei, beta particles are streams of high-speed electrons, and gamma rays are high-energy radiations similar to X-rays. The emission of these three types of radiation is depicted below.

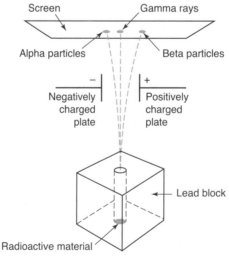

Deflection of Radioactive Emissions

The important characteristics of each type of radiation can be summarized as follows:

Alpha particle (helium nucleus $^4_2$He) Positively charged, +2

1. Ejection reduces the atomic number by 2 and the atomic weight by 4 amu.
2. High energy, relatively low velocity.
3. Range: about 5 cm in air.
4. Shielding needed: stopped by the thickness of a sheet of paper, skin.
5. Interactions: produces about 100,000 ionizations per centimeter; repelled by the positively charged nucleus; attracts electrons but does not capture them until its speed is much reduced.
6. An example: Thorium-230 has an unstable nucleus and undergoes radioactive decay through alpha emission. The nuclear equation that describes this reaction is

$$^{230}_{90}\text{Th} \rightarrow ^4_2\text{He} + ^{226}_{88}\text{Ra}$$

In a decay reaction like this, the initial element (thorium-230) is called the parent nuclide and the resulting element (radium-226) is called the daughter nuclide.

Beta particle (fast electron) Negatively charged, −1

1. Ejected when a *neutron* decays into a proton and an electron.
2. High velocity, low energy.
3. Range: about 12 m.
4. Shielding needed: stopped by 1 cm of aluminum or the thickness of the average book.
5. Interactions: weak due to high velocity but produces about 100 ionizations per centimeter.
6. An example: Protactinium-234 is a radioactive nuclide that undergoes beta emission. The nuclear equation is

$$^{234}_{91}\text{Pa} \rightarrow ^{234}_{92}\text{U} + ^0_{-1}\text{e}$$

Gamma radiation (electromagnetic radiation identical with light; high energy) No charge

1. Beta particles and gamma rays are usually emitted together; after a beta is emitted, a gamma ray follows.
2. Arrangement in nucleus is unknown. Same velocity as visible light.
3. Range: no specific range.
4. Shielding needed: about 13 cm of lead.
5. Interactions: weak of itself; gives energy to electrons, which then perform the ionization.

METHODS OF DETECTION OF ALPHA, BETA, AND GAMMA RAYS

All methods of detection of these radiations rely on their ability to ionize. There are a number of methods in common use.

Photographic plate The fogging of a photographic emulsion led to the discovery of radioactivity. If this emulsion is viewed under a high-power microscope, it is seen that beta and gamma rays cause the silver bromide grains to develop in a scattered fashion. However, alpha particles, owing to the dense ionization they produce, leave a definite track of exposed grains in the emulsion. Hence, not only is the alpha particle detected, but also its range (in the emulsion) can be measured. Special emulsions are capable of showing the beta-particle tracks.

Film pages are used to measure radiation exposure of people working in a radiation environment.

Scintillation counter A fluorescent screen (e.g., ZnS) will show the presence of electrons and X-rays, as we have already seen. If the screen is viewed with a magnifying eyepiece, small flashes of light, called scintillations, will be observed. By observing the scintillations, one not only can detect the presence of alpha particles, but also can actually count them.

The cloud chamber One of the most useful instruments for detecting and measuring radiation is the Wilson cloud chamber. Its operation depends on the well-known fact that moisture tends to condense around ions (the probable explanation for some formation of clouds in the sky). If an enclosed region of air is saturated with water vapor (this is always the case if water is present) and the air is cooled suddenly, it becomes supersaturated, that is, it contains for the instant more water vapor than it can hold permanently, and a fog of water droplets develops around the ions in the chamber.

For example, an alpha particle traveling through such a supersaturated atmosphere supplies a trail of ions on which water droplets will condense. This trail is thus made visible. These trails are usually photographed by a camera that operates whenever a piston moves downward causing the air to expand and cool.

Cloud Chamber

The bubble chamber This device utilizes a liquid such as ether, ethyl alcohol, pentane, or propane, which is superheated to well above the boiling point. When the pressure is released quickly, the liquid is in a highly unstable condition, ready to boil violently. An ionizing particle passing through the liquid at this instant leaves a trail of tiny bubbles that may be photographed.

The Geiger counter This instrument is perhaps the most widely used at the present time for measuring individual radiation. It consists of a fine wire of tungsten mounted along the axis of a tube that contains a gas at reduced pressure. A difference of potential of about 1000 V is applied in such a way as to make the metal tube negative with respect to the wire. The voltage is high enough so that the electrons produced are accelerated by the electric field. Near the wire, where the field is strongest, the accelerated particles produce more ions—positive ones going to the walls and negative ones being collected by the wire. Any particle that will produce an ion gives rise to the same avalanche of ions, and so the type of particle cannot be identified. However, each individual particle can be detected.

DECAY SERIES AND TRANSMUTATIONS

The nuclei of uranium, radium, and other radioactive elements are continually disintegrating. It should be emphasized that spontaneous disintegration produces the gas known as radon. The time it takes for half of the material to disintegrate is called its **half-life**.

For example, for radium, we know that on the average half of all the radium nuclei present will have disintegrated to radon in 1590 years. In another 1590 years, half of this remainder will decay, and so on. When a radium atom disintegrates, it loses an alpha particle, which upon gaining two electrons eventually becomes a neutral helium atom. The remainder of the atom becomes radon.

Such a conversion of an element to a new element (because of a change in the number of protons) is called a **transmutation**. This transmutation can be produced artificially by bombarding the nuclei of a substance with various particles from a particle accelerator such as the cyclotron.

The following uranium-radium disintegration series shows how a radioactive atom can change when it loses each kind of particle. Note that the atomic number is shown by a subscript ($_{92}$U), and the isotopic mass by a superscript (^{238}U). The alpha particle is shown by the Greek symbol α, and the beta particle by β.

$$^{238}_{92}\text{U} \xrightarrow{-\alpha} {}^{234}_{90}\text{Th} \xrightarrow{-\beta} {}^{234}_{91}\text{Pa} \xrightarrow{-\beta} {}^{234}_{92}\text{U} \xrightarrow{-\alpha} {}^{230}_{90}\text{Th} \xrightarrow{-\alpha}$$

$$^{226}_{88}\text{Ra} \xrightarrow{-\alpha} {}^{222}_{86}\text{Rn} \xrightarrow{-\alpha} {}^{218}_{84}\text{Po} \xrightarrow{-\alpha} {}^{214}_{82}\text{Pb} \xrightarrow{-\beta} {}^{214}_{83}\text{Bi} \xrightarrow{-\beta}$$

$$^{214}_{84}\text{Po} \xrightarrow{-\alpha} {}^{214}_{82}\text{Pb} \xrightarrow{-\beta} {}^{210}_{83}\text{Bi} \xrightarrow{-\beta} {}^{210}_{84}\text{Po} \xrightarrow{-\alpha} {}^{206}_{82}\text{Pb(stable)}$$

This can be shown graphically in the radioactive decay series below.

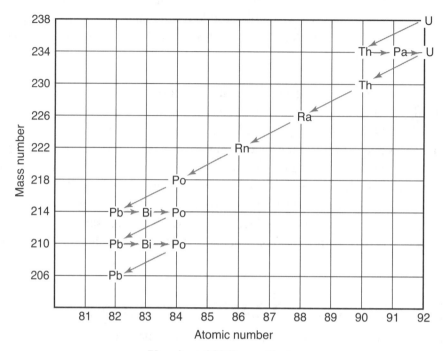

Uranium-238 Decay Series

The stability of an atom seems to be related to binding energy. The binding energy is the amount of energy released when a nucleus is formed from its component particles. If you add the mass of the components and compare this sum to the actual mass of the nucleus formed, there will be a small difference in these figures. This difference in mass can be converted to its energy equivalent using Einstein's equation $E = mc^2$, where E is the energy, m is the mass, and c is the velocity of light. It is this energy that is called the binding energy.

It has been found that the lightest and heaviest elements have the smallest binding energy per nuclear particle and thus are less stable than the elements of intermediate atomic masses, which have the greatest binding energy.

The relationship of an even or odd number of protons to the number of neutrons affects the stability of a nucleus. Many stable nuclei have even numbers of protons and neutrons, while stability is less frequent in nuclei that have an even number of protons and an odd number of neutrons, or vice versa. Only a few stable nuclei are known that have odd numbers of protons and neutrons.

The changes that occur in radioactive reactions are summarized in the following chart.

RADIOACTIVE DECAY AND NUCLEAR CHANGE

Change in Type of Decay	Change in Decay Particle	Particle Mass	Charge	Nucleon Number	Atomic Number
Alpha decay	α	4	2+	Decreases by 4	Decreases by 2
Beta decay	β	0	1−	No change	Increases by 1
Gamma radiation	γ	0	0	No change	No change
Positron emission	β^+	0	1+	No change	Decreases by 1
Electron capture	e^-	0	1−	No change	Decreases by 1

NUCLEAR SYMBOLS FOR SUBATOMIC PARTICLES

Particle	Symbols	Nuclear Symbols
Proton	p	$_1^1 p$ or $_1^1 H$
Neutron	n	$_0^1 n$
Electron	e^- or β	$_{-1}^0 e$ or $_{-1}^0 \beta$
Positron	e^+ or β^+	$_{+1}^0 e$ or $_{+1}^0 \beta$
Alpha particle	α	$_2^4 He$ or $_{+2}^4 \alpha$
Beta particle	β or β^-	$_{-1}^0 e$ or $_{-1}^0 \beta$
Gamma ray	γ	$_0^0 \gamma$

RADIOACTIVE DATING

A helpful adaptation of radioactive decay is its use in determining the age of substances such as rocks and relics. Because ^{14}C has a half-life of about 5700 years and occurs in the remains of organic materials, it has been useful in dating these materials. A small percentage of CO_2 in the atmosphere contains ^{14}C. The stable isotope of carbon is ^{12}C. ^{14}C is a beta emitter and decays to form ^{14}N:

$$_6^{14}C \rightarrow \, _7^{14}N + \, _{-1}^0 e$$

In any living organism, the ratio of ^{14}C to ^{12}C is the same as it is in the atmosphere because of the constant interchange of materials between organism and surroundings. When an organism dies, this interaction stops and the ^{14}C gradually decays to nitrogen. By comparing the relative amounts of ^{14}C and ^{12}C in the remains, the age of the organism can be established. Since ^{14}C has a half-life of 5700 years, if there was 5 g of ^{14}C at first and now there is only half as much, 2.5 g, its age would be 5700 years. In other words, old wood emits half as much beta radiation per gram of carbon as that emitted by living plant tissues. This method was used to determine the age of the Dead Sea Scrolls (about 1900 years) and has been found to be in agreement with several other dating techniques that have been employed.

Nuclear Energy

Since Einstein predicted that matter could be converted to energy over a century ago, scientists have tried to unlock this energy source. Einstein's prediction was verified in 1932 by Sir John Cockcroft and Ernest Walton, who produced small quantities of helium and a tremendous amount of energy from the reaction

$$_3^7\text{Li} + _1^1\text{H} \rightarrow 2_2^4\text{He} + \text{energy}$$

In fact, the energy released was almost exactly the amount calculated from Einstein's equation.

In 1942, Enrico Fermi and his co-workers discovered that a sustained **chain reaction** of **fission** could be controlled to produce large quantities of energy. This led to the development of the nuclear fission bomb, which brought World War II to an end.

CONDITIONS FOR FISSION

When fissionable material like ^{235}U is bombarded with a "slow" neutron, fission occurs, giving off different fission products. An example of such a reaction is shown in Figure 42. As long as one of the released neutrons produces another reaction, the chain reaction will continue. If each fission starts more than one reaction, the process becomes tremendously powerful in a very short time. This occurs in the atomic bomb.

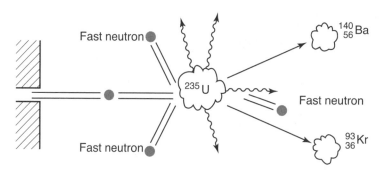

FIGURE 42. Fission chain reaction.

To obtain the neutron "trigger" for this reaction, neutrons must be slowed down so that they do not pass through the nucleus without effect. This slowing down or moderating is best done by letting fast neutrons collide with many relatively light atoms such as hydrogen, deuterium, and carbon. Graphite, paraffin, ordinary water, and "heavy water" (containing deuterium instead of ordinary hydrogen) are all suitable moderators.

Nuclear fission can be made to occur in an uncontrolled explosion or in a controlled nuclear reactor. In both cases, enough fissionable material must be present so that, once the reaction starts, it can at least sustain itself. This amount of fissionable material is called the **critical mass**. In the bomb, the number of reactions increases tremendously, whereas in a reactor the rate of fissions is controlled.

The typical nuclear reactor or "pile" is made up of the following kinds of material:

1. Fissionable material—sustains the chain reaction.
2. Moderator—slows down fission neutrons.
3. Control rods (cadmium or boron steel rods)—absorb excess neutrons and control rate of reaction.
4. Concrete encasement—provides shielding from radiation.

See the model diagram of a nuclear reactor below.

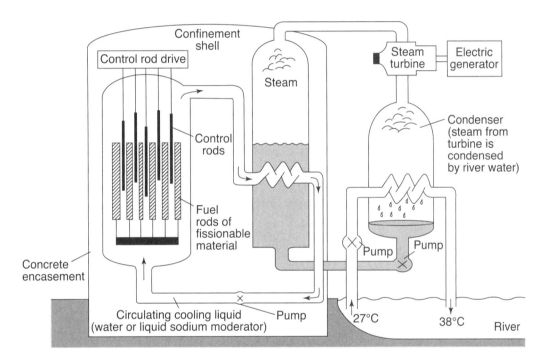

Nuclear Reactor

There are many variations of this basic reactor as reactors become more efficient. Many U.S. cities now receive some of their electrical power from a nuclear power station.

METHODS OF OBTAINING FISSIONABLE MATERIAL

The natural abundance of ^{235}U is extremely small: about 0.0005% of the Earth's crust, with 0.7% natural uranium. Therefore the problem is to separate this isotope from the rest of the more abundant isotope. Two methods have been used to isolate ^{235}U: the electromagnetic process and gaseous diffusion.

The electromagnetic process is based on the principle of the mass spectrograph. Gaseous ions with similar charges but different masses are passed between the poles of a magnet. The lighter ones are deflected more than the heavier ones and are thus collected in separate compartments. These large mass separators are called calutrons and were used extensively to collect the first uranium-235 isotopes.

The diffusion process was found to be faster and less expensive than the electromagnetic process. This process makes use of the gas diffusion principle that light molecules diffuse faster than heavier ones. When uranium is converted into gaseous uranium hexafluoride, UF_6, and allowed to diffuse through miles of porous partitions under reduced pressure, the lighter $^{235}UF_6$ will diffuse faster than the $^{238}UF_6$. Thus, the gas coming through first will be richer in ^{235}U than when it started. The ^{235}U is then retrieved from the fluoride compound for use in atomic devices or reactors.

FUSION

Fusion is the opposite of fission in that instead of a nucleus being split, two nuclei join to form a new nucleus with a great amount of released energy. The sun's energy results from such a fusion reaction, in which four hydrogen protons are eventually made into one helium nucleus with the liberation of a large amount of energy. The reactions are

$$^1_1H + ^1_1H \rightarrow ^2_1H + ^0_1e$$

$$^1_1H + ^2_1H \rightarrow ^3_2He$$

$$^3_2He + ^3_2He \rightarrow ^4_2He + 2^1_1H$$

$$^3_2He + ^1_1H \rightarrow ^4_2He + ^0_1e$$

So far, nuclear fusion has been produced only in an uncontrolled explosion, the hydrogen bomb. Scientists in every major country of the world are attempting to find a means of controlling a sustained fusion reaction.

In the meantime, nuclear fission is being adapted to produce energy for electrical power stations, merchant ships, aircraft carriers, and submarines. Experimentation is proceeding on the nuclear aircraft engine and a nuclear rocket engine.

Radiation Exposure

Nuclear radiation can transfer its energy to the electrons of atoms or molecules and cause ionization. The **roentgen** is a unit used to measure nuclear radiation. It is equal to the amount of radiation that produces 2×10^9 ion pairs when it passes through 1 cm^3 of dry air. Ionization can damage living tissue. Radiation damage to human tissue is measured in **rems** (roentgen equivalent, man). One rem is the quantity of ionizing radiation that does as much damage to human tissue as is done by 1 roentgen of high-voltage X-rays. Cancer and genetic effects caused by DNA mutations are long-term radiation damage to living tissue. DNA can be mutated directly by interaction with radiation or indirectly by interaction with previously ionized molecules.

Everyone is exposed to environmental background radiation. Average exposure for people living in the United States is estimated to be about 0.1 rem per year. At high altitudes, people have increased exposure because of increased cosmic-ray levels.

Radon-222 trapped inside home basements may also cause increased exposure. Because it is a gas, it can move up from the soil into homes through cracks and holes in the foundation.

Chapter Summary

The following terms summarize all the concepts and ideas that were introduced in this chapter. You should be able to explain their meaning and how you would use them in chemistry. They appear in boldface type in this chapter to draw your attention to them. The boldface type also makes the terms easier for you to look up if you need to. You could also use a search engine on your computer to get a quick and expanded explanation of these terms, laws, and formulas.

TERMS YOU SHOULD KNOW

alpha particle
beta particle
bubble chamber
chain reaction
cloud chamber
critical mass
fission
fluorescence
fusion
gamma ray

Geiger counter
half-life
ionization
radioactive dating
radioactivity
radon-22
rem
roentgen
scintillation counter
transmutation

Chapter 15 Review Exercises

1. Radioactive changes differ from ordinary chemical changes because radioactive changes

    (A) involve changes in the nucleus
    (B) are explosive
    (C) absorb energy
    (D) release energy

2. Isotopes of uranium have different

    (A) atomic numbers
    (B) atomic masses
    (C) numbers of planetary electrons
    (D) numbers of protons

3. Pierre and Marie Curie discovered

    (A) oxygen
    (B) hydrogen
    (C) chlorine
    (D) radium

4. The number of protons in the nucleus of an atom of atomic number 32 is

    (A) 4
    (B) 32
    (C) 42
    (D) 73

5. Atoms of ^{235}U and ^{238}U differ in structure by three

    (A) electrons
    (B) isotopes
    (C) neutrons
    (D) protons

6. In a hydrogen bomb, hydrogen is converted into

    (A) uranium
    (B) helium
    (C) barium
    (D) plutonium

7. The use of radioactive isotopes has already produced promising results in the treatment of certain types of

   (A) cancer
   (B) heart disease
   (C) pneumonia
   (D) diabetes

8. The emission of a beta particle results in a new element with the atomic number

   (A) increased by 1
   (B) increased by 2
   (C) decreased by 1
   (D) decreased by 2

9. A substance used as a moderator in a nuclear reactor is

   (A) marble
   (B) hydrogen
   (C) tritium
   (D) graphite

10. A self-sustaining nuclear fission chain reaction depends upon the release of

    (A) protons
    (B) neutrons
    (C) electrons
    (D) alpha particles

11. Einstein's formula for the Law of Conservation of Mass and Energy states that

    $E =$ _____ .

12. The energy of the sun is thought to be produced by the fusion of hydrogen atoms into

    _____ .

13. Could a mass spectrograph (or calutron) be used to separate ^{54}Mn from ^{54}Cr?

    _____ .

14. The release of energy from the combination of two nuclei into one is called

   _____ .

15. Write the equation for the first observed artificial transmutation by Rutherford:

   _____ .

MATCHING

Choose an answer from this list and match it with the appropriate numbered statement below.

A. Crooke	H. gamma ray
B. alpha particle	I. gaseous diffusion
C. beta particle	J. neptunium
D. Becquerel	K. plutonium
E. Einstein	L. ^{235}U
F. chain reaction	M. ^{238}U
G. deuteron	

16. Helium nucleus: _____

17. Stable isotope of uranium: _____

18. One method of separating isotopes: _____

19. Most penetrating ray from radioactive decay: _____

20. Mass-energy conversion: _____

Answers and Explanations

1. **(A)** Radioactive changes differ because they involve changes in the nucleus.

2. **(B)** Isotopes differ in their atomic mass due to differences in the number of neutrons.

3. **(D)** The Curies discovered radium.

4. **(B)** The atomic number, 32, gives the number of protons in the nucleus.

5. **(C)** These two isotopes, ^{235}U and ^{238}U, differ in their atomic mass by 3 neutrons.

6. **(B)** The nuclear fusion of hydrogen isotopes in the hydrogen bomb produces helium.

7. **(A)** Radioactive isotopes have been successful in the treatment of certain cancers.

8. **(A)** A beta particle emission causes an increase of 1 in the atomic number.

9. **(D)** A moderator slows down emission particles. Graphite is a moderator used in nuclear reactors.

10. **(B)** The release of neutrons, slow enough to be captured by other nuclei, self-sustains a fission chain reaction.

11. mc^2

12. He

13. No. (The mass spectrograph depends on differences in mass for separation.)

14. fusion

15. $^{14}_{7}\text{N} + ^{4}_{2}\text{He} \rightarrow ^{17}_{8}\text{O} + ^{1}_{1}\text{H}$

16. **(B)** A helium nucleus, two protons, and two neutrons is the makeup of an alpha particle.

17. **(M)** The stable isotope of uranium is ^{238}U.

18. **(I)** One method of separating isotopes is gaseous diffusion. Since the isotopes differ in mass, they pass through the diffusion mechanism at different rates and therefore separate.

19. **(H)** The gamma ray is the most penetrating ray of radioactivity because it has no mass.

20. **(E)** Einstein's famous equation $E = mc^2$ is used to calculate the mass converted to energy in a nuclear reaction.

REPRESENTATIVE LABORATORY SETUPS

Laboratory Safety Rules

THE TEN COMMANDMENTS OF LAB SAFETY

(A summary of rules you should be well aware of in your own chemistry lab)

1. **Dress appropriately for the lab.** Wear safety goggles and a lab apron or coat. Tie back long hair.

2. **Know what safety equipment is available and how to use it.** This includes the eyewash fountain, fire blanket, fire extinguisher, and emergency shower.

3. **Know the dangers of the chemicals in use and read labels carefully.** Do not taste or sniff chemicals.

4. **Dispose of chemicals according to instructions.** Use designated disposal sites, and follow the rules. Never return unneeded chemicals to their original containers.

5. **Always slowly add acid to water to avoid splattering.** This is especially important when using strong acids that can generate significant heat to form steam and splash out of the container. Never add water to acid. This rule also applies to adding bases to water.

6. **Never point heating test tubes at yourself or others.** Be aware of reactions that are occurring so that you can remove the test tubes from the heat if necessary before the contents "shoot" out of them.

7. **Do not pipette anything by mouth!** Never use your mouth as a suction pump, not even at home with toxic or flammable liquids.

8. **Use the fume hood when dealing with toxic fumes!** If you can smell them, you are exposing yourself to a dose that can harm you.

9. **Do not eat or drink in the lab!** Ingesting some dangerous substance is too easy.

10. **Follow all directions.** Never haphazardly mix chemicals. Pay attention to the order in which chemicals are to be added to each other, and do not deviate!

Technology in the Laboratory

Laboratory setups vary from school to school depending on whether the lab is equipped with macro- or microscale equipment. Microlabs use specialized equipment that allows lab work to be done on a much smaller scale. The basic principles are the same as when using full-sized equipment, but microscale equipment lowers the cost of materials, results in less waste, and poses less danger. The examples in this book are of macroscale experiments.

Along with learning to use microscale equipment, most labs require a student to learn how to use technological tools to assist in experiments. The most common are

Gravimetric balance with direct readings to thousandths of a gram instead of a triple-beam balance

pH meters that give pH readings directly instead of using indicators

Spectrophotometer that measures the percentage of light transmitted at specific frequencies so that the molarity of a sample can be determined without doing a titration

Computer-assisted labs that use probes to take readings, for example, temperature and pressure, so that programs available for computers can print out a graph of the relationship of readings taken over time

Some Basic Setups

Figures 43 through 54 show some of the basic laboratory setups used in beginning chemistry. The purpose of these diagrams is to review the basic techniques of assembling equipment with regard to some knowledge about the reactants and products involved.

In the equations, the letters in parentheses have the following meanings:

$$(s) = \text{solid}$$
$$(\ell) = \text{liquid}$$
$$(g) = \text{gas}$$

Preparation of a gaseous product, nonsoluble in water, by water displacement from solid reactants.

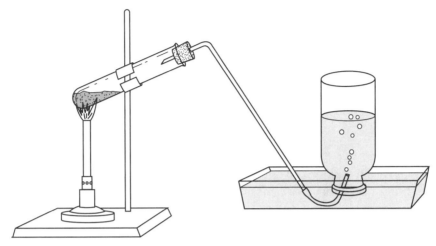

FIGURE 43. Preparation of oxygen.

Preparation of oxygen (O_2).

$$2KClO_3(s) + MnO_2(s) \rightarrow 2KCl + 3O_2(g) + MnO_2$$

Preparation of a gaseous product, nonsoluble in water, by water displacement from at least one reactant in solution.

Note: Purpose of the thistle tube shown in Figure 44:
    a. Introduction of more liquid without "opening" the reacting vessel.
    b. Safety valve to indicate blocked delivery tube by the rise of liquid in the thistle tube.

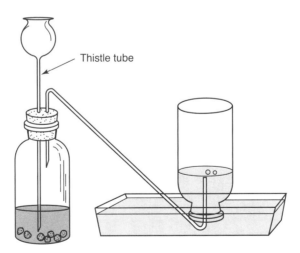

Thistle tube

FIGURE 44. Preparation of carbon dioxide, nitric oxide, and hydrogen.

Preparation of carbon dioxide (CO_2).

$$CaCO_3(s) + 2HCl(\ell) \rightarrow CaCl_2 + H_2O + CO_2(g)$$

Preparation of nitric oxide (NO).

$$3Cu(s) + 8 \text{ dilute } HNO_3(\ell) \rightarrow 3Cu(NO_3)_2 + 4H_2O + 2NO(g)$$

Preparation of hydrogen (H_2).

$$Zn(s) + 2HCl(\ell) \rightarrow ZnCl_2 + H_2(g)$$

3 | **Preparation of a gaseous product heavier than air that can best be collected by the upward displacement of air.**

Using a thistle tube with a stopcock to control the flow of a liquid reactant:

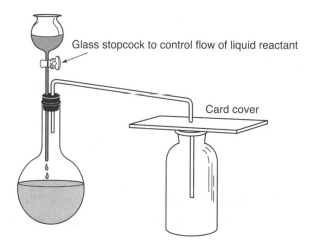

Glass stopcock to control flow of liquid reactant

Card cover

FIGURE 45. Preparation of sulfur dioxide and hydrogen sulfide.

Preparation of sulfur dioxide (SO_2).

$$Na_2SO_3(s) + H_2SO_4(\ell) \rightarrow Na_2SO_4 + H_2O + SO_2(g)$$

strong, irritating odor

Preparation of hydrogen sulfide (H_2S).

$$2HCl(\ell) + FeS(s) \rightarrow FeCl_2 + H_2S(g)$$

rotten egg odor

4 Preparation of a gaseous product that is soluble in water and lighter than air by the downward displacement of air.

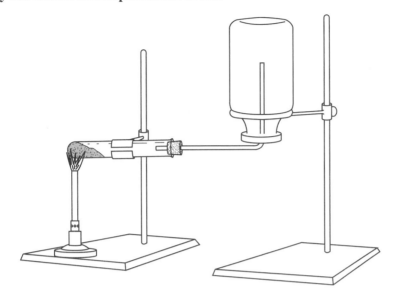

FIGURE 46. Preparation of ammonia.

Preparation of ammonia (NH_3).

$$2NH_4Cl(s) + Ca(OH)_2(s) \rightarrow CaCl_2 + 2H_2O + 2NH_3(g)$$

5 Distillation of a liquid.

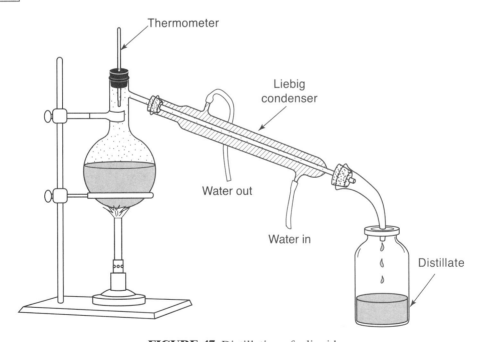

FIGURE 47. Distillation of a liquid.

Removes dissolved impurities, which remain in the flask. Does not remove volatile materials, which vaporize and pass over into the distillate.

6 **Preparation of a gaseous product, not dissolved by water, by means of electrolysis.**

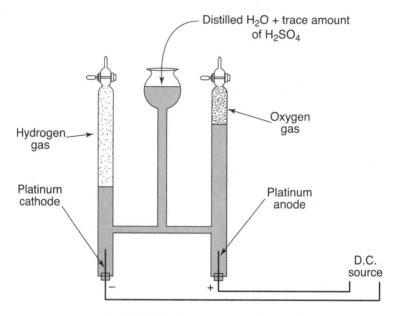

FIGURE 48. Electrolysis setup.

Anode reaction: $H_2O(\ell) \rightarrow \frac{1}{2}O_2(g) + 2H^+ + 2e^-$

Cathode reaction: $2H_2O(\ell) + 2e^- \rightarrow H_2(g) + 2OH^-$

Cell reaction: $3H_2O(\ell) \rightarrow \frac{1}{2}O_2(g) + H_2(g) + 2H_2O(\ell)$

or $H_2O(\ell) \rightarrow \frac{1}{2}O_2(g) + H_2(g)$

7 | **Separation of a mixture by chromatography.**

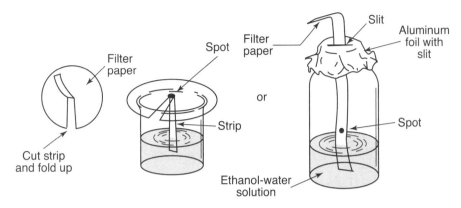

FIGURE 49. Chromatography setup.

Chromatography is a process used to separate parts of a mixture. The component parts separate as the solvent carrier moves past the spot by capillary action. Because of variations in solubility, attraction to the filter paper, and density, each fraction moves at a different rate. Once separation occurs, the fractions are either identified by their color or removed for other tests. A usual example is the use of Shaeffer Skrip Ink No. 32, which separates into yellow, red, and blue streaks of dyes.

8 | **Measuring potentials in electrochemical cells.**

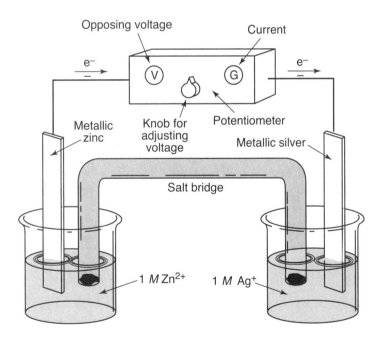

FIGURE 50. Potentiometer setup for measuring potential.

The voltmeter in this zinc-silver electrochemical cell reads approximately 1.56 V. This means that the Ag to Ag^+ half-cell has 1.56 V more electron-attracting ability than the Zn to Zn^{2+} half-cell. If the potential of the zinc half-cell is known, the potential of the silver half-cell can be determined by adding 1.56 V to the potential of the zinc half-cell. In a setup like this, only the difference in potential between two half-cells can be measured.

<table>
<tr><td>9</td></tr>
</table>

9 **Titration setup.**

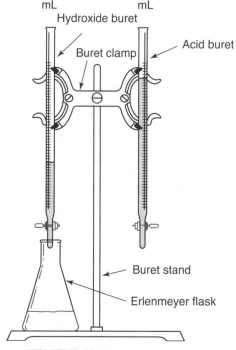

FIGURE 51. Titration using burets.

The titration of a NaOH solution of unknown concentration in one buret with 0.1 M HCl (standard solution) in the other.

Introduce approximately 15 mL of NaOH into the flask and add an indicator such as litmus or phenolphthalein. Add the HCl slowly with constant swirling. When a color change occurs and is retained, record the amount used. To find the molarity of the NaOH use the formula

$$M_{acid} \times V_{acid} = M_{base} \times V_{base}$$

10 | Replacement of hydrogen by a metal.

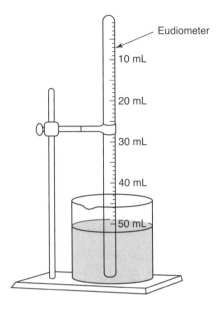

FIGURE 52. Eudiometer apparatus.

Measure the mass of a strip of magnesium with an analytical balance to the nearest 0.001 g. Using a strip with a mass of about 0.040 g produces about 40 mL of H_2. Pour 5 mL of concentrated HCl into the eudiometer and slowly fill the remainder with water. Try to minimize mixing. Lower the coil of Mg strip into the tube, invert it, and lower it to the bottom of the battery jar. After the reaction is complete, measure the volume of the gas released and calculate the mass of hydrogen replaced by the magnesium. (Refer to Chapter 5 for a discussion of gas laws.)

11   **Drying gases to remove water vapor.**

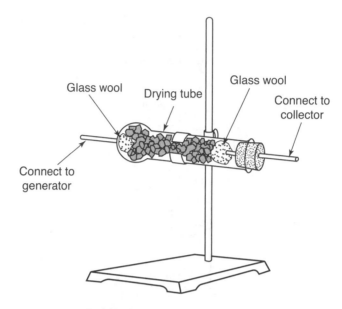

FIGURE 53. CaCl$_2$ drying tube.

Calcium chloride is hydroscopic and absorbs water vapor as the gas passes through this tube.

12   **Measuring temperature for phase change and heating curves.**

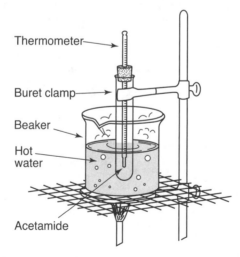

FIGURE 54. Heating curve and phase change setup.

This setup can be used with either acetamide or paradichlorobenzene in the test tube. By keeping a careful time and temperature observation chart, you can obtain data to plot a time-temperature graph and can note the effect of a phase change on this plot.

Summary of Qualitative Tests

I. IDENTIFICATION OF GASES

Gas	Test	Result
Ammonia NH_3	1. Smell cautiously. 2. Test with litmus. 3. Expose to HCl fumes.	1. Sharp odor. 2. Red litmus turns blue. 3. White fumes form (NH_4Cl).
Carbon dioxide CO_2	1. Pass through limewater, $Ca(OH)_2$.	1. White precipitate (ppt.) forms, $CaCO_3$.
Carbon monoxide CO	1. Burn it and pass product through limewater, $Ca(OH)_2$.	1. White ppt. forms, $CaCO_3$.
Hydrogen H_2	1. Allow it to mix with some air and then ignite. 2. Burn it and then trap product.	1. Gas explodes. 2. Burns with blue flame— product H_2O turns cobalt chloride paper from blue to pink.
Hydrogen chloride HCl	1. Smell cautiously. 2. Exhale over the gas. 3. Dissolve in water and test with litmus. 4. Add $AgNO_3$ to the solution.	1. Choking odor. 2. Vapor fumes form. 3. Blue litmus turns red. 4. Forms white ppt.
Hydrogen sulfide H_2S	1. Smell cautiously. 2. Test with moist lead acetate paper.	1. Rotten egg odor. 2. Turns brown-black (PbS).
Nitric oxide NO	1. Expose to the air.	1. Colorless gas turns reddish brown.
Nitrous oxide N_2O	1. Insert glowing splint. 2. Add nitric oxide gas.	1. Bursts into flame. 2. Remains colorless.
Oxygen O_2	1. Insert glowing splint. 2. Add nitric oxide gas.	1. Bursts into flame. 2. Turns reddish brown.
Sulfur dioxide SO_2	1. Smell cautiously. 2. Allow it to bubble into purple potassium permanganate solution.	1. Choking odor. 2. Solution becomes colorless.

II. IDENTIFICATION OF NEGATIVE IONS

Ion	Test	Result
Acetate $C_2H_3O_2^-$	Add conc. H_2SO_4 and warm gently.	Odor of vinegar released.
Bromide Br^-	Add chlorine water and some CCl_4; shake.	Reddish-brown color concentrated in CCl_4 layer.
Carbonate CO_{2-}	Add HCl acid; pass released gas through limewater.	White cloudy ppt. forms.
Chloride Cl^-	1. Add silver nitrate solution. 2. Then add nitric acid, later followed by ammonium hydroxide.	1. White ppt. forms. 2. Ppt. insoluble in HNO_3 but dissolves in NH_4OH.
Hydroxide OH^-	Test with red litmus paper.	Turns blue.
Iodide I^-	Add chlorine water and some CCl_4; shake.	Purple color concentrated in CCl_4 layer.
Nitrate NO_3^-	Add freshly made ferrous sulfate sol., and then conc. H_2SO_4 carefully down the side of the tilted tube.	Brown ring forms at junction of layers.
Nitrite NO_2^-	Add dilute H_2SO_4.	Brown fumes (NO_2) released.
Sulfate SO^{2-}	Add sol. of $BaCl_2$, then HCl.	White ppt. forms; insoluble in HCl.
Sulfide S^{2-}	Add HCl and test gas released with lead acetate paper.	Gas, with rotten egg odor, turns paper brown-black.
Sulfite SO_3^{2-}	Add HCl and pass gas into purple $KMnO_4$ sol.	Solution turns colorless.

III. IDENTIFICATION OF POSITIVE IONS

Ion	Test	Result
Ammonium NH_4^+	Add strong base (NaOH); heat gently.	Odor of ammonia.
Ferrous Fe^{2+}	Add sol. of potassium ferricyanide, $K_3Fe(CN)_6$.	Dark blue ppt. forms (Turnball's blue).
Ferric Fe^{3+}	Add sol. of potassium ferrocyanide, $K_4Fe(CN)_6$.	Dark blue ppt. forms (Prussian blue).
Hydrogen H^+	Test with blue litmus paper.	Turns red.

IV. QUALITATIVE TESTS FOR METALS

FLAME TESTS

Carefully clean a platinum wire by dipping it into dilute HNO_3 and heating in a Bunsen flame. Repeat until the flame is colorless. Dip heated wire into the substance being tested (either solid or solution) and then hold it in the hot outer part of the Bunsen flame.

Compound of	Color of Flame
Na	Yellow
K	Violet (use cobalt-blue glass to screen out Na impurities)
Li	Crimson
Ca	Orange-red
Ba	Green
Sr	Bright red

HYDROGEN SULFIDE TESTS

Bubble hydrogen sulfide gas through the solution of a salt of the metal being tested. Check color of the precipitate formed.

Compound of	Color of Sulfide Precipitate
Lead (Pb)	Brown-black (PbS)
Copper (Cu)	Black (CuS)
Silver (Ag)	Black (Ag_2S)
Mercury (Hg)	Black (HgS)
Nickel (Ni)	Black (NiS)
Iron (Fe)	Black (FeS)
Cadmium (Cd)	Yellow (CdS)
Arsenic (As)	Light yellow (As_2S_3)
Antimony (Sb)	Orange (Sb_2S_3)
Zinc (Zn)	White (ZnS)
Bismuth (Bi)	Brown (Bi_2S_3)

1.

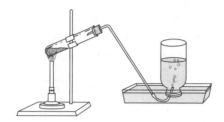

In the reaction setup shown above, which of the following is true?

I.   This setup can be used to prepare a soluble gas by water displacement.
II. This setup involves a decomposition reaction if the substance heated is potassium chlorate.
III. This setup can be used to prepare an insoluble gas by water displacement.

(A) I only
(B) II only
(C) I and III
(D) II and III
(E) I, II, and III

Questions 2–4 refer to the following diagram.

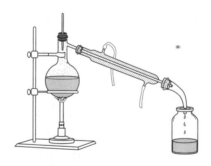

(A) around the thermometer
(B) in the condenser
(C) in the circulating water
(D) in the heated flask
(E) in the distillate

2. In this laboratory setup for distillation, where does the vaporization take place?

3. If the liquid being distilled contains dissolved magnesium chloride, where will the magnesium chloride be found after distillation is completed?

4. If the liquid being distilled contains dissolved ammonia gas, where will the ammonia be bound after distillation is completed?

5. If the flame used to heat the flask is an orange color and blackens the bottom of the flask, what correction should you make to solve this problem?

   (A) Move the flask farther from the flame.
   (B) Move the flask closer to the flame.
   (C) Allow less air into the collar of the burner.
   (D) Allow more air into the collar of the bruner.
   (E) The problem is in the gas supply, and you cannot fix it.

Questions 6–8 refer to the following diagram.

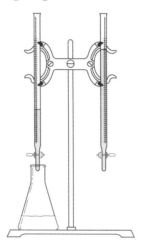

6. In the above titration setup, if you introduce 15 mL of the NaOH with an unknown molarity into the flask and then add 5 drops of phenolphthalein indicator, what will you observe?

   (A) A pinkish color will appear throughout the solution.
   (B) A blue color will appear throughout the solution.
   (C) There will be a temporary pinkish color that will dissipate.
   (D) There will be a temporary blue color that will dissipate.
   (E) There will not be a color change.

7. If the HCl is 0.1 M standard solution and you must add 30 mL to reach the end-point, what is the molarity of the NaOH?

(A) 0.1 M
(B) 0.2 M
(C) 0.3 M
(D) 1 M
(E) 2 M

8. When is the endpoint reached and the volume of the HCl recorded in this reaction?

   I. When the color first appears in the flask.
  II. When equal amounts of HCl and NaOH are in the flask.
 III. When the color in the flask disappears and does not return.

(A) I only
(B) III only
(C) I and III
(D) II and III
(E) I, II, and III

Questions 9–11 refer to the following diagram.

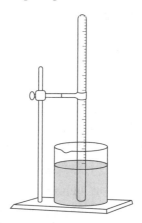

In this setup, a clean strip of magnesium with a mass of 0.040 g was introduced into the bottom of the tube that contained a dilute solution of HCl and allowed to react completely. The hydrogen gas formed was collected and the following data recorded:

Air pressure in the room = 730 mm Hg
Temperature of the water solution = 302 K
Vapor pressure of water at 302 K = 30.0 mm Hg

The gas collected did not fill the eudiometer. The height of the meniscus above the level of the water was 40.8 mm.

9. What is the theoretical yield (in mL) at STP of hydrogen gas produced when the 0.040 g of Mg reacted completely?

   (A) 10 mL
   (B) 25 mL
   (C) 37 mL
   (D) 46 mL
   (E) 51 mL

10. What is the correction to the atmospheric pressure due to the 40.8 mm height of the solution up in the tube and above the level in the beaker?

    (A) 3.0 mm Hg
    (B) 6.0 mm Hg
    (C) 13.6 mm Hg
    (D) 27.2 mm Hg
    (E) 40.8 mm Hg

11. What is the pressure of the collected gas once you have also corrected for vapor pressure of water?

    (A) 730 mm Hg
    (B) 727 mm Hg
    (C) 30.0 mm Hg
    (D) 697 mm Hg
    (E) 760 mm Hg

Questions 12–14 use the following choices.

   (A) The rule is to add concentrated acid to water slowly.
   (B) The rule is to add water to the concentrated acid slowly.
   (C) Carefully replace unused or excess chemicals into the properly labeled containers from which they came.
   (D) Flush eyes with water at the eyewash fountain for at least 15 minutes, and then report the accident for further help.
   (E) Dispose of chemicals in the proper places, and follow posted procedures.

12. Which of the above choices is the proper way to dilute a concentrated acid?

13. How do you properly dispose of chemicals not needed in the experiment?

14. What should you do if a chemical splatters into your eye?

15. What instrument is being used in chemistry labs to measure the molarity of a colored solution by measuring the light transmitted through it?

    (A) electronic gravimetric balance
    (B) pH meter
    (C) spectrophotometer
    (D) computer-assisted probes
    (E) galvanometer

Answers and Explanations

1. **(D)** This setup can be used to prepare an insoluble gas but not a soluble one. If the substance is potassium chlorate, it does decompose into potassium chloride and oxygen.

2. **(D)** Vaporization occurs in the heated flask.

3. **(D)** The magnesium chloride will be left behind in the heated flask since it is not volatile as the liquid boils off.

4. **(E)** Since the dissolved ammonia is volatile below the boiling point of water, it will be found in the distillate. Some will also escape as a gas.

5. **(D)** One of the basic adjustments to a burner is to assure enough air is mixing with the gas to form a blue cone-shaped flame. Insufficient air will cause carbon deposits on the flask and an orange flame.

6. **(A)** Phenolphthalein indicator is colorless below a pH of 8.3 but is red to pink in basic solutions above this pH. In this NaOH solution, it will be red to pink.

7. **(B)** The calculation of the molarity is found by using the formula:

$$M_{acid} \times V_{acid} = M_{base} \times V_{base}$$
$$0.1\ M \times 30\ \text{mL} = M_{base} \times 15\ \text{mL}$$
$$M_{base} = 0.2\ M$$

    Notice that as long as the units of volume are the same, they cancel out of the equation.

8. **(B)** The endpoint is reached when the color of the indicator first appears and stays visible. The color will first appear temporarily before the endpoint but will dissipate with stirring.

9. **(C)** The theoretical yield at STP can be found from the chemical equation of the reaction:

$$
\begin{array}{ccc}
0.040\ \text{g Mg} & & x\ \text{L} \\
\text{Mg} + 2\text{HCl} & \rightarrow & \text{MgCl}_2 + \text{H}_2\ (g) \\
24.0\ \text{g} & & 22.4\ \text{L}
\end{array}
$$

    Solving for x L = 0.037 L or 37 mL

10. **(A)** The 40.8 mm of water being held up in the tube by atmospheric pressure can be changed to its equivalent height of mercury by dividing by 13.6 since 1 mm of Hg = 13.6 mm of water:

$$40.8 \ \cancel{mm \ H_2O} \times \frac{1 \ mm \ Hg}{13.6 \ \cancel{mm \ H_2O}} = 3 \ mm \ Hg$$

By correcting the atmospheric pressure, we get 730 mm Hg – 3 mm Hg = 727 mm Hg.

11. **(D)** Correcting the previous pressure by subtracting the given amount for the vapor pressure of water at 302 K gives 727 mm Hg – 30 mm Hg vapor pressure = 697 mm Hg as the final pressure.

12. **(A)** The correct and safe way to dilute concentrated acids is to add water slowly down the side of the beaker and be aware of any heat buildup.

13. **(E)** You never return chemicals or solutions to their original containers for fear of contaminating the original sources.

14. **(D)** It is essential to get your eyes washed of any chemicals. Know where the eyewash fountains are, and know how to use them.

15. **(C)** The spectrophotometer uses the absorption of light waves to do qualitative and quantitative investigations in the lab.

PART VIII

TESTING YOUR ACHIEVEMENT IN CHEMISTRY

PRACTICE TESTS IN CHEMISTRY

General Information

This section comprises three tests. These test the same body of material usually included in a full-year college-preparatory chemistry course.

Practice Test 1 and Practice Test 2 are directly modeled after the SAT Subject Test in Chemistry. The directions are similar to those used on the present SAT Subject Test in Chemistry.

Students take the subject tests to demonstrate to colleges their mastery of particular topics, including chemistry. The tests' content evolves to reflect current trends in high school curricula, but the types of questions change little from year to year.

Many colleges use the subject tests for admission, for course placement, and to advise students about course selection. When used in combination with other background information (your high school record, scores from other tests, and so on), subject tests provide a dependable measure of your academic achievement. They are also a good predictor of future performance. A good score in chemistry may help to get you into the college of your choice.

Some colleges specify the subject tests they require for admission or placement. Others allow applicants to choose which tests to take. Check the catalogs of the colleges you are interested in for this information.

Study the descriptions of the test and the directions in this book, and visit the SAT Subject Tests Preparation Center at *www.collegeboard.com* for more information about the test, registration, test dates, and scoring.

Practice Test 3 covers the same content and skills but uses only a variety of multiple-choice questions. Notably, it does not use the true/false questions that are distinctively used in the first two tests.

The following is a chart of the content covered on these tests and the approximate distribution of questions in each area.

BASIC TOPICS AND ABILITIES TESTED

	Percent of Test (Approx.)	Number of Questions (Approx.)
Structure of Matter	25	21
I. Atomic theory and structure, energy levels, quantum numbers, orbitals, electron configurations, periodic trends, nucleonics		
II. Molecular structures, shapes, Lewis structures, polarity		
III. Bonding (ionic, covalent, metallic), relationships to properties and structures, intermolecular forces, hydrogen bonding, London dispersion forces, dipole-dipole forces		
States of Matter	16	14
IV. Gases, Kinetic Molecular Theory, gas law relationships, molar volume, density and related problems		
V. Liquids, solids, forces in liquids and solids, types of solids, phase diagrams, phase changes		
VI. Solutions; molarity; molality; percent by mass; solubility factors for solids, liquids, and gases; colligative properties		
Reaction Types	14	12
VII. Acids and bases, acid and base theories, strong and weak forms acids and bases, pH, titration problems, indicators		
VIII. Oxidation-reduction, combustion, oxidation numbers, activity series		
Stoichiometry	14	12
IX. Mole concept, molar mass, Avogadro's number, empirical and molecular formulas, chemical equations, balancing equations, solving related problems, determining yield, determining limiting factors		
Equilibrium and Reaction Rates	5	4
X. Equilibrium systems, factors affecting equilibrium, Le Châtelier's principle in gaseous and aqueous reactions, constants, rates of reactions, factors affecting rates, activation energies, reaction diagrams		
Thermochemistry	6	5
XI. Conservation of energy, calorimetry, specific heat, thermal curves, enthalpy changes, entropy changes		
Descriptive Chemistry	12	10
XII. Physical and chemical properties; nomenclature of elements, and ions; properties related to the compounds, periodic table; examples of basic organic compounds; environmental concerns		
Laboratory	8	7
XIII. Laboratory safety rules; nomenclature; use of equipment; observations; data to analyze, interpret graphical data, and draw conclusions		**Total Questions (85)**

Note: Each test contains approximately five questions on equation balancing and/or predicting products of chemical reactions. These are distributed among the various content categories.

Thinking Skills Tested	Percent of Test (approx.)
Recalling fundamental concepts, specific pieces of information, and basic terminology (low-level skill)	20
Showing a *comprehension of the basics* and the *ability to apply this information* in a rather straightforward manner to questions, situations, and the solution of qualitative or quantitative problem-oriented questions (medium-level skill)	45
Using the ability to *analyze* information and/or situations and to *synthesize* the knowledge learned to *evaluate* how and what ideas or relationships should be used to draw conclusions or to solve problems (high-level skill)	35

The chart on page 386 gives you a general overview of the content of the tests. Your knowledge of the topics and your skills in recalling, applying, and synthesizing this knowledge are evaluated through 85 multiple-choice questions. This is the material that is generally covered in an introductory course in chemistry at a level suitable for college preparation. While every test covers the topics listed, different aspects of each topic are stressed from year to year. Add to this the differences that exist in high school courses with respect to the percentage of time devoted to each major topic and to the specific subtopics covered, and you may find that there are questions on topics with which you have little or no familiarity.

Each of the sample tests in this book is constructed to match closely the distribution of topics shown in the above chart so that you will gain a feel for the makeup of a chemistry test on a full year course. After each test, a chart will show you which questions relate to each topic. This will be very helpful to you in planning your review because you can identify the areas on which you need to concentrate in your studies. Another chart enables you to see which chapters correspond to the various topic areas.

What Types of Questions Appear on the Tests?

There are three general types of questions on the SAT Subject Test in Chemistry. They are matching questions, true/false and relationship questions, and multiple-choice questions. This section will discuss each type of question and give specific examples of how to answer each type. You should learn the directions for each type so that you will not waste time becoming familiar with them on the test day. The directions in this section are similar to those that are on the test. Test 3 is not divided into parts. It has 85 questions that are all varieties of multiple-choice questions.

TYPE 1: MATCHING QUESTIONS IN PART A

In each of these questions, you are given five lettered choices to be used to answer all the questions in that set. The choices may be in the form of statements, pictures, graphs, experimental findings, equations, or specific situations. Answering a question may be as simple as recalling information or as difficult as analyzing the information given to establish what you need to do qualitatively or quantitatively to synthesize your answer. The directions for this type of question specifically state that a choice can be used once, more than once, or not at all in each set.

PART A

Directions: Each set of lettered choices below refers to the numbered statements or formulas immediately following it. Select the one lettered choice that best fits each statement or formula and then fill in the corresponding oval on the answer sheet. A choice may be used once, more than once, or not at all in each set.

Example

Questions 1–3 refer to the following graphs:

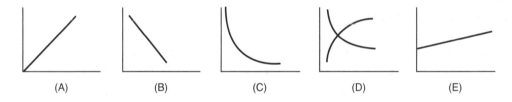

1. The graph that best shows the relationship of volume to temperature for an ideal gas while the pressure is held constant

2. The graph that best shows the relationship of volume to pressure for an ideal gas while the temperature is held constant

3. The graph that best shows the relationship of the number of grams of solute that is soluble in 100 grams of water at varying temperatures if the solubility begins as a small quantity and increases slowly as the temperature is increased.

These three questions require you to recall the basic gas laws and the graphic depiction of the relationship expressed in each law as well as how solubility can be shown graphically.

To answer the first question, you must recognize that the relationship between gas volume and changes in temperature is a direct relationship that is depicted by graphing Charles' Law, $V_1/T_1 = V_2/T_2$. The only graph that shows this type of direct relationship with the appropriate slope is (A).

To answer the second question you need to understand that Boyle's Law states that the pressure of a gas specimen is inversely proportional to its volume at constant temperature. Mathematically this means that pressure (P) times volume (V) is constant, or $P_1V_1 = P_2V_2$. This inversely proportional relationship is accurately depicted in (C). Although (B) shows the x-axis increasing as the y-axis decreases, it is not the graph for an inverse proportion.

The third question requires that you have knowledge about solubility curves and can translate the solubility relationship given in words to the graph shown as choice (E).

TYPE 2: TRUE/FALSE AND RELATIONSHIP QUESTIONS IN PART B (ON TESTS 1 AND 2 ONLY)

On the actual Chemistry Test, this type of question must be answered in a special section of your answer sheet labeled "chemistry" at the lower left-hand corner of page 2. These questions will be numbered beginning with 101. They consist of a statement or assertion in column I and, on the other side of the word BECAUSE, another statement or assertion in column II. Your first task is to determine if each of the statements is true or false and record your answer for each respectively in the answer blocks for column I and column II by darkening either the Ⓣ oval or the Ⓕ oval. Then you must use your reasoning skills and your understanding of the topic to determine if there is a cause-and-effect relationship between the two statements.

Here are the directions and two examples of a relationship analysis question.

PART B

Directions: Each question below consists of two statements, I in the left-hand column and II in the right-hand column. For each question, determine whether statement I is true or false and if statement II is true or false and fill in the corresponding T or F ovals on your answer sheet. Fill in oval CE only if statement II is a correct explanation of statement I.

Sample Answer Grid:

CHEMISTRY* Fill in oval CE only if
II is a correct explanation of I.

	I	II	CE*
101.	Ⓣ Ⓕ	Ⓣ Ⓕ	◯

Example 1

101. When 2 liters of oxygen gas react completely with 2 liters of hydrogen gas, the limiting factor is the volume of the oxygen

BECAUSE the coefficients in the balanced equation of a gaseous reaction give the volume relationship of the reacting gases.

The reaction that takes place is

$$2H_2 + O_2 \rightarrow 2H_2O$$

The coefficients of this gaseous reaction show that 2 liters of hydrogen react with 1 liter of oxygen. This leaves 1 liter of unreacted oxygen. The limiting factor is the quantity of hydrogen.

Knowing how to solve this quantitative relationship shows that statement I is not true. However, statement II is a true statement of the relationship of coefficients in a balanced equation for a gaseous chemical reaction. So the answer blocks should be marked like this:

	I	II	CE*
101.	T (F)	(T) F	◯

Example 2

101. Water is a good solvent of   BECAUSE   the water molecule has polar
ionic and polar compounds   properties caused by the factors
                                                         involved in the bonding of the
                                                         hydrogen and oxygen atoms.

Statement I is true because water is known to be such a good solvent that it is sometimes referred to as the universal solvent. This property is attributed mostly to its polar structure. Your knowledge of the polar covalent bond between the oxygen and hydrogen atoms and the angular orientation of the hydrogens with 105° between them contribute to the establishment of a permanent dipole moment. This feature also gives rise to a high degree of hydrogen bonding. These properties combine to make water a powerful solvent for both polar and ionic compounds. Knowing these concepts and how substances go into solution, you know that statement I is true and that the assertion in statement II explains statement I. There is a cause-and-effect relationship between the two statements. Therefore the answers should be marked like this:

	I	II	CE*
101.	(T) F	(T) F	●

TYPE 3: MULTIPLE-CHOICE QUESTIONS IN PART C

These five-choice questions are usually written as questions but are sometimes written as incomplete statements. You are given five suggested answers or completions. You must select the one that is best in each case and record your choice in the appropriate oval. In some questions you are asked to select the one inappropriate answer. Such questions contain a word in capital letters such as NOT, LEAST, or EXCEPT.

In some of these questions, you may be asked to make an association between a graphic, pictorial, or mathematical representation and a word explanation or problem. The solution may involve solving a scientific problem by interpreting these representations. In some cases, the same representation may be used for a series of two or more questions. In no case, however, is the answer to one question necessary for answering a subsequent question correctly. Each question in the set is independent of the others.

PART C

Directions: Each of the questions or incomplete statements below is followed by five suggested answers or completions. Select the one that is best in each case and then fill in the corresponding oval on the answer sheet.

Example 1

40. In this graphic representation of a chemical reaction, which arrow depicts the activation energy?

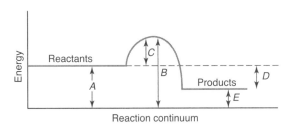

To answer this question you need to know how to interpret the energy levels in this graphic representation of energy level changes along the time continuum of the reaction. The activation energy is the minimum energy required for a chemical reaction to take place. The reactant molecules come together, and chemical bonds are stretched, broken, and formed in producing the products. During this process, the energy of the system increases to a maximum and then decreases to the energy of the products. The activation energy is the difference between the maximum energy and the energy of the reactants. Choice (C) in the graphic depiction shows this energy barrier that has to be overcome for the reaction to proceed. This oval should be darkened.

Example 2

41. If the molecular mass of NH_3 is 17, what is the density of this compound at STP?

    (A) 0.25 g/L
    (B) 0.76 g/L
    (C) 1.25 g/L
    (D) 3.04 g/L
    (E) 9.11 g/L

The solution of this quantitative problem depends on the application of several principles. One principle is that the molar mass of a gas expressed in grams will occupy 22.4 liters at standard temperature and pressure (STP). The other is that the density of a gas at STP is the mass of one liter of the gas. Therefore, 17 grams of ammonia occupy 22.4 liters, and 1 liter of this is 17 g/22.4 L, or 0.76 g/L. The correct mathematical answer is (B).

Example 3

Some questions in this part will give you a question followed by three or four bits of information labeled by Roman numerals I through III or IV. One or more of these statements may correctly answer the question. You must select from the five lettered choices the one that best answers the question.

45. Which bond(s) is(are) ionic?

      I. $H-Cl(g)$
      II. $S-Cl(g)$
      III. $Cs-F(g)$

      (A) I, II, and III
      (B) I and II only
      (C) II and III only
      (D) I only
      (E) III only

To determine the type of bonding that exists in these three substances, you must use your knowledge of ionic bonds and of how they are formed instead of other types of bonds. You must also use your knowledge of the relationship between the electronegativity of individual elements and the position of an element in the periodic table. The first two choices (I and II) are compounds formed between elements that do not differ enough in electronegativity to cause the formation of an ionic bond. This can be confirmed by checking the position of the elements in the periodic table and how electronegativity varies with respect to the position of the elements in the periodic table. Because cesium fluoride, choice III, consists of elements that appear in the lower right corner and the upper left corner of the periodic table, they have a sufficient difference in electronegativity values that an ionic bond can be predicted between them. Of the choices given, only (E) would be a correct answer.

Taking the Practice Tests in This Book

- **Use the answer sheet** provided for each test.

- **Keep the periodic table that is provided handy for reference.**

- **Complete each test in *one hour*.** Set an alarm clock to allow yourself one hour to take each test. Place it in view so that you can pace yourself.

- **Answer the easy questions first.**

- **Mark only one answer for each question.** Answer the questions you know first, and come back later to those you did not know.

- **Mark your answer in the correct column.** The answer sheet has five circles for each question. Some questions, though, have only three or four answers.

- **Use your test pages for scratch work,** but transfer your answers to your answer sheet. On machine-scored tests, you will receive credit ONLY for what is written in the answer blocks.

- **Guess only when you know a little about the topic.** If you can rule out one or more answers as wrong, your chances of guessing correctly among the remaining choices improve.

- **Omit questions only when you have no idea how to answer them.**

- **For all questions involving solutions, assume that the solvent is water unless otherwise noted.**

Reminder: You may NOT use a calculator on these tests.

The following symbols have the meanings listed unless otherwise noted.	
H = enthalpy	atm = atmosphere
M = molar	g = gram(s)
n = number of moles	J = joules(s)
P = pressure	kJ = kilojoules
R = molar gas constant	L = liter(s)
S = entropy	mL = milliliter(s)
T = temperature	mm = millimeter(s)
V = volume	V = volt(s)

After you have taken each test, follow the instructions for diagnosing your strengths and weaknesses and how to plan to improve in those areas.

Material in the following table may be useful in answering the questions in this examination

Periodic Table of the Elements

1 H 1.0078																	2 He 4.0026
3 Li 6.941	4 Be 9.012											5 B 10.811	6 C 12.011	7 N 14.007	8 O 16.00	9 F 19.00	10 Ne 20.179
11 Na 22.99	12 Mg 24.30											13 Al 26.98	14 Si 28.09	15 P 30.974	16 S 32.06	17 Cl 35.453	18 Ar 39.948
19 K 39.10	20 Ca 40.08	21 Sc 44.96	22 Ti 47.90	23 V 50.94	24 Cr 52.00	25 Mn 54.938	26 Fe 55.85	27 Co 58.93	28 Ni 58.69	29 Cu 63.55	30 Zn 65.39	31 Ga 69.72	32 Ge 72.59	33 As 74.92	34 Se 78.96	35 Br 79.90	36 Kr 83.80
37 Rb 85.47	38 Sr 87.62	39 Y 88.91	40 Zr 91.22	41 Nb 92.91	42 Mo 95.94	43 Tc (98)	44 Ru 101.1	45 Rh 102.91	46 Pd 106.42	47 Ag 107.87	48 Cd 112.41	49 In 114.82	50 Sn 118.71	51 Sb 121.75	52 Te 127.60	53 I 126.91	54 Xe 131.29
55 Cs 132.91	56 Ba 137.33	57 La 138.91	72 Hf 178.49	73 Ta 180.95	74 W 183.85	75 Re 186.21	76 Os 190.2	77 Ir 192.2	78 Pt 195.08	79 Au 196.97	80 Hg 200.59	81 Tl 204.38	82 Pb 207.2	83 Bi 208.98	84 Po (209)	85 At (210)	86 Rn (222)
87 Fr (223)	88 Ra 226.02	89 †Ac 227.03	104 Rf (261)	105 Db (262)	106 Sg (263)	107 Bh (262)	108 Hs (265)	109 Mt (266)	110 § (269)	111 § (272)	112 § (277)						

§ Not yet named

*Lathanides Series

58 Ce 140.12	59 Pr 140.91	60 Nd 144.24	61 Pm (145)	62 Sm 150.4	63 Eu 151.97	64 Gd 157.25	65 Tb 158.93	66 Dy 162.50	67 Ho 164.93	68 Er 167.26	69 Tm 168.93	70 Yb 173.04	71 Lu 174.97

†Actinides Series

90 Th 232.04	91 Pa 231.04	92 U 238.03	93 Np 237.05	94 Pu (244)	95 Am (243)	96 Cm (247)	97 Bk (247)	98 Cf (251)	99 Es (252)	100 Fm (257)	101 Md (258)	102 No (259)	103 Lr (260)

Answer Sheet for Practice Test 1

Determine the correct answer for each question. Then, using a No. 2 pencil, blacken completely the oval containing the letter of your choice.

1. Ⓐ Ⓑ Ⓒ Ⓓ Ⓔ
2. Ⓐ Ⓑ Ⓒ Ⓓ Ⓔ
3. Ⓐ Ⓑ Ⓒ Ⓓ Ⓔ
4. Ⓐ Ⓑ Ⓒ Ⓓ Ⓔ
5. Ⓐ Ⓑ Ⓒ Ⓓ Ⓔ
6. Ⓐ Ⓑ Ⓒ Ⓓ Ⓔ
7. Ⓐ Ⓑ Ⓒ Ⓓ Ⓔ
8. Ⓐ Ⓑ Ⓒ Ⓓ Ⓔ
9. Ⓐ Ⓑ Ⓒ Ⓓ Ⓔ
10. Ⓐ Ⓑ Ⓒ Ⓓ Ⓔ
11. Ⓐ Ⓑ Ⓒ Ⓓ Ⓔ
12. Ⓐ Ⓑ Ⓒ Ⓓ Ⓔ
13. Ⓐ Ⓑ Ⓒ Ⓓ Ⓔ
14. Ⓐ Ⓑ Ⓒ Ⓓ Ⓔ
15. Ⓐ Ⓑ Ⓒ Ⓓ Ⓔ
16. Ⓐ Ⓑ Ⓒ Ⓓ Ⓔ
17. Ⓐ Ⓑ Ⓒ Ⓓ Ⓔ
18. Ⓐ Ⓑ Ⓒ Ⓓ Ⓔ
19. Ⓐ Ⓑ Ⓒ Ⓓ Ⓔ
20. Ⓐ Ⓑ Ⓒ Ⓓ Ⓔ
21. Ⓐ Ⓑ Ⓒ Ⓓ Ⓔ
22. Ⓐ Ⓑ Ⓒ Ⓓ Ⓔ
23. Ⓐ Ⓑ Ⓒ Ⓓ Ⓔ

ON THE ACTUAL CHEMISTRY TEST, THE FOLLOWING TYPE OF QUESTION MUST BE ANSWERED ON A SPECIAL SECTION (LABELED "CHEMISTRY") AT THE LOWER LEFT-HAND CORNER OF PAGE 2 OF YOUR ANSWER SHEET. THESE QUESTIONS WILL BE NUMBERED BEGINNING WITH 101 AND MUST BE ANSWERED ACCORDING TO THE FOLLOWING DIRECTIONS.

CHEMISTRY* Fill in oval CE only if II is a correct explanation of I.

	I		II		CE*
101.	Ⓣ	Ⓕ	Ⓣ	Ⓕ	◯
102.	Ⓣ	Ⓕ	Ⓣ	Ⓕ	◯
103.	Ⓣ	Ⓕ	Ⓣ	Ⓕ	◯
104.	Ⓣ	Ⓕ	Ⓣ	Ⓕ	◯
105.	Ⓣ	Ⓕ	Ⓣ	Ⓕ	◯
106.	Ⓣ	Ⓕ	Ⓣ	Ⓕ	◯
107.	Ⓣ	Ⓕ	Ⓣ	Ⓕ	◯
108.	Ⓣ	Ⓕ	Ⓣ	Ⓕ	◯
109.	Ⓣ	Ⓕ	Ⓣ	Ⓕ	◯
110.	Ⓣ	Ⓕ	Ⓣ	Ⓕ	◯
111.	Ⓣ	Ⓕ	Ⓣ	Ⓕ	◯
112.	Ⓣ	Ⓕ	Ⓣ	Ⓕ	◯
113.	Ⓣ	Ⓕ	Ⓣ	Ⓕ	◯
114.	Ⓣ	Ⓕ	Ⓣ	Ⓕ	◯
115.	Ⓣ	Ⓕ	Ⓣ	Ⓕ	◯
116.	Ⓣ	Ⓕ	Ⓣ	Ⓕ	◯

ON THE ACTUAL CHEMISTRY TEST, THE REMAINING QUESTIONS MUST BE ANSWERED BY RETURNING TO THE SECTION OF YOUR ANSWER SHEET YOU STARTED FOR CHEMISTRY.

24. Ⓐ Ⓑ Ⓒ Ⓓ Ⓔ
25. Ⓐ Ⓑ Ⓒ Ⓓ Ⓔ
26. Ⓐ Ⓑ Ⓒ Ⓓ Ⓔ
27. Ⓐ Ⓑ Ⓒ Ⓓ Ⓔ
28. Ⓐ Ⓑ Ⓒ Ⓓ Ⓔ
29. Ⓐ Ⓑ Ⓒ Ⓓ Ⓔ
30. Ⓐ Ⓑ Ⓒ Ⓓ Ⓔ
31. Ⓐ Ⓑ Ⓒ Ⓓ Ⓔ
32. Ⓐ Ⓑ Ⓒ Ⓓ Ⓔ
33. Ⓐ Ⓑ Ⓒ Ⓓ Ⓔ
34. Ⓐ Ⓑ Ⓒ Ⓓ Ⓔ
35. Ⓐ Ⓑ Ⓒ Ⓓ Ⓔ
36. Ⓐ Ⓑ Ⓒ Ⓓ Ⓔ
37. Ⓐ Ⓑ Ⓒ Ⓓ Ⓔ
38. Ⓐ Ⓑ Ⓒ Ⓓ Ⓔ
39. Ⓐ Ⓑ Ⓒ Ⓓ Ⓔ

40. Ⓐ Ⓑ Ⓒ Ⓓ Ⓔ
41. Ⓐ Ⓑ Ⓒ Ⓓ Ⓔ
42. Ⓐ Ⓑ Ⓒ Ⓓ Ⓔ
43. Ⓐ Ⓑ Ⓒ Ⓓ Ⓔ
44. Ⓐ Ⓑ Ⓒ Ⓓ Ⓔ
45. Ⓐ Ⓑ Ⓒ Ⓓ Ⓔ
46. Ⓐ Ⓑ Ⓒ Ⓓ Ⓔ
47. Ⓐ Ⓑ Ⓒ Ⓓ Ⓔ
48. Ⓐ Ⓑ Ⓒ Ⓓ Ⓔ
49. Ⓐ Ⓑ Ⓒ Ⓓ Ⓔ
50. Ⓐ Ⓑ Ⓒ Ⓓ Ⓔ
51. Ⓐ Ⓑ Ⓒ Ⓓ Ⓔ
52. Ⓐ Ⓑ Ⓒ Ⓓ Ⓔ
53. Ⓐ Ⓑ Ⓒ Ⓓ Ⓔ
54. Ⓐ Ⓑ Ⓒ Ⓓ Ⓔ

55. Ⓐ Ⓑ Ⓒ Ⓓ Ⓔ
56. Ⓐ Ⓑ Ⓒ Ⓓ Ⓔ
57. Ⓐ Ⓑ Ⓒ Ⓓ Ⓔ
58. Ⓐ Ⓑ Ⓒ Ⓓ Ⓔ
59. Ⓐ Ⓑ Ⓒ Ⓓ Ⓔ
60. Ⓐ Ⓑ Ⓒ Ⓓ Ⓔ
61. Ⓐ Ⓑ Ⓒ Ⓓ Ⓔ
62. Ⓐ Ⓑ Ⓒ Ⓓ Ⓔ
63. Ⓐ Ⓑ Ⓒ Ⓓ Ⓔ
64. Ⓐ Ⓑ Ⓒ Ⓓ Ⓔ
65. Ⓐ Ⓑ Ⓒ Ⓓ Ⓔ
66. Ⓐ Ⓑ Ⓒ Ⓓ Ⓔ
67. Ⓐ Ⓑ Ⓒ Ⓓ Ⓔ
68. Ⓐ Ⓑ Ⓒ Ⓓ Ⓔ
69. Ⓐ Ⓑ Ⓒ Ⓓ Ⓔ

Practice Test 1

Note: For all questions involving solutions and/or chemical equations, assume that the system is in water unless otherwise stated.

PART A

Directions: Each set of lettered choices refers to the numbered statements or formulas immediately following it. Select the one lettered choice that best fits each statement or formula and then fill in the corresponding oval on the answer sheet. A choice may be used once, more than once, or not at all in each set.

Questions 1–9

PERIODIC TABLE (ABBREVIATED)

^{3}Li			. . .		(D)	^{10}Ne
	(A)		. . .	(C)		
(B)	^{20}Ca		. . .			(E)

1. The most electronegative element

2. The element with a possible oxidation number of −2

3. The element that would react in a one-to-one ratio with (D)

4. The element with the smallest ionic radius

5. The element with the smallest first ionization potential

6. The element with a complete *p* orbital as its outermost energy level

7. A member of the alkali-metals family

8. A noble gas

9. The element that would react most actively when placed in water to form a strong base

Questions 10–12 refer to the following heating curve for water:

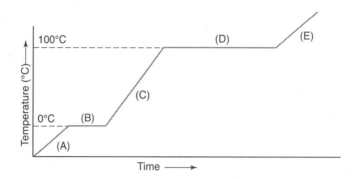

10. In which part of the curve is the state of H_2O only a solid?

11. When is the heat to change the state of H_2O greater?

12. Where is the temperature of H_2O changing at 1 K/4.18 J/g 1°C/cal/g?

Questions 13–15 refer to the following diagram:

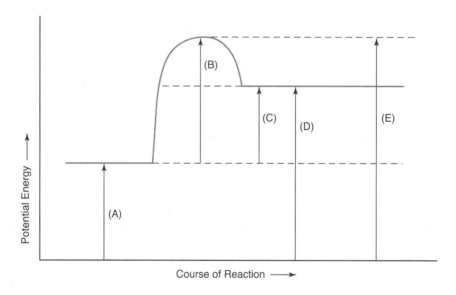

13. Indicates the forward activation energy of this reaction

14. Indicates the portion of the curve that would be directly affected by the addition of a catalyst

15. Indicates the difference between the activation energies for the reverse and forward reactions and equals the energy change in the reaction.

Questions 16–18

(A) Iron
(B) Gold
(C) Sodium
(D) Argon
(E) Uranium

16. A metal element that resists reaction with acids

17. A monatomic element with filled *p* orbitals

18. A transition element that occurs when the inner $3d$ orbital is partially filled

Questions 19–20

(A) Rhombic sulfur
(B) Monoclinic sulfur
(C) Sulfur trioxide
(D) Phosphate
(E) Sulfite

19. A substance that exhibits a resonance structure

20. A product formed from a base reacting with sulfurous acid

Questions 21–23

(A) Dilute
(B) Concentrated
(C) Unsaturated
(D) Saturated
(E) Supersaturated

21. The condition, unrelated to quantities, that indicates that the rate going into solution is equal to the rate coming out of solution

22. The condition that exists when a water solution that has been at equilibrium and is saturated is heated to a higher temperature with a higher solubility but no additional solute is added

23. The descriptive term that indicates there is a large quantity of solute, compared to the amount of solvent, in a solution

ON THE ACTUAL CHEMISTRY TEST, THE FOLLOWING TYPE OF QUESTION MUST BE ANSWERED ON A SPECIAL SECTION (LABELED "CHEMISTRY") AT THE LOWER LEFT-HAND CORNER OF PAGE 2 OF YOUR ANSWER SHEET. THESE QUESTIONS WILL BE NUMBERED BEGINNING WITH 101 AND MUST BE ANSWERED ACCORDING TO THE FOLLOWING DIRECTIONS.

Directions: Each question below consists of two statements, I in the left-hand column and II in the right-hand column. For each question, determine whether statement I is true or false <u>and</u> if statement II is true or false and fill in the corresponding T or F ovals on your answer sheet. <u>Fill in oval CE only if statement II is a correct explanation of statement I.</u>

Sample Answer Grid:

CHEMISTRY *Fill in oval CE only if
II is a correct explanation of I.

	I	II	CE*
101.	T F	T F	⬭

101. Nonmetallic oxides are usually acid anhydrides   BECAUSE   nonmetallic oxides form acids when placed in water.

102. When HCl gas and NH_3 gas come into contact, a white smoke forms   BECAUSE   the NH_3 and HCl react to form a white solid, ammonium chlorate.

103. The reaction of barium chloride and sodium sulfate does not go to completion   BECAUSE   the compound barium formed as an insoluble sulfate is precipitate.

104. When two elements react exothermically to form a compound, the compound should be relatively stable   BECAUSE   the release of energy from a combustion reaction indicates that the compound formed is at a lower energy level than the reactants and thus relatively stable.

105. The ion of a nonmetallic atom is larger in radius than the atom BECAUSE when a nonmetallic ion is formed, it gains electrons in the outer orbital and thus increases the size of the electron cloud around the nucleus.

106. Oxidation and reduction occur together BECAUSE in redox reactions electrons must be gained and lost.

107. Decreasing the atmospheric pressure on a pot of boiling water causes it to stop boiling BECAUSE changes in pressure are directly related to the boiling point of water.

108. The reaction of hydrogen with oxygen to form water is an exothermic reaction BECAUSE water molecules have polar covalent bonds.

109. Atoms of different elements can have the same mass number BECAUSE the atoms of each element have a characteristic number of protons in the nucleus.

110. The transmutation decay of $^{238}_{92}U$ can be shown as $^{238}_{92}U \rightarrow {}^{234}_{90}Th + {}^{4}_{2}He$ BECAUSE the transmutation of $^{238}_{92}U$ is accompanied by the release of a beta particle.

111. $^{12}_{6}C$ and $^{13}_{6}C$ are isotopes of the element carbon BECAUSE isotopes of an element have the same number of protons in the nucleus but a different number of neutrons.

112. The Cu^{2+} ion needs to be oxidized to form Cu metal BECAUSE oxidation is a gain of electrons.

113. The volume of a gas at 373 K and 600 mm Hg pressure will be decreased at STP BECAUSE decreasing temperature and increasing pressure will cause the volume to decrease because $V_2 = V_1 \times \dfrac{P_1}{P_2} \times \dfrac{T_2}{T_1}$.

114. The pH of a 0.01 M solution of HCl will be 2 BECAUSE dilute HCl dissociates into two essentially ionic particles.

115. Nuclear fusion on the sun converts hydrogen to helium with a release of energy

    BECAUSE some mass is converted to energy in a solar fusion.

116. The "bullet" usually used to initiate the fusion of ^{235}U is a neutron

    BECAUSE capture of the neutron by the ^{235}U nucleus causes an unstable condition that leads to its disintegration.

PART C

Directions: Each of the questions or incomplete statements below is followed by five suggested answers or completions. Select the one that is best in each case and then fill in the corresponding oval on the answer sheet.

24. What is the approximate formula mass of $Ca(NO_3)_2$?

    (A) 70
    (B) 82
    (C) 102
    (D) 150
    (E) 164

25. In this reaction: $XClO_3 + A \rightarrow XCl + O_2(g) + A$, which substance is the catalyst?

    (A) X
    (B) $XClO_3$
    (C) A
    (D) XCl
    (E) O_2

26. The normal electron configuration for ethyne (acetylene) is

    (A) H:C::C:H
    (B) H:C̈:C̈:H
    (C) H·C:::C·H
    (D) H:C::C:H
    (E) H:Ċ:Ċ:H

27. According to the Kinetic Molecular Theory, molecules increase in kinetic energy when they

    (A) are mixed with other molecules at lower temperature
    (B) are frozen into a solid
    (C) are condensed into a liquid
    (D) are heated to a higher temperature
    (E) collide with each other in a container at lower temperature

28. How many atoms are represented in the formula $Ca_3(PO_4)_2$?

   (A) 5
   (B) 8
   (C) 9
   (D) 12
   (E) 13

29. All of the following have covalent bonds EXCEPT

   (A) HCl
   (B) CCl_4
   (C) H_2O
   (D) CsF
   (E) CO_2

30. Which of the following is(are) the weakest attractive force?

   (A) dipole-dipole forces
   (B) coordinate covalent bonding
   (C) covalent bonding
   (D) polar covalent bonding
   (E) ionic bonding

31. Which of these resembles the molecular structure of the water molecule?

   (A)

   (B)

   (C)

   (D)

   (E)

32. The two most important considerations in deciding whether a reaction will occur spontaneously are

    (A) the stability and state of the reactants
    (B) the energy gained and the heat evolved
    (C) a negative value for ΔH and a positive value for ΔS
    (D) a positive value for ΔH and a negative value for ΔS
    (E) the endothermic energy and the structure of the products

33. The reaction of an acid like HCl and a base like NaOH always

    (A) forms a precipitate
    (B) forms a volatile product
    (C) forms an insoluble salt and water
    (D) forms a sulfate salt and water
    (E) forms a salt and water

34. The oxidation number of sulfur in H_2SO_4 is

    (A) +2
    (B) +3
    (C) +4
    (D) ⏐6
    (E) +8

35. Balance this redox reaction by using the half-reaction method of electron exchange:

$$KMnO_4 + H_2SO_3 \rightarrow K_2SO_4 + MnSO_4 + H_2SO_4 + H_2O$$

Which of the following partial equations is the correct reduction half-reaction for the *balanced* equation?

    (A) $5SO_3^{2-} + 5H_2O \rightarrow 5SO_4^{2-} + 10H^+ + 10e^-$
    (B) $2MnO_4^- + 16H^+ + 10e^- \rightarrow 2Mn^{2+} + 8H_2O$
    (C) $SO_3^{2-} \rightarrow SO_4^{2-} + 2e^-$
    (D) $SO_3^{2-} + 2H^+ \rightarrow SO_4^{2-} + H_2O + 2e^-$
    (E) $Mn^{7+} \rightarrow Mn^{2+} + 5e^-$

36. Which of the following, when placed into water, will test as an acid solution?

    I. $HCl(g) + H_2O$
    II. Excess $H_3O^+ + H_2O$
    III. $CuSO_4(s) + H_2O$

    (A) I only
    (B) III only
    (C) I and II only
    (D) II and III only
    (E) I, II, and III

37. The property of matter that is independent of its surrounding conditions and position is

(A) volume
(B) density
(C) mass
(D) weight
(E) state

38. Where are the highest ionization energies found in the periodic table?

(A) upper left corner
(B) lower left corner
(C) upper right corner
(D) lower right corner
(E) middle of transition elements

39. Which of the following pairs of compounds can be used to illustrate the Law of Multiple Proportions?

(A) NO and NO_2
(B) CH_4 and CO_2
(C) ZnO_2 and $ZnCl_2$
(D) NH_3 and NH_4Cl
(E) H_2O and HCl

40. In the equilibrium reaction $A + B \rightleftharpoons AB + heat$ (in a closed container), how could the forward reaction rate be increased?

I. By increasing the concentration of AB
II. By increasing the concentration of A
III. By removing some of product AB

(A) I only
(B) III only
(C) I and III only
(D) II and III only
(E) I, II, and III

41. In the reaction of sodium with water, the balanced equation using smallest whole numbers has which of the following coefficients?

    I.   1
    II. 2
    III. 3

    (A) I only
    (B) III only
    (C) I and II only
    (D) II and III only
    (E) I, II, and III

42. If 10 L of CO gas react with sufficient oxygen to completely react, how many liters of CO_2 gas are formed?

    (A) 5
    (B) 10
    (C) 15
    (D) 20
    (E) 40

43. If 49 g of H_2SO_4 react with 80 g of NaOH, how much reactant will be left over after the reaction is complete?

    (A) 24.5 g H_2SO_4
    (B) none of either compound
    (C) 20 g NaOH
    (D) 40 g NaOH
    (E) 60 g NaOH

44. If the molar concentration of Ag^+ ions in 1 liter of a saturated water solution of silver chloride is 1.38×10^{-5} mol/L, what is the K_{sp} of this solution?

    (A) 0.34×10^{-10}
    (B) 0.69×10^{-5}
    (C) 1.90×10^{-10}
    (D) 2.76×10^{-5}
    (E) 2.76×10^{-10}

45. If the density of a diatomic gas at STP is 1.43 g/L, what is its molar mass?

    (A) 16.3
    (B) 32.0
    (C) 48.0
    (D) 64.3
    (E) 128

46. From 2 mol of $KClO_3$ how many liters of O_2 can be produced at STP by decomposition of all the $KClO_3$?

(A) 11.2
(B) 22.4
(C) 33.6
(D) 44.8
(E) 67.2

47. Which value best determines whether a reaction is spontaneous?

(A) change in Gibbs free energy, ΔG
(B) change in entropy, ΔS
(C) change in kinetic energy, ΔKE
(D) change in enthalpy, ΔH
(E) change in heat of formation, ΔH^0

Questions 48–52 refer to the following experimental setup and data:

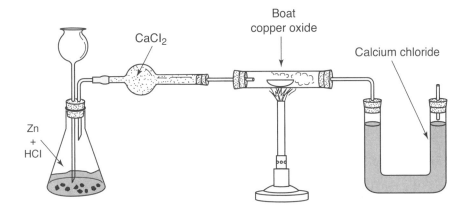

Recorded data:

Before: CuO + porcelain boat = 62.869 g
    $CaCl_2$ + U-tube   = 80.483 g

After: Porcelain boat + contents = 54.869 g
    $CaCl_2$ + U-tube   = 89.483 g

Reactions: $Zn + 2HCl \rightarrow ZnCl_2 + H_2$
        then $H_2 + CuO \rightarrow H_2O + Cu$

48. What type of reaction occurred in the porcelain boat?

    (A) electrolysis
    (B) double displacement
    (C) reduction and oxidation
    (D) decomposition or analysis
    (E) combination or synthesis

49. Why was the $CaCl_2$ tube placed between the generator and the tube containing the porcelain boat?

    (A) to absorb evaporated HCl
    (B) to absorb evaporated H_2O
    (C) to slow down the gases released
    (D) to absorb the evaporated Zn particles
    (E) to remove the initial air that passes through the tube

50. How many grams of hydrogen were used in the formation of the water that was a product?

    (A) 1
    (B) 2
    (C) 4
    (D) 8
    (E) 9

51. What conclusion can you draw from this experiment?

    (A) Hydrogen diffuses faster than oxygen.
    (B) Hydrogen is lighter than oxygen.
    (C) The molar mass of oxygen is 32 g/mol.
    (D) Water is a triatomic molecule with polar characteristics.
    (E) Water is formed from hydrogen and oxygen in a ratio of 1 : 8 by mass.

52. Which of the following is an observation rather than a conclusion?

    (A) A substance is an acid if it changes litmus paper from blue to red.
    (B) A gas is lighter than air if it escapes from a bottle left mouth upward.
    (C) The gas H_2 forms an explosive mixture with air.
    (D) Air is mixed with hydrogen gas and ignited; it explodes.
    (E) An oil liquid is immiscible with water because it separates into a layer above the water.

53. When HCl fumes and NH_3 fumes are introduced into opposite ends of a long, dry glass tube, a white ring forms in the tube. Which answer explains this phenomenon?

   (A) NH_4Cl forms.
   (B) HCl diffuses faster.
   (C) NH_3 diffuses faster.
   (D) The ring occurs closer to the end into which HCl was introduced.
   (E) The ring occurs in the middle of the tube.

54. The correct formula for calcium hydrogen sulfate is

   (A) CaH_2SO_4
   (B) $CaHSO_4$
   (C) $Ca(HSO_4)_2$
   (D) Ca_2HSO_4
   (E) $Ca_2H_2SO_4$

55. Ten grams of sodium hydroxide is dissolved in enough water to make 1.0 L of solution. What is the molarity of the solution?

   (A) $0.25\ M$
   (B) $0.5\ M$
   (C) $1\ M$
   (D) $1.5\ M$
   (E) $4\ M$

56. For a saturated solution of salt in water, which statement is true?

   (A) All dissolving has stopped.
   (B) Crystals begin to grow.
   (C) An equilibrium has been established.
   (D) Crystals of the solute will visibly continue to dissolve.
   (E) The solute is exceeding its solubility.

57. In which of the following series is the pi bond present in the bonding structure?

   I. Alkane
   II. Alkene
   III. Alkyne

   (A) I only
   (B) III only
   (C) I and III only
   (D) II and III only
   (E) I, II, and III

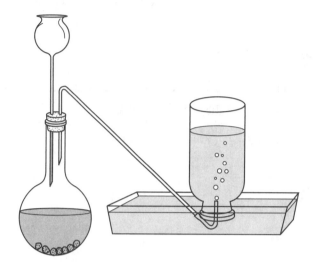

58. Why could you NOT use this setup for preparing H_2 if the generator contained Zn and vinegar?

    (A) Hydrogen would not be produced.
    (B) The setup of the generator is improper.
    (C) The generator must be heated with a burner.
    (D) The delivery tube setup is wrong.
    (E) The gas cannot be collected over water.

59. In a proper laboratory setup for collecting a gas by water displacement, which of these gases could NOT be collected over H_2O because of its solubility?

    (A) CO_2
    (B) NO
    (C) O_2
    (D) NH_3
    (E) CH_4

60. What is the approximate percentage of oxygen in the formula mass of $Ca(NO_3)_2$?

    (A) 28
    (B) 42
    (C) 58
    (D) 68
    (E) 84

61. For the reaction $N_2O_4(g) \rightleftharpoons 2NO_2(g)$, the K_{eq} expression is

(A) $K_{eq} = \dfrac{[N_2O_4]}{[NO_2]}$

(B) $K_{eq} = \dfrac{[N_2O_4]}{[NO_2]^2}$

(C) $K_{eq} = \dfrac{[NO_2]}{[N_2O_4]}$

(D) $K_{eq} = \dfrac{[NO_2]^2}{[N_2O_4]}$

(E) $K_{eq} = \dfrac{[N_2O_4]^2}{[NO_2]}$

62. What is the K_{eq} for the above reaction in question 61 if at equilibrium the concentration of N_2O_4 is 4×10^{-2} mol/L and that of NO_2 is 2×10^{-2} mol/L?

(A) 1×10^{-2}
(B) 2×10^{-2}
(C) 3×10^{-2}
(D) 4×10^{-4}
(E) 8×10^{-2}

63. How much water, in liters, must be added to 0.5 L of 6 M HCl to make it 2 M?

(A) 0.33
(B) 0.5
(C) 1
(D) 1.5
(E) 2

64. Four grams of hydrogen are ignited with 4 g of oxygen. How many grams of water can be formed?

(A) 0.5
(B) 2.5
(C) 4.5
(D) 8
(E) 36

65. Which structure is an ester?

(A)
$$H-\underset{\underset{H}{|}}{\overset{\overset{H}{|}}{C}}-O-\underset{\underset{H}{|}}{\overset{\overset{H}{|}}{C}}-H$$

(B)
$$H-\underset{\underset{H}{|}}{\overset{\overset{H}{|}}{C}}-\underset{\underset{H}{|}}{\overset{\overset{H}{|}}{C}}-\underset{\underset{H}{|}}{\overset{\overset{H}{|}}{C}}-OH$$

(C)
$$H-\underset{\underset{H}{|}}{\overset{\overset{H}{|}}{C}}-\underset{\underset{H}{|}}{\overset{\overset{H}{|}}{C}}-C\underset{\diagdown\ O-H}{\overset{\diagup\!\!\diagup\ O}{}}$$

(D)
$$H-\underset{\underset{H}{|}}{\overset{\overset{H}{|}}{C}}-\underset{\underset{H}{|}}{\overset{\overset{H}{|}}{C}}-C\underset{\diagdown\ H}{\overset{\diagup\!\!\diagup\ O}{}}$$

(E)
$$H-\underset{\underset{H}{|}}{\overset{\overset{H}{|}}{C}}-\underset{\underset{H}{|}}{\overset{\overset{H}{|}}{C}}-\overset{\overset{O}{\|}}{C}-O-\underset{\underset{H}{|}}{\overset{\overset{H}{|}}{C}}-H$$

66. What piece of apparatus can be used to introduce more liquid into a reaction and also serve as a pressure valve?

(A) stopcock
(B) pinchcock
(C) thistle tube
(D) flask
(E) condenser

67. Which formulas could represent the empirical formula and the molecular formula of a given compound?

(A) CH_2O and $C_4H_6O_4$
(B) CHO and $C_6H_{12}O_6$
(C) CH_4 and C_5H_{12}
(D) CH_2 and C_3H_6
(E) CO and CO_2

68. The reaction Fe $\rightarrow$ Fe^{2+} + 2e$^-$ (+0.44 volt) would occur spontaneously with which of the following?

   I.   Pb $\rightarrow$ Pb^{2+} + 2e$^-$ (+0.13 volt)
  II.   Cu $\rightarrow$ Cu^{2+} + 2e$^-$ (−0.34 volt)
 III.   2Ag$^+$ + 2e$^-$ $\rightarrow$ 2Ag^0 (+0.80 volt)

(A) I only
(B) III only
(C) I and III only
(D) II and III only
(E) I, II, and III

69. 2Na(s) + Cl$_2$(g) $\rightarrow$ 2NaCl(s) + 822 kJ

How much heat is released by the above reaction if 0.5 mole of sodium reacts completely with chlorine?

(A) 205 kJ
(B) 411 kJ
(C) 822 kJ
(D) 1644 kJ
(E) 3288 kJ

STOP
IF YOU FINISH BEFORE ONE HOUR IS UP, YOU MAY GO BACK
TO CHECK YOUR WORK OR COMPLETE UNANSWERED QUESTIONS.

(Total questions 85)
By Sections (1–23) (101–116) (24–69)
23 + 16 + 46 = 85

Answers and Explanations for Test 1

1. **(D)** The most electronegative element (F) would be found in the upper right corner; the noble gases are exceptions at the far right.

2. **(C)** Elements in the group with (C) have a possible oxidation number of –2.

3. **(B)** Elements in the group with (B) react in a 1 : 1 ratio with elements in the group with (D) since one has an electron to lose and the other one needs an electron to complete its outer energy level.

4. **(A)** (A) loses two electrons to form an ion whose remaining electrons, being close to the nucleus, are pulled in closer because of the unbalanced 2+ charge.

5. **(B)** Since (B) has only one electron in the outer $4s$ orbital, it can more easily be removed than can an electron from the $3s$ orbital of (A), which is closer to the positive nucleus.

6. **(E)** All Group VIII elements have a complete p orbital as the outer energy level. This explains why these elements are "inert."

7. **(B)** The Group IA elements comprise the alkali-metals family.

8. **(E)** The Group VIII elements are the inert or noble gases.

9. **(B)** The alkali metals react with water to form a strong base.

10. **(A)** H_2O is ice in part A.

11. **(D)** H_2O changes state at parts B and D. The heat of vaporization at D (2.26×10^3 J/g) is greater than the heat of fusion (3.34×10^2 J/g) at B.

12. **(C)** Water is heating at 1 K/4.18 J/g in part C.

13. **(B)** This is the energy needed to start the reaction. It is called the activation energy.

14. **(B)** The addition of a catalyst affects only the part indicated by B, by lowering the energy needed to commence the reaction.

15. **(C)** The net energy released is the endothermic quantity indicated by C.

16. **(B)** Gold is known as a noble metal because of its resistance to acids. Aqua regia, a mixture of HNO_3 and HCl, will react with gold.

17. **(D)** Argon is the only element that is monatomic in the molecular form.

18. **(A)** Iron has five electrons in the d orbitals, which are partially filled.

19. **(C)** Only sulfur trioxide has a resonance structure (as shown here):

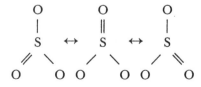

20. **(E)** Sulfurous acid reacts with a base to form a sulfite salt.

21. **(D)** The condition described is the equilibrium that exists at saturation.

22. **(C)** With the increased temperature more solute may go into solution; therefore, the solution is now unsaturated.

23. **(B)** The term *concentrated* means that there is a large amount of solute in the solvent.

101. **(T, T, CE)** Nonmetallic oxides are usually acid anhydrides, and they form acids in water.

102. **(T, F)** The white smoke formed is ammonium chloride, not ammonium chlorate.

103. **(F, T)** The reaction does go to completion since barium sulfate is a precipitate.

104. **(T, T, CE)** The product of an exothermic reaction is relatively stable because it is at a lower energy level than the reactants.

105. **(T, T, CE)** Both statements are true, and the reason explains the assertion.

106. **(T, T, CE)** Both the statements are true, and the reason explains the assertion.

107. **(F, T)** The statement is false while the reason is true. Decreasing the pressure on a boiling pot will only cause it to boil more vigorously.

108. **(T, T)** The statements are true, but the reason doesn't explain the assertion.

109. **(T, T)** The statements are true, but the reason doesn't explain the assertion.

110. **(T, F)** The transmutation equation is true and shows the release of an alpha particle, not a beta particle.

111. **(T, T, CE)** The two configurations of carbon are isotopes because they have the same atomic number, which means that they have the same number of protons in the nucleus.

112. **(F, F)** Both the statement and the reason are false.

113. **(T, T, CE)** To go from 100°C to 0°C decreases the volume as the gas gets colder; therefore, the temperature fraction expressed in kelvins must be $\frac{273}{373}$ to decrease the volume. To go from 600 mm to 760 mm of pressure increases the pressure, thus causing the gas to contract. The fraction must then be $\frac{600}{760}$ to cause the volume to decrease. You could use the formula

$$V_2 = V_1 \times \frac{T_2}{T_1} \times \frac{P_1}{P_2}$$

114. **(T, T)** Since HCl is a strong acid and ionizes completely in a dilute solution of water, the $[H^+]$ (H_3O^+ is the same thing) or molar concentration of a 0.01 M concentration is 1×10^{-2} mol/L.

pH = $-\log[H^+]$
pH = $-\log[1 \times 10^{-2}]$
pH = $-(-2) = 2$

The pH is 2, but the reason, although true, does not explain the statement.

115. **(T, T, CE)** Both statements are true, and the reason explains the assertion.

116. **(F, T)** The statement should refer to fission, not fusion. The reason is true.

24. **(E)** The total formula mass is

$$
\begin{array}{ll}
\text{Ca} & = 40 \\
\text{2N} & = 28 \\
\underline{\text{6O}} & = \underline{96} \\
\text{Total} & 164
\end{array}
$$

25. **(C)** The catalyst, by definition, is not consumed in the reaction and ends up in its original form as one of the products.

26. **(D)** The ethyne molecule is the first member of the acetylene series with a general formula of C_nH_{2n-2}. It contains a triple bond between the two C atoms: H:C⦂⦂C:H.

27. **(D)** Heating molecules increases their kinetic energy.

28. **(E)** $3Ca + 2P + 8O = 13$ atoms.

29. **(D)** Cesium and fluorine are from the most electropositive and electronegative portions, respectively, of the periodic chart and thus form an ionic bond by cesium giving an electron to fluorine to form the respective ions.

30. **(A)** Dipole-dipole forces are the weak attraction of the nuclear positive charge of one atom to the negative electron field of an adjacent atom. They are much weaker than the others named.

31. **(D)** The molecular structure of water is a polar covalent compound with the hydrogens 105° apart.

32. **(C)** The most important considerations for a spontaneous reaction are (1) that the reaction is exothermic with a $-\Delta H$ so that once started it tends to continue on its own because of the energy released and (2) that reactions tend to go to the highest state of randomness shown by a $+\Delta S$ value.

33. **(E)** Normal H^+ acids and OH^- bases form water and a salt—not necessarily a soluble salt.

34. **(D)** All compounds have a charge of 0. H usually has +1, and O usually has −2, so

$$
\begin{array}{ll}
H_2 & = +2 \\
O_4 & = -8 \\
\underline{S} & = \underline{x} \\
\text{Total} & = 0 \\
\end{array}
$$
$$
\begin{array}{l}
+2 - 8 + x = 0 \\
x = -2 + 8 = +6
\end{array}
$$

35. **(B)** In the balanced equation the two half reactions are

$$
\begin{aligned}
5SO_3{}^{2-} + 5H_2O &\rightarrow 5SO_4{}^{2-} + 10H^+ + 10e^- \\
\underline{2MnO_4{}^- + 16H^+ + 10e^-} &\underline{\rightarrow 2Mn^{2+} + 8H_2O} \\
2KMnO_4 + 5H_2SO_3 &\rightarrow \\
K_2SO_4 + 2MnSO_4 &+ 2H_2SO_4 + 3H_2O
\end{aligned}
$$

36. **(E)** HCl and H_3O^+ give acid solutions, as does $CuSO_4$ (the salt of a weak base and a strong acid), when it hydrolyzes in water. I, II, and III are correct.

37. **(C)** Mass is a constant and is not dependent on position or surrounding conditions.

38. **(C)** The complete outer energy level of electrons of the smallest inert gases has the highest ionization potential.

39. **(A)** Only NO and NO_2 fit the definition of the Law of Multiple Proportions, in which one substance stays the same and the other varies in units of whole integers.

40. **(D)** Increasing the concentration of one or both of the reactants and removing some of the product formed would cause the forward reaction to increase in rate to try to regain the equilibrium condition. II and III are correct.

41. **(C)** I and II are correct. The equation is $2Na + 2H_2O \rightarrow 2NaOH + H_2(g)$. The coefficients include a 1 and a 2.

42. **(B)** $2CO + O_2 \rightarrow 2CO_2$ indicates 2 vol of CO react with 1 vol of O_2 to form 2 vol of CO_2. Therefore, 10 L of CO form 10 L of CO_2.

43. **(D)** $H_2SO_4 + 2NaOH \rightarrow 2H_2O + Na_2SO_4$ is the equation for this reaction. 1 mol $H_2SO_4 = 98$ g. 1 mol NaOH $= 40$ g.

Then 49 g of $H_2SO_4 = \frac{1}{2}$ mol H_2SO_4.

The equation shows that 1 mol of sulfuric reacts with 2 mol of sodium hydroxide or a ratio of 1 : 2. Therefore, $\frac{1}{2}$ mol of sulfuric reacts with 1 mol of sodium hydroxide in this reaction. We found 1 mol of NaOH to equal 40 g. Since 80 g of NaOH is given, 40 g of it will remain after the reaction has gone to completion.

44. **(C)** The $K_{sp} = [Ag^+][Cl^-] = 1.9 \times 10^{-10}$. The molar concentration of both Ag^+ and Cl^- is 1.38×10^{-5} mol/L.

$$K_{sp} = [Ag^+][Cl^-]$$
$$= [1.38 \times 10^{-5}][1.38 \times 10^{-5}]$$
$$= 1.9 \times 10^{-10}$$

45. **(B)** If 1.43 g is the weight of 1 L, then the weight of 22.4 L, which is the molar volume of a gas at STP, will give the molar mass. So 22.4 L $\times$ 1.43 g/L $= 32$ g, the gram-molecular mass.

46. **(E)** The equation is

$$2KClO_3 \rightarrow 2KCl + 3O_2$$

This shows that 2 mol of $KClO_3$ yields 3 mol of O_2. Three mol of $O_2 = 3 \times 22.4$ L $= 67.2$ L of O_2.

47. **(A)** Gibbs free energy combines the overall energy changes and the entropy change. The formula is $\Delta G = \Delta H - T\Delta S$. Only if ΔG is negative will the reaction be spontaneous in the forward direction.

48. **(C)** The CuO was reduced while the H_2 was oxidized, forming $H_2O +$ Cu. The reaction is

$$CuO + H_2 \rightarrow H_2O + Cu$$

49. **(B)** The purpose of this $CaCl_2$ tube is to absorb any water evaporated from the generator. If it were not present, some water vapor would pass through to the final drying tube and cause the weight of water gained there to be larger than it should be from the reaction only.

50. **(A)** Since 8 g of O_2 was lost by the CuO, the weight of water gained by the U-tube came from 8 g of O_2 and 1 g of H_2, to make 9 g of water that it absorbed.

51. **(E)** The weight ratio of water is 1 g of H_2 to 8 g of O_2 or 1 : 8. The other statements are true but are not conclusions from this experiment.

52. **(D)** This is the only observation of the group. All the others are conclusions. Remember that an observation is only what you see, smell, taste, or measure with a piece of equipment.

53. **(A)** The phenomenon is the formation of the white ring, which is NH_4Cl. Although measuring the distance traveled by each gas could be used to verify Graham's Law of Gaseous Diffusion, this was not asked. The relationship is that the diffusion rate is inversely proportional to the square root of the gas's molar mass.

54. **(C)** Since Ca^{2+} and HSO_4^- combine, the formula is $Ca(HSO_4)_2$.

55. **(A)** NaOH is 40 g/mol. 10 g is 10/40 or 0.25 mol in 1 L or 0.25 M.

56. **(C)** A saturated solution represents a condition where the solute is going into solution as rapidly as some solute is coming out of the solution.

57. **(D)** II and III are correct since the double-bonded carbons in the alkene and the triple-bonded carbons in the alkyne series have pi bonds.

58. **(B)** The thistle tube is not below the level of the liquid in the generator, and the gas would escape into the air. (Vinegar is an acid and would produce hydrogen.)

59. **(D)** NH_3 is very soluble and could not be collected in this manner. All others are not sufficiently soluble to hamper this method of collection.

60. **(C)** The formula mass is the total of (Ca = 40) + (2N = 28) + (6O = 96), or 164. Since oxygen is 96 of the 164 total, the percentage is $96/164 \times 100\% = 58.5\%$.

61. **(D)** The K_{eq} expression is made up of the concentration(s) of the products over those of the reactants, with the coefficients becoming exponents. So

$$K_{eq} = \frac{[NO_2]^2}{[N_2O_4]}$$

62. **(A)**

$$K_{eq} = \frac{[2 \times 10^{-2}]^2}{[4 \times 10^{-2}]} = \frac{4 \times 10^{-4}}{4 \times 10^{-2}} = 1 \times 10^{-2}$$

63. **(C)** In dilution problems, the expression $M_1 \times V_1 = M_2 \times V_2$ can be used. Substituting $(6\ M)(0.5\ L) = (2\ M)(x\ liters)$ gives $x = 1.5$ L total volume. Since there was 0.5 L to begin with, an additional 1 L must be added.

64. **(C)** The reaction equation and information given can be set up like this:

Given   Given

   4 g   4 g x grams

$2H_2$ + O_2 $\rightarrow$ $2H_2O$

   4 g   32 g   36 g

Studying this shows that the limiting element will be the 4 g of oxygen since 4 g of H_2 would require 32 g of O_2. The solution setup is

$$\frac{4 \text{ g } O_2}{32 \text{ g } O_2} = \frac{x \text{ grams } H_2O}{36 \text{ g } H_2O}$$
$$\therefore = 4.5 \text{ g } H_2O$$

65. **(E)** The functional group is $R - \overset{\displaystyle O}{\overset{\displaystyle \|}{C}} - R_1$. This appears only in (E).

66. **(C)** The thistle tube serves both these purposes.

67. **(D)** The empirical formula is a representation of the elements in their simplest ratio. Therefore, CH_2 is the simplest ratio of the molecular formula C_3H_6.

68. **(E)** I, II, and III would occur since their reduction reactions would be −0.13, +0.34, and +0.80 volt, respectively. These numbers added to +0.44 volt separately to give a positive E^0 for the reaction with Fe.

69. **(A)** In the equation, 2 moles of Na will release 822 kJ of heat. If only 0.5 mole of Na is consumed, only one-fourth as much heat will be released or $\frac{1}{4} \times 822 \text{ kJ} = 205 \text{ kJ}$.

Diagnosing Your Needs

After taking Practice Test 1, check your answers against the correct ones. Then fill in the chart below.

In the space under each equation number, place a check if you answered that question correctly.

Example

If your answer to question 5 was correct, place a check in the appropriate box.

Next, total the check marks for each section and insert the number in the designated block. Now do the arithmetic indicated and insert your percent for each area.

SUBJECT AREA

(✓) QUESTIONS ANSWERED CORRECTLY

I.	**Atomic Theory and Structure**, including periodic relationships	6	8	105	109	110	111	31	38

☐ No. of checks ÷ 8 × 100 = _____ %

II.	**Nucleonics**							115	116

☐ No. of checks ÷ 2 × 100 = _____ %

III.	**Chemical Bonding and Molecular Structure**	3	17	18	19	29	30	39	54	57

☐ No. of checks ÷ 9 × 100 = _____ %

IV.	**States of Matter and Kinetic Molecular Theory of Gases**	10	11	12	107	108	113	27

☐ No. of checks ÷ 7 × 100 = _____ %

V.	**Solutions**, including concentration units, solubility, and colligative properties	21	103	60	55	63

☐ No. of checks ÷ 5 × 100 = _____ %

SUBJECT AREA

VI. Acids and Bases	20	101	102	114	33	36	53

☐ No. of checks ÷ 7 × 100 = _____%

VII. Oxidation-Reduction and Electrochemistry	106	112	34	35	48	68

☐ No. of checks ÷ 6 × 100 = _____%

VIII. Stoichiometry	24	28	41	42	43	45	46	50	60	64

☐ No. of checks ÷ 10 × 100 = _____%

IX. Reaction Rates	25	32

☐ No. of checks ÷ 2 × 100 = _____%

X. Equilibrium	22	40	56	61	62

☐ No. of checks ÷ 5 × 100 = _____%

XI. Thermodynamics: energy changes in chemical reactions, randomness, and criteria for spontaneity	13	14	15	47	104	69

☐ No. of checks ÷ 6 × 100 = _____%

XII. Descriptive Chemistry: physical and chemical properties of elements and their familiar compounds; organic chemistry; periodic properties	1	2	4	5	7	9
	16	23	26	37	65	67

☐ No. of checks ÷ 12 × 100 = _____%

XIII. Laboratory: equipment, procedures, observations, safety, calculations, and interpretation of results	49	51	52	58	59	66

☐ No. of checks ÷ 6 × 100 = _____%

Planning Your Study

The percentages give you an idea of how you have done on the various major areas of the test. Because of the limited number of questions on some parts, these percentages may not be as reliable as the percentages for parts with larger numbers of questions. However, you should now have at least a rough idea of the areas in which you have done well and those in which you need more study.

Start your study with the areas in which you are weakest. The corresponding chapters are indicated below.

Subject Area	Chapters to Review
I. Atomic Theory and Structure, including periodic relationships	2
II. Nuclear Reactions	15
III. Chemical Bonding and Molecular Structure	3, 4
IV. States of Matter and Kinetic Molecular Theory of Gases	1
V. Solutions, including concentration units, solubility, and colligative properties	7
VI. Acids and Bases	11
VII. Oxidation-Reduction and Electrochemistry	12
VIII. Stoichiometry	5, 6
IX. Reaction Rates	9
X. Equilibrium	10
XI. Thermodynamics, including energy changes in chemical reactions, randomness, and criteria for spontaneity	8
XII. Descriptive Chemistry: physical and chemical properties of elements and their familiar compounds; organic chemistry; periodic properties	1, 2, 13, 14
XIII. Laboratory: equipment, procedures, observations, safety, calculations, and interpretation of results	All lab diagrams, 16

Answer Sheet for Practice Test 2

Determine the correct answer for each question. Then, using a No. 2 pencil, blacken completely the oval containing the letter of your choice.

1. Ⓐ Ⓑ Ⓒ Ⓓ Ⓔ
2. Ⓐ Ⓑ Ⓒ Ⓓ Ⓔ
3. Ⓐ Ⓑ Ⓒ Ⓓ Ⓔ
4. Ⓐ Ⓑ Ⓒ Ⓓ Ⓔ
5. Ⓐ Ⓑ Ⓒ Ⓓ Ⓔ
6. Ⓐ Ⓑ Ⓒ Ⓓ Ⓔ
7. Ⓐ Ⓑ Ⓒ Ⓓ Ⓔ
8. Ⓐ Ⓑ Ⓒ Ⓓ Ⓔ
9. Ⓐ Ⓑ Ⓒ Ⓓ Ⓔ
10. Ⓐ Ⓑ Ⓒ Ⓓ Ⓔ
11. Ⓐ Ⓑ Ⓒ Ⓓ Ⓔ
12. Ⓐ Ⓑ Ⓒ Ⓓ Ⓔ
13. Ⓐ Ⓑ Ⓒ Ⓓ Ⓔ
14. Ⓐ Ⓑ Ⓒ Ⓓ Ⓔ
15. Ⓐ Ⓑ Ⓒ Ⓓ Ⓔ
16. Ⓐ Ⓑ Ⓒ Ⓓ Ⓔ
17. Ⓐ Ⓑ Ⓒ Ⓓ Ⓔ
18. Ⓐ Ⓑ Ⓒ Ⓓ Ⓔ
19. Ⓐ Ⓑ Ⓒ Ⓓ Ⓔ
20. Ⓐ Ⓑ Ⓒ Ⓓ Ⓔ
21. Ⓐ Ⓑ Ⓒ Ⓓ Ⓔ
22. Ⓐ Ⓑ Ⓒ Ⓓ Ⓔ
23. Ⓐ Ⓑ Ⓒ Ⓓ Ⓔ

ON THE ACTUAL CHEMISTRY TEST, THE FOLLOWING TYPE OF QUESTION MUST BE ANSWERED ON A SPECIAL SECTION (LABELED "CHEMISTRY") AT THE LOWER LEFT-HAND CORNER OF PAGE 2 OF YOUR ANSWER SHEET. THESE QUESTIONS WILL BE NUMBERED BEGINNING WITH 101 AND MUST BE ANSWERED ACCORDING TO THE FOLLOWING DIRECTIONS.

CHEMISTRY* Fill in oval CE only if II is a correct explanation of I.

	I		II		CE*
101.	Ⓣ	Ⓕ	Ⓣ	Ⓕ	◯
102.	Ⓣ	Ⓕ	Ⓣ	Ⓕ	◯
103.	Ⓣ	Ⓕ	Ⓣ	Ⓕ	◯
104.	Ⓣ	Ⓕ	Ⓣ	Ⓕ	◯
105.	Ⓣ	Ⓕ	Ⓣ	Ⓕ	◯
106.	Ⓣ	Ⓕ	Ⓣ	Ⓕ	◯
107.	Ⓣ	Ⓕ	Ⓣ	Ⓕ	◯
108.	Ⓣ	Ⓕ	Ⓣ	Ⓕ	◯
109.	Ⓣ	Ⓕ	Ⓣ	Ⓕ	◯
110.	Ⓣ	Ⓕ	Ⓣ	Ⓕ	◯
111.	Ⓣ	Ⓕ	Ⓣ	Ⓕ	◯
112.	Ⓣ	Ⓕ	Ⓣ	Ⓕ	◯
113.	Ⓣ	Ⓕ	Ⓣ	Ⓕ	◯
114.	Ⓣ	Ⓕ	Ⓣ	Ⓕ	◯
115.	Ⓣ	Ⓕ	Ⓣ	Ⓕ	◯
116.	Ⓣ	Ⓕ	Ⓣ	Ⓕ	◯

ON THE ACTUAL CHEMISTRY TEST, THE REMAINING QUESTIONS MUST BE ANSWERED BY RETURNING TO THE SECTION OF YOUR ANSWER SHEET YOU STARTED FOR CHEMISTRY.

24. Ⓐ Ⓑ Ⓒ Ⓓ Ⓔ
25. Ⓐ Ⓑ Ⓒ Ⓓ Ⓔ
26. Ⓐ Ⓑ Ⓒ Ⓓ Ⓔ
27. Ⓐ Ⓑ Ⓒ Ⓓ Ⓔ
28. Ⓐ Ⓑ Ⓒ Ⓓ Ⓔ
29. Ⓐ Ⓑ Ⓒ Ⓓ Ⓔ
30. Ⓐ Ⓑ Ⓒ Ⓓ Ⓔ
31. Ⓐ Ⓑ Ⓒ Ⓓ Ⓔ
32. Ⓐ Ⓑ Ⓒ Ⓓ Ⓔ
33. Ⓐ Ⓑ Ⓒ Ⓓ Ⓔ
34. Ⓐ Ⓑ Ⓒ Ⓓ Ⓔ
35. Ⓐ Ⓑ Ⓒ Ⓓ Ⓔ
36. Ⓐ Ⓑ Ⓒ Ⓓ Ⓔ
37. Ⓐ Ⓑ Ⓒ Ⓓ Ⓔ
38. Ⓐ Ⓑ Ⓒ Ⓓ Ⓔ
39. Ⓐ Ⓑ Ⓒ Ⓓ Ⓔ

40. Ⓐ Ⓑ Ⓒ Ⓓ Ⓔ
41. Ⓐ Ⓑ Ⓒ Ⓓ Ⓔ
42. Ⓐ Ⓑ Ⓒ Ⓓ Ⓔ
43. Ⓐ Ⓑ Ⓒ Ⓓ Ⓔ
44. Ⓐ Ⓑ Ⓒ Ⓓ Ⓔ
45. Ⓐ Ⓑ Ⓒ Ⓓ Ⓔ
46. Ⓐ Ⓑ Ⓒ Ⓓ Ⓔ
47. Ⓐ Ⓑ Ⓒ Ⓓ Ⓔ
48. Ⓐ Ⓑ Ⓒ Ⓓ Ⓔ
49. Ⓐ Ⓑ Ⓒ Ⓓ Ⓔ
50. Ⓐ Ⓑ Ⓒ Ⓓ Ⓔ
51. Ⓐ Ⓑ Ⓒ Ⓓ Ⓔ
52. Ⓐ Ⓑ Ⓒ Ⓓ Ⓔ
53. Ⓐ Ⓑ Ⓒ Ⓓ Ⓔ
54. Ⓐ Ⓑ Ⓒ Ⓓ Ⓔ

55. Ⓐ Ⓑ Ⓒ Ⓓ Ⓔ
56. Ⓐ Ⓑ Ⓒ Ⓓ Ⓔ
57. Ⓐ Ⓑ Ⓒ Ⓓ Ⓔ
58. Ⓐ Ⓑ Ⓒ Ⓓ Ⓔ
59. Ⓐ Ⓑ Ⓒ Ⓓ Ⓔ
60. Ⓐ Ⓑ Ⓒ Ⓓ Ⓔ
61. Ⓐ Ⓑ Ⓒ Ⓓ Ⓔ
62. Ⓐ Ⓑ Ⓒ Ⓓ Ⓔ
63. Ⓐ Ⓑ Ⓒ Ⓓ Ⓔ
64. Ⓐ Ⓑ Ⓒ Ⓓ Ⓔ
65. Ⓐ Ⓑ Ⓒ Ⓓ Ⓔ
66. Ⓐ Ⓑ Ⓒ Ⓓ Ⓔ
67. Ⓐ Ⓑ Ⓒ Ⓓ Ⓔ
68. Ⓐ Ⓑ Ⓒ Ⓓ Ⓔ
69. Ⓐ Ⓑ Ⓒ Ⓓ Ⓔ

Practice Test 2

Note: For all questions involving solutions and/or chemical equations, assume that the system is in water unless otherwise stated.

PART A

Directions: Each set of lettered choices refers to the numbered statements or formulas immediately following it. Select the one lettered choice that best fits each statement or formula and then fill in the corresponding oval on the answer sheet. A choice may be used once, more than once, or not at all in each set.

Questions 1–4

    (A)   Law of Definite Composition
    (B)   High dielectric constant
    (C)   van der Waals forces
    (D)   Graham's Law of Diffusion (Effusion)
    (E)   Triple point

1. At a particular temperature and pressure, three states of a substance may coexist.

2. Water is a good solvent.

3. The rate of movement of hydrogen gas compared to oxygen gas is 4 : 1.

4. The molecules of nitrogen monoxide and nitrogen dioxide differ by a multiple of the mass of one oxygen.

Questions 5–7 refer to the following diagram:

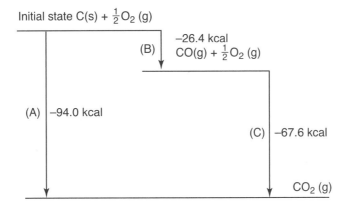

5. The ΔH of the reaction to form CO from $C + O_2$

6. The ΔH of the reaction to form CO_2 from $CO + O_2$

7. The ΔH of the reaction to form CO_2 from $C + O_2$

Questions 8–11

    (A) Hydrogen bond
    (B) Ionic bond
    (C) Polar covalent bond
    (D) Pure covalent bond
    (E) Metallic bond

8. The type of bond between atoms of potassium and chloride in a crystal of potassium chloride

9. The type of bond between the atoms in a nitrogen molecule

10. The type of bond between the atoms in a molecule of CO_2 (electronegativity difference = 1)

11. The type of bond between the atoms of calcium in a crystal of calcium

Questions 12–14 refer to the following diagram which is a phase diagram for CO_2.

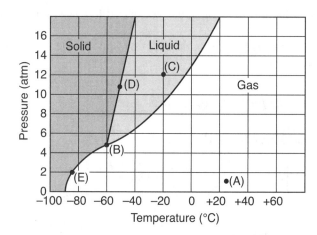

12. The point at which all three states of CO_2 can exist

13. The point at which CO_2 can exist only as a liquid

14. The point at which CO_2 can exist as a solid and a gas under 2 atmospheres of pressure

Questions 15–23 refer to the following graph, which shows the variation of the first ionization potential with respect to increasing atomic numbers.

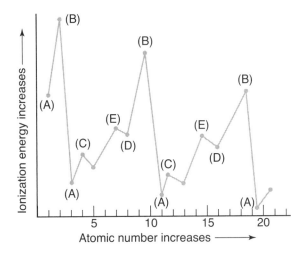

15. The atoms likely to react with water to release hydrogen

16. Nonmetals that are all found in the gaseous state at STP

17. The noble gases

18. The alkali metals

19. The half-filled condition of the *p* orbitals

20. The filled *s* orbitals with the exception of He

21. The beginning of pairing in the *p* orbitals

22. The most active metals

23. The filled *p* orbitals

ON THE ACTUAL CHEMISTRY TEST, THE FOLLOWING TYPE OF QUESTION MUST BE ANSWERED ON A SPECIAL SECTION (LABELED "CHEMISTRY") AT THE LOWER LEFT-HAND CORNER OF PAGE 2 OF YOUR ANSWER SHEET. THESE QUESTIONS WILL BE NUMBERED BEGINNING WITH 101 AND MUST BE ANSWERED ACCORDING TO THE FOLLOWING DIRECTIONS.

Directions: Each question below consists of two statements, I in the left-hand column and II in the right-hand column. For each question, determine whether statement I is true or false <u>and</u> if statement II is true or false and fill in the corresponding T or F ovals on your answer sheet. <u>Fill in oval CE only if statement II is a correct explanation of statement I</u>.

Sample Answer Grid:

CHEMISTRY *Fill in oval CE only if
II is a correct explanation of I.

	I	II	CE*
101.	T F	T F	◯

101. The structure of SO_3 is shown by using more than one structural formula   BECAUSE SO_3 is very unstable and resonates between these possible structures.

102. When the ΔG value of a reaction at a given temperature is negative, the reaction occurs spontaneously   BECAUSE   when ΔG is negative, the ΔH is also negative.

103. One mole of CO_2 has a greater mass than one mole of H_2O BECAUSE   the molecular mass of CO_2 is greater than the molecular mass of H_2O.

104. Hydrosulfuric acid is often used in qualitative tests   BECAUSE $H_2S(aq)$ reacts with many metallic ions to produce colorful precipitates.

105. Crystals of sodium chloride go into solution in water as ions   BECAUSE   the sodium ion has a 1+ charge and the chloride ion has a 1− charge and they are hydrated by the water molecules.

106. In an equilibrium reaction, if the concentration of the reactants is increased, the reaction will increase its forward rate

BECAUSE when a stress is applied to a reaction in equilibrium, the equilibrium shifts in the direction that opposes the stress.

107. The $\Delta H_{reaction}$ of a particular reaction can be arrived at by the summation of the $\Delta H_{reaction}$ values of two or more reactions that, added together, give the $\Delta H_{reaction}$ of the particular reaction

BECAUSE Hess's Law conforms to the First Law of Thermodynamics, which states that the total energy of the universe is a constant.

108. In a reaction that has both a forward and a reverse reaction, $A + B \rightleftharpoons AB$, when only A and B are introduced into a reacting vessel, the forward reaction rate is the highest at the beginning and begins to decrease from that point until equilibrium is reached

BECAUSE the reverse reaction does not begin until equilibrium is reached.

109. At equilibrium, the forward reaction and reverse reaction stop

BECAUSE the reactants and products have reached the equilibrium concentrations.

110. The hybrid orbital form of carbon in acetylene is believed to be the *sp* form

BECAUSE C_2H_2 is a linear compound with a triple bond between the carbons.

111. The weakest of the bonds between molecules are coordinate covalent bonds

BECAUSE coordinate covalent bonds represent the weak attractive force of the electrons of one molecule for the positively charged nucleus of another.

112. A saturated solution is not necessarily concentrated

BECAUSE *dilute* and *concentrated* are terms that relate only to the relative amount of solute dissolved in the solvent.

113. Lithium is the most active metal in the first group of the periodic chart

BECAUSE lithium has only one electron in the outer energy level.

114. The anions migrate to the cathode in an electrolytic cell | BECAUSE | positively charged ions are attracted to the negatively charged cathode.

115. The atomic number of a neutral atom that has a mass of 39 and has 19 electrons is 19 | BECAUSE | the number of protons in a neutral atom is equal to the number of electrons.

116. For an element with an atomic number of 17, the most probable oxidation number is +1 | BECAUSE | the outer energy level of the halogen family has a tendency to add one electron to itself.

PART C

Directions: Each of the questions or incomplete statements below is followed by five suggested answers or completions. Select the one that is best in each case and then fill in the corresponding oval on the answer sheet.

24. All of the following involve a chemical change EXCEPT

  (A) the formation of HCl from H_2 and Cl_2
  (B) the color change when NO is exposed to air
  (C) the formation of steam from burning H_2 and O_2
  (D) the solidification of Crisco at low temperatures
  (E) the odor of NH_3 when NH_4Cl is rubbed together with $Ca(OH)_2$ powder

25. When most fuels burn, the products include carbon dioxide and

  (A) hydrocarbons
  (B) hydrogen
  (C) water
  (D) hydroxide
  (E) hydrogen peroxide

26. In the metric system, the prefix *kilo-* means

  (A) 10^0
  (B) 10^{-1}
  (C) 10^{-2}
  (D) 10^2
  (E) 10^3

27. How many atoms are in 1 mol of water?

    (A) 3
    (B) 54
    (C) 6.02×10^{23}
    (D) $2(6.02 \times 10^{23})$
    (E) $3(6.02 \times 10^{23})$

28. Which of the following elements normally exists as a monatomic molecule?

    (A) Cl
    (B) H
    (C) O
    (D) N
    (E) He

29. Which method is often employed in the separation of the hydrocarbons found in petroleum?

    (A) alkylation
    (B) hydrogenation
    (C) catalytic cracking
    (D) fractional distillation
    (E) polymerization

30. The complete loss of an electron of one atom to another atom with the consequent formation of electrostatic charges is said to be

    (A) a covalent bond
    (B) a polar covalent bond
    (C) an ionic bond
    (D) a coordinate covalent bond
    (E) a pi bond between p orbitals

31. In the electrolysis of water, the cathode reaction is

    (A) $2H_2O(\ell) + 2e^- \rightarrow H_2(g) + 2OH^- + O_2(g)$
    (B) $2H_2O(\ell) \rightarrow \frac{1}{2}O_2(g) + 2H^+ + 2e^-$
    (C) $2OH^- + 2e^- \rightarrow O^2(g) + H_2(g)$
    (D) $2H^+ + 2e^- \rightarrow H_2(g)$
    (E) $2H_2O(\ell) + 4e^- \rightarrow O_2(g) + 2H_2(g)$

Question 32 refers to a solution of 0.100 M acetic acid.

32. What is the H_3O^+ concentration? ($K_a = 1.8 \times 10^{-5}$)

(A) 1.8×10^{-5}
(B) 1.8×10^{-4}
(C) 1.8×10^{-3}
(D) 1.3×10^{-3}
(E) 0.9×10^{-3}

33. If a radioactive element with a half-life of 100 years is found to have transmutated so that only 25% of the original sample remains, what is the age of the sample?

(A) 25 years
(B) 50 years
(C) 100 years
(D) 200 years
(E) 400 years

34. What is the pH of an acetic acid solution if the $[H_3O^+] = 1 \times 10^{-4}$ mol/L?

(A) 1
(B) 2
(C) 3
(D) 4
(E) 5

35. The polarity of water is useful in explaining which of the following?

  I. The solution process
 II. The ionization process
III. The high conductivity of distilled water

(A) I only
(B) II only
(C) I and II only
(D) II and III only
(E) I, II, and III

36. When sulfur dioxide is bubbled through water, the solution will contain

(A) sulfurous acid
(B) sulfuric acid
(C) hyposulfuric acid
(D) persulfuric acid
(E) anhydrous sulfuric acid

37. Four grams of hydrogen gas at STP contain

   (A) 6.02×10^{23} atoms
   (B) 12.04×10^{23} atoms
   (C) 12.04×10^{46} atoms
   (D) 1.2×10^{22} molecules
   (E) 12.04×10^{23} molecules

38. Analysis of a gas gave C = 85.7% and H = 14.3%. If the formula mass of this gas is 42 amu, what are the empirical formula and the true formula?

   (A) CH; C_4H_4
   (B) CH_2; C_3H_6
   (C) CH_3; C_3H_9
   (D) C_2H_2; C_3H_6
   (E) C_2H_4; C_3H_6

39. Which fraction would be used to correct a given volume of gas at 30°C to its new volume when it is heated to 60°C and the pressure is kept constant?

   (A) $\dfrac{30}{60}$

   (B) $\dfrac{60}{30}$

   (C) $\dfrac{273}{333}$

   (D) $\dfrac{303}{333}$

   (E) $\dfrac{333}{303}$

40. What would be the predicted freezing point of a solution that has 684 g of sucrose (1 mol = 342 g) dissolved in 1500. g of water?

   (A) 1.86°C or 271.1 K
   (B) −0.93°C or 272.1 K
   (C) −1.39°C or 271.6 K
   (D) −2.47°C or 270.5 K
   (E) −2.79°C or 269.3 K

41. What is the approximate pH of a 0.005 M solution of H_2SO_4?

   (A) 1
   (B) 2
   (C) 5
   (D) 9
   (E) 13

42. How many grams of NaOH are needed to make 100 g of a 5% solution?

   (A) 2
   (B) 5
   (C) 20
   (D) 40
   (E) 95

43. For the Haber process, $N_2 + 3H_2 \rightleftharpoons 2NH_3$ + heat (at equilibrium), which of the following statements concerning the reaction rate are true?

   I. The reaction to the right increases when the pressure is increased.
   II. The reaction to the right decreases when the temperature is increased.
   III. The reaction to the right decreases when NH_3 is removed from the chamber.

   (A) I only
   (B) II only
   (C) I and II only
   (D) II and III only
   (E) I, II, and III

44. If you titrate a 1 M H_2SO_4 solution against 50 mL of 1 M NaOH solution, what volume of H_2SO_4, in milliliters, will be needed for neutralization?

   (A) 10
   (B) 25
   (C) 40
   (D) 50
   (E) 100

45. How many grams of CO_2 can be prepared from 150 g of calcium carbonate reacting with an excess of hydrochloric acid solution?

   (A) 11
   (B) 22
   (C) 33
   (D) 44
   (E) 66

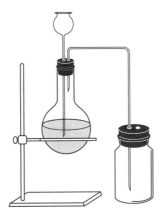

46. The above diagram represents a setup that may be used to prepare and collect

    (A) NH_3
    (B) NO
    (C) H_2
    (D) SO_3
    (E) CO_2

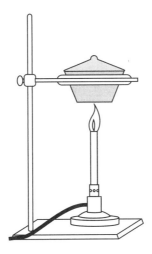

47. The lab setup shown above was used for the gravimetric analysis of the empirical formula of MgO. In synthesizing MgO from a Mg strip in the crucible, which of the following is NOT true?

    (A) The initial strip of Mg should be cleaned.
    (B) The lid of the crucible should fit tightly to exclude oxygen.
    (C) The heating of the covered crucible should continue until the Mg is fully reacted.
    (D) Cool the crucible, lid, and contents to room temperature before measuring their mass.
    (E) When the Mg appears to be fully reacted, partially remove the crucible lid and continue heating.

Questions 48–50 refer to the following experimental setup and data.

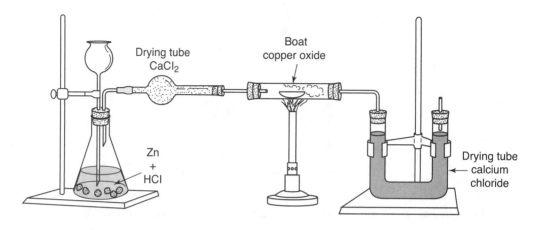

Recorded data:

Weight of U-tube .... 20.36 g
Weight of U-tube and calcium chloride before........ 39.32 g
Weight of U-tube and calcium chloride after .... 57.32 g
Weight of boat and copper oxide before.................. 30.23 g
Weight of boat and contents after .... 14.23 g
Weight of boat.. 5.00 g

48. What is the reason for the first $CaCl_2$ drying tube?

   (A) generate water
   (B) absorb hydrogen
   (C) absorb water that evaporates from the flask
   (D) decompose the water from the flask
   (E) act as a catalyst for the combination of hydrogen and oxygen

49. What conclusion can be derived from the data collected?

   (A) Oxygen was lost from the $CaCl_2$.
   (B) Oxygen was generated in the U-tube.
   (C) Water was formed from the reaction.
   (D) Hydrogen was absorbed by the $CaCl_2$.
   (E) CuO was formed in the decomposition.

50. What is the ratio of the mass of water formed to the mass of hydrogen used in the formation of water?

   (A) 1 : 8
   (B) 1 : 9
   (C) 8 : 1
   (D) 9 : 1
   (E) 8 : 9

51. What is the mass, in grams, of 1 mol of $KAl(SO_4)_2 \cdot 12H_2O$?

   (A) 132
   (B) 180
   (C) 394
   (D) 474
   (E) 516

52. What mass of aluminum will be completely oxidized by 44.8 L of oxygen at STP?

   (A) 18 g
   (B) 37.8 g
   (C) 50.4 g
   (D) 72.0 g
   (E) 100.8 g

53. In general, when metal oxides react with water, they form solutions that are

   (A) acidic
   (B) basic
   (C) neutral
   (D) unstable
   (E) colored

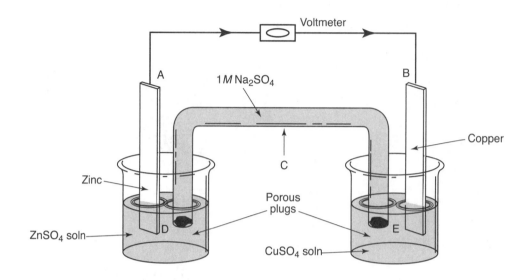

54. The oxidation reaction will occur at

    (A) A
    (B) B
    (C) C
    (D) D
    (E) E

55. The apparatus at C is called the

    (A) anode
    (B) cathode
    (C) salt bridge
    (D) ion bridge
    (E) osmotic bridge

56. The standard potentials of the metals are:

    $$Zn^{2+} + 2\ e^- = Zn \qquad\qquad E_0 = -0.76 \text{ volt}$$
    $$Cu^0 = Cu^{2+} + 2e^- \qquad\qquad E_0 = -0.34 \text{ volt}$$

    What will be the voltmeter reading for this reaction?

    (A) +1.10
    (B) −1.10
    (C) +0.42
    (D) −0.42
    (E) −1.52

57. How many liters of oxygen (STP) can you prepare from the decomposition of 212 g of sodium chlorate (1 mol = 106 g)?

   (A) 11.2
   (B) 22.4
   (C) 44.8
   (D) 67.2
   (E) 78.4

58. In the equation $Al(OH)_3 + H_2SO_4 \rightarrow Al_2(SO_4)_3 + H_2O$, the coefficients of the balanced equation are

   (A) 1, 3, 1, 2
   (B) 2, 3, 2, 6
   (C) 2, 3, 1, 6
   (D) 2, 6, 1, 3
   (E) 1, 3, 1, 6

59. What is $\Delta H_{reaction}$ for the decomposition of sodium chlorate? [ΔH_f^0 values are $NaClO_3(s) = -358.2$ kJ/mol, $NaCl(s) = -410.5$ kJ/mol, $O_2(g) = 0$ kJ/mol].

   (A) −12.8 kJ
   (B) −25.7 kJ
   (C) −52.3 kJ
   (D) −102.6 kJ
   (E) −153.9 kJ

60. Isotopes of an element are related because which of the following is(are) the same in these isotopes?

   I.   Atomic mass
   II. Atomic number
   III. Arrangement of orbital electrons

   (A) I only
   (B) II only
   (C) I and II only
   (D) II and III only
   (E) I, II, and III

61. In the reaction of Zn with dilute HCl to form H_2, which of the following will increase the reaction rate?

    I.   Increasing the temperature
    II. Increasing the exposed surface of Zn
    III. Using a more concentrated solution of HCl

    (A) I only
    (B) II only
    (C) I and III only
    (D) II and III only
    (E) I, II, and III

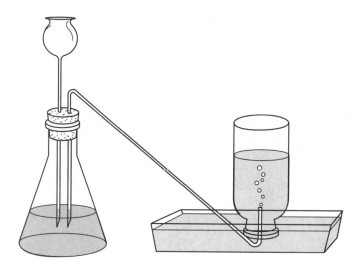

62. The laboratory setup shown above can be used to prepare a

    (A) gas less dense than air and soluble in water
    (B) gas more dense than air and soluble in water
    (C) gas soluble in water
    (D) gas insoluble in water
    (E) gas that reacts with water

63. In the reaction $CaCO_3 + 2HCl \rightarrow CaCl_2 + H_2O + CO_2$, if 4 mol of HCl is available to the reaction with an unlimited supply of $CaCO_3$, how many moles of CO_2 can be produced at STP?

    (A) 1
    (B) 1.5
    (C) 2
    (D) 2.5
    (E) 3

64. A saturated solution of $BaSO_4$ at 25°C contains 3.9×10^{-5} mol/L of Ba^{2+} ions. What is the K_{sp} of this salt?

   (A) 3.9×10^{-5}
   (B) 3.9×10^{-6}
   (C) 2.1×10^{-7}
   (D) 1.5×10^{-8}
   (E) 1.5×10^{-9}

65. If 0.1 mol of K_2SO_4 were added to the solution in question 64, what would happen to the Ba^{2+} concentration?

   (A) It would increase.
   (B) It would decrease.
   (C) It would remain the same.
   (D) It would first increase and then decrease.
   (E) It would first decrease and then increase.

66. Which of the following will definitely cause the volume of a gas at STP to increase?

   I.   Decrease the pressure with the temperature held constant.
   II. Increase the pressure with a temperature decrease.
   III. Increase the temperature with a pressure increase.

   (A) I only
   (B) II only
   (C) I and III only
   (D) II and III only
   (E) I, II, and III

67. In a neutralization reaction of an Arrhenius acid and base, what are the products?

   I. H_2O + a sulfate
   II. H_2O + a chloride
   III. H_2O + a salt

   (A) I only
   (B) II only
   (C) I and III only
   (D) II and III only
   (E) III only

68. Which of the following particles has the least mass?

    (A) alpha particle
    (B) beta particle
    (C) proton
    (D) neutron
    (E) gamma ray

69. Which emission from a radioactive source is an electromagnetic wave and is not affected by an electric field?

    (A) alpha particle
    (B) beta particle
    (C) proton
    (D) neutron
    (E) gamma ray

STOP
IF YOU FINISH BEFORE ONE HOUR IS UP, YOU MAY GO BACK
TO CHECK YOUR WORK OR COMPLETE UNANSWERED QUESTIONS.

(Total questions 85)
By Sections (1–23) (101–116) (24–69)
23 + 16 + 46 = 85

Answers and Explanations for Test 2

1. **(E)** The triple point of a phase diagram shows that all three states can exist at this point.

2. **(B)** The high dielectric constant of water is due to the polar nature of the water molecules. It is this property that is largely responsible for the ability of water to dissolve so many solutes.

3. **(D)** According to Graham's Law of Gaseous Diffusion (of Effusion), the rate of diffusion is inversely proportional to the square root of the molecular weight. Therefore,

$$\frac{\text{Rate}_H}{\text{Rate}_O} = \frac{\sqrt{\text{Mol. mass } O_2}}{\sqrt{\text{Mol. mass } H_2}} = \frac{\sqrt{32}}{\sqrt{2}} = \sqrt{\frac{16}{1}} = \frac{4}{1}$$

4. **(A)** The Law of Definite Composition states that when compounds form, they always form in the same ratio by mass. Water, for instance, always forms in a ratio of 1 : 8 of hydrogen to oxygen by mass. For nitrogen monoxide (NO) and nitrogen dioxide (NO_2), the difference in molecular mass is one atomic mass of oxygen.

5. **(B)** The first step is the ΔH for $C + \frac{1}{2}O_2 \rightarrow CO$. It releases 26.4 kcal of heat. This is written as −26.4 kcal.

6. **(C)** This is the second step on the diagram.

7. **(A)** To arrive at the ΔH of this, take the total drop (−94.0 kcal) or algebraically add these reactions:

$$C + \tfrac{1}{2}O_2 \quad \rightarrow CO \; -26.4 \text{ kcal}$$
$$\underline{CO + \tfrac{1}{2}O_2 \rightarrow CO_2 \; -67.6 \text{ kcal}}$$
$$C + O_2 \quad\;\; \rightarrow CO_2 \; -94.0 \text{ kcal}$$

8. **(B)** Potassium and chlorine have a large enough difference in their electronegativities to form ionic bonds. The respective positions of these two elements in the periodic chart also are indicative of the large difference in their electronegativity values.

9. **(D)** Two atoms of an element that forms a diatomic molecule always have a pure nonpolar covalent bond between them since the electron attraction or electronegativity of each atom is the same.

10. **(C)** Electronegativity differences between 0.4 and 1.7 are usually indicative of polar covalent bonding. CO_2 is an interesting example of a *nonpolar molecule* with polar covalent bonds since the bonds are symmetrical in the molecule.

11. **(E)** Calcium is a metal and forms a metallic bond between atoms.

12. **(B)** The point at which all three states can exist is called the triple point.

13. **(C)** The liquid state can exist only at (C).

14. **(E)** The only point at which CO_2 can exist as both a solid and a gas under 2 atmospheres of pressure is (E).

15. **(A)** These are the alkali metals.

16. **(B)** Because of their complete outer energy levels these elements have very low boiling points.

17. **(B)** The complete outer energy level makes it difficult to remove an electron; thus the high peaks would be the noble gases.

18. **(A)** With only one electron in the outer energy level, the alkali metals lose an electron relatively easily.

19. **(E)** There is a slight increase in the stability of the p orbitals when each has one electron—thus the little peaks at (E).

20. **(C)** The filled s orbital shows a slight increase in stability—thus the little peaks at C.

21. **(D)** The dip at the D areas on the chart shows the first pairing of electrons in the p orbitals and indicates that this first paired electron has a lower ionization potential than in the atom which has only one electron in each p orbital.

22. **(A)** The most active metals, the alkalis, have the lowest ionization potentials.

23. **(B)** The filled p orbitals occur in the noble gases.

101. **(T, F)** Sulfur trioxide is shown as three structural formulas because each bond is a "hybrid" of a single and a double bond. Resonance in chemistry does not mean that the bonds resonate between the structures shown in the structural drawing.

102. **(T, F)** When ΔG is negative in the Gibbs equation, the reaction is spontaneous. However, the total equation determines this, not just ΔH. The Gibbs equation is

$$\Delta G = \Delta H - T\Delta S$$

103. **(T, T, CE)** One mole of each of the gases contains 6.022×10^{23} molecules, but their molecular masses are different. CO_2 is found by adding one C = 12 and two O = 32, or a total of 44 amu. H_2O is found by adding two H = 2 and one O = 16, or a total of 18 amu. So it is true that 1 mole of CO_2 is heavier than 1 mole of H_2O.

104. **(T, T, CE)** Hydrosulfuric acid is used in qualitative tests because of the distinctly colored precipitates of sulfides that it forms with many metallic ions.

105. **(T, T, CE)** Sodium chloride is an ionic crystal and its ions are hydrated by the polar water molecules.

106. **(T, T, CE)** This statement and its reason fit the rules of Le Châtelier's Principle.

107. **(T, T, CE)** The statement is true, and the reason is also true and explains the statement.

108. (**T, F**) The statement is true, but not the reason. In an equilibrium reaction, concentrations can be shown to progress like this:

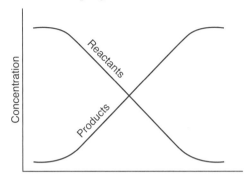

until equilibrium is reached. Then the concentrations stabilize.

109. (**F, T**) The forward and reverse reactions are occurring at equal rates when equilibrium is reached. The reactions do not stop. The concentrations remain the same at this point.

110. (**T, T, CE**) Since acetylene is known to be a linear molecule with a triple bond between the two carbons, the *sp* orbitals along the central axis with the hydrogens bonded on either end fit the experimental evidence.

111. (**F, F**) The weakest bonds between molecules are called van der Waals forces.

112. (**T, T, CE**) The terms *dilute* and *concentrated* merely indicate a relatively large amount of solvent and a small amount of solvent, respectively. You can have a dilute saturated solution if the solute is only slightly soluble.

113. (**F, T**) Cs is the most active Group I metal because it has (a) the largest atomic radius, thus making it easier to lose its outer energy level electron, and (b) the intermediate electrons help screen the positive attraction of the nucleus, also increasing the ease with which the outer electron is lost.

114. (**F, T**) The cations are positively charged ions and migrate to the cathode, while the anions are negatively charged and migrate to the anode.

115. (**T, T, CE**) There are as many electrons as there are protons in a neutral atom, and the atomic number represents the number of each.

116. (**F, T**) The first two principal energy levels fill up at 2 and 8 electrons, respectively. That leaves 7 electrons to fill the $3s$ and $3p$ orbitals like this: $3s^2 \, 3p^5$. With only one electron missing in the $3p$ orbitals, the most likely oxidation number is -1.

24. (**D**) The solidification of vegetable oil is merely a physical change just like the formation of ice from liquid water at lower temperatures. All the others involve actual recombinations of the atoms and thus are chemical changes.

25. (**C**) Water is formed because most common fuels contain hydrogen in their structure.

26. **(E)** The other choices, in order, would be 1, $\frac{1}{10}$ or deci-, $\frac{1}{100}$ or centi-, and 100 or hecto-.

27. **(E)** One mole of any substance contains 6.02×10^{23} molecules. Since each water molecule is triatomic, there would be $3(6.02 \times 10^{23})$ atoms present.

28. **(E)** The noble gases are all monatomic because of their complete outer energy level. A rule to help you remember diatomic gases is: Gases ending in *-gen* or *-ine* usually form diatomic molecules.

29. **(D)** Fractional distillation separates the hydrocarbons by taking advantage of the difference in boiling points. All other processes mentioned are used to modify the molecular structure of the hydrocarbons.

30. **(C)** The complete loss and gain of electrons is an ionic bond. All other bonds indicated are "sharing of electrons" type bonds or some form of covalent bonding.

31. **(D)** The cathode reaction releases only H_2 gas. The half-reaction is as given in reaction D.

32. **(D)** $K_a = 1.8 \times 10^{-5} = \dfrac{\left[H_3O^+\right]\left[Ac^-\right]}{\left[HAc\right]} = \dfrac{(x)(x)}{0.1 - x}$. Since acetic acid is a weak acid, the 0.1 M concentration is much greater than the amount ionized (x), and so $0.1 - x$ is essentially equal to 0.1. Therefore $1.8 \times 10^{-5} = \dfrac{(x)(x)}{0.1} = \dfrac{x^2}{0.1}$ and $x = 1.3 \times 10^{-3}$.

33. **(D)** The sample went from 25% to 50% going back 100 years. If you go back another 100 years, it would be back to 100%. That is a total of 200 years.

34. **(D)** $pH = -\log[H_3O^+] = -\log[1 \times 10^{-4}] = -(-4) = 4$.

35. **(C)** Only I and II are true. Distilled water does not significantly conduct an electric current. The polarity of the water molecule is helpful in ionization and in causing substances to go into solution.

36. **(A)** SO_2 is the acid anhydride of H_2SO_3 or sulfurous acid. $H_2O + SO_2 \rightarrow H_2SO_3$.

37. **(E)** Four grams of hydrogen gas at STP represent 2 mol of hydrogen since 2 g is the gram-molecular mass of hydrogen. Each mole of a gas contains 6.02×10^{23} molecules, and 2 mol would contain $2 \times 6.02 \times 10^{23}$ or 12.04×10^{23} molecules.

38. **(B)** To solve percent composition problems, first divide the percentage given by the atomic mass:

$$12\overline{)85.7} \qquad 1\overline{)14.3}$$
$$7.14 \qquad\quad 14.3$$

Then divide by the smallest quotient to get small whole numbers:

$$7.14\overline{)7.14} \quad 7.14\overline{)14.3}$$
$$1.0 \qquad\quad 2.0$$

So the empirical formula is CH_2. Since the molecular mass is 42 and the empirical formula has a molecular mass of 14, the true formula must be 3 times the empirical formula, or C_3H_6.

39. **(E)** Because the temperature (in kelvins) increases from 303 to 333 K, the volume of the gas should increase with the pressure held constant. The correct fraction is $\frac{333}{303}$.

40. **(D)** One mole of dissolved substance (which does not ionize) causes a 1.86 drop (Celsius) in the freezing point of a 1 m solution. Since 1500 g of water was used, the solution has $\frac{684}{342}$ or 2 mol in 1500 g of water. Then

$$\frac{2 \text{ mol}}{1500 \text{ g}} = \frac{x \text{ moles}}{1000 \text{ g}}$$

$$x = \frac{2000}{1500} = 1.33 \text{ mol/1000 g}$$

So the freezing point is depressed $1.33 \times -1.86° = -2.48°C$ or 270.5 K.

41. **(B)** The pH is $-\log[H^+]$. A 0.005 M solution of H_2SO_4 ionizes in a dilute solution to release two H^+ ions per molecule of H_2SO_4. Therefore, the molar concentration of H^+ ion is 2×0.005 mol/L or 0.010 mol/L. Substituting this in the formula gives pH $= -\log(0.010) = -\log(1 \times 10^{-2})$. The log of a number is the exponent to which the base 10 is raised to express that number in exponential form:

$-\log(1 \times 10^{-2}) = -(-2) = 2$

$-\log(6 \times 10^{-3}) = -\log(10^7 \times 10^{-3})$

$= -[0.7 + (-3)] = -(-2.3) = 2.3$ (approx.)

42. **(B)** If the solution is to be 5% sodium hydroxide, then 5% of 100 g is 5 g. Percent is always by mass unless otherwise specified.

43. **(C)** Since this equation is exothermic, higher temperatures will decrease the reaction to the right and increase the reaction to the left; so II is true. Also, statement I is true because with an increase in pressure the reaction will try to relieve that pressure by going in the direction that has the least volume. In this reaction, that would be to the right. Statement III is false because removing product in this reaction would increase the forward reaction. So I and II are true.

44. **(B)** The reaction is

$H_2SO_4 + 2NaOH \rightarrow Na_2SO_4 + H_2O$

1 mol   2 mol
acid   base

1 mol acid $= \frac{1}{2}$ mol of base

Since molarity $\times$ vol (L) = moles, then

$$M_a V_a = \tfrac{1}{2} M_b V_b$$

$$(1 M)(x \text{ liters}) = \tfrac{1}{2}(1 M)(0.05 \text{ L})$$

$$x \text{ L} = \tfrac{1}{2}(0.05 \text{ L})$$

$$x \text{ L} = 0.025 \text{ L or } 25 \text{ mL}$$

45. **(E)** The reaction is 150 g

$$CaCO_3 + 2HCl \rightarrow CaCl_2 + H_2O + CO_2$$

The gram-molecular mass of calcium is 100 g. Then $150 \text{ g} = \frac{150}{100}$ or 1.5 mol of calcium carbonate. According to the equation, 1 mol of calcium carbonate yields 1 mol of carbon dioxide, and so 1.5 mol of calcium carbonate would yield 1.5 mol of carbon dioxide.

The gram-molecular mass of $CO_2 = 44$ g:

$$1.5 \text{ mol of } CO_2 = 1.5 \times 44 \text{ g} = 66 \text{ g } CO_2$$

46. **(E)** The other choices are wrong because
    (A) is lighter than air
    (B) reacts with air
    (C) is lighter than air
    (D) needs heat to be evolved

47. **(B)** The Mg needs oxygen to form MgO; so the lid cannot be tightly sealed. All other choices are true.

48. **(C)** To ensure that all the water vapor collected in the U-tube comes from the reaction, the first drying tube is placed in the path of the hydrogen to absorb any evaporated water.

49. **(C)** Calcium chloride is deliquescent, and its weight gain of water indicates that water was formed from the reaction.

50. **(D)**

$CaCl_2$ and U-tube after	$= 57.32$ g
$CaCl_2$ and U-tube before	$= 39.32$ g
Mass of water formed	$= 18.00$ g
CuO and boat before	$= 30.23$ g
CuO and boat after	$= 14.23$ g
Mass of oxygen reacted	$= 16.00$ g
So, mass of water	$= 18$ g
mass of oxygen	$= 16$ g
Mass of hydrogen	$= 2$ g

The ratio of mass of water to mass of hydrogen is 18 : 2 or 9 : 1.

51. **(D)** $1K = 39$, $1Al = 27$, $2SO_4 = 2(32 + 16 \times 4) = 192$, and $12H_2O = 12(2 + 16) = 216$. This totals 474 g.

    x grams   44.8 L

52. **(D)**

$$4Al + 3O_2 \rightarrow 2Al_2O_3$$
$$\quad 180 \text{ g} \quad 67.2 \text{ L}$$

$$\frac{x}{108} = \frac{44.8}{67.2}, \text{ and so } x = 72 \text{ g}$$

or, using the mole method: 44.8 L = 2 mol

x mol   2 mol
$$4Al + 3O_2 \rightarrow 2Al_2O_3$$
4 mol   3 mol

This shows that: $\dfrac{x \text{ mol Al}}{4 \text{ mol Al}} = \dfrac{2 \text{ mol O}_2}{3 \text{ mol O}_2}$

$3x = 8$

$x = 8/3$ mol Al

8/3 ~~mol~~ Al $\times$ 27 g Al/1 ~~mol Al~~ = 72 g Al

53. **(B)** Metal oxides are generally basic anhydrides.

54. **(A)** Since A is marked +, the oxidation (or loss of electrons) will occur at this pole.

55. **(C)** *Salt bridge* is the correct terminology.

56. **(A)** $Zn \rightarrow Zn^{2+} + e^-$ $E^0 = +0.76$ V
     $Cu^{2+} + 2e^- \rightarrow Cu^0$ $\underline{E^0 = +0.34 \text{ V}}$

Adding these $= +1.10$ V

The voltmeter will read +1.10 V

          213 g x liters
57. **(D)** $2NaClO_3 \rightarrow 2NaCl + 3O_2$

$$\frac{213}{213} = \frac{x}{67.2}, \text{ and so } x = 67.2 \text{ L}$$

58. **(C)** The balanced equation has the coefficients 2, 3, 1, and

$$6 : 2Al(OH)_3 + 3H_2SO_4 \rightarrow Al_2(SO_4)_3 + 6H_2O$$

59. **(E)** The reaction is $NaClO_3 \rightarrow NaCl + \frac{3}{2}O_2$.

$\Delta H_{\text{reaction}} = \Delta H_{f(\text{products})} - H_{f(\text{reactants})}$

$\Delta H_{\text{reaction}} = -410.5 + 0 - (-358.7)$

$\Delta H_{\text{reaction}} = -52.3$ kJ

60. **(D)** II and III are identical; isotopes differ only in the number of neutrons in the nucleus, and this affects only the atomic mass.

61. **(E)** I, II, and III would affect this reaction.

62. **(D)** This setup depends on water displacement of an insoluble gas.

63. **(C)** The coefficients give the molar relations, so 2 mol of HCl give off 1 mol of CO_2. Given 4 mol of HCl, you have

$$\frac{4}{2} = \frac{x}{1}, \text{ then } x = \frac{4}{2} = 2 \text{ mol}$$

64. **(E)** $K_{\text{sp}} = [B_2^{2+}] \, [SO_4^{2-}] = (3.9 \times 10^{-5})(3.9 \times 10^{-5}) = 1.5 \times 10^{-9}$

These two will be the same since

$$BaSO_4 \rightarrow Ba^{2+} + SO_4^{2-}$$

65. **(B)** The introduction of the common ion SO_4^{2-} at 0.1 *M* forces the equilibrium to shift to the left and reduce the Ba^{2+} concentration.

66. **(A)** According to the gas laws, only I will cause an increase in the volume of a confined gas.

67. **(E)** This type of neutralization always involves the formation of H_2O and a salt. I and II would be true only if the acids used were sulfuric and hydrochloric, respectively.

68. **(B)** The beta particle is a high-speed electron and has the smallest mass of the first four choices. Gamma rays are electromagnetic waves.

69. **(E)** A gamma ray is not affected by an electric field because it is an electromagnetic wave. The neutron is not an electromagnetic wave; it is a particle with no charge but with the approximate mass of a proton.

Diagnosing Your Needs

After taking Practice Test 2, check your answers against the correct ones. Then fill in the chart below.

In the space under each equation number, place a check if you answered that question correctly.

Example

If your answer to question 5 was correct, place a check in the appropriate box.

Next, total the check marks for each section and insert the number in the designated block. Now do the arithmetic indicated and insert your percent for each area.

SUBJECT AREA

(✓) QUESTIONS ANSWERED CORRECTLY

I.	**Atomic Theory and Structure**, including periodic relationships	19	20	21	23	101	115	28	60

☐ No. of checks ÷ 8 × 100 = _____ %

II.	**Nuclear Reactions**	33	68	69

☐ No. of checks ÷ 3 × 100 = _____ %

III.	**Chemical Bonding and Molecular Structure**	4	8	9	10	11	110	111	62

☐ No. of checks ÷ 8 × 100 = _____%

IV.	**States of Matter and Kinetic Molecular Theory of Gases**	3	12	13	14	27	38	39	66

☐ No. of checks ÷ 8 × 100 = _____%

V.	**Solutions**, including concentration units, solubility, and colligative properties	2	105	112	40	44

☐ No. of checks ÷ 5 × 100 = _____%

VI.	**Acids and Bases**	32	34	36	41	53	67

☐ No. of checks ÷ 6 × 100 = _____%

VII.	**Oxidation-Reduction and Electrochemistry**	114	116	63	35	54	55	56

☐ No. of checks ÷ 7 × 100 = _____%

VIII.	**Stoichiometry**	37	42	45	50	51	52	57	58	63	103

☐ No. of checks ÷ 10 × 100 = _____%

IX.	**Reaction Rates**	43	77

☐ No. of checks ÷ 2 × 100 = _____%

X.	**Equilibrium**	106	108	109	64	65

☐ No. of checks ÷ 5 × 100 = _____%

XI. **Thermodynamics:** energy changes in chemical reactions, randomness, and criteria for spontaneity	5	6	7	107	102	59

☐ No. of checks ÷ 6 × 100 = _____%

XII. **Descriptive Chemistry:** physical and chemical properties of elements and their familiar compounds; organic chemistry; periodic properties	1	15	16	17	18	22
	104	113	24	25	26	29

☐ No. of checks ÷ 12 × 100 = _____%

XIII. **Laboratory:** equipment, procedures, observations, safety, calculations, and interpretation of results	46	47	48	49	62

☐ No. of checks ÷ 5 × 100 = _____%

Planning Your Study

The percentages give you an idea of how you have done on the various major areas of the test. Because of the limited number of questions on some parts, these percentages may not be as reliable as the percentages for parts with larger numbers of questions. However, you should now have at least a rough idea of the areas in which you have done well and those in which you need more study.

Start your study with the areas in which you are weakest. The corresponding chapters are indicated below.

Subject Area	Chapters to Review
I. Atomic Theory and Structure, including periodic relationships	2
II. Nuclear Reactions	15
III. Chemical Bonding and Molecular Structure	3, 4
IV. States of Matter and Kinetic Molecular Theory of Gases	1
V. Solutions, including concentration units, solubility, and colligative properties	7
VI. Acids and Bases	11
VII. Oxidation-Reduction and Electrochemistry	12
VIII. Stoichiometry	5, 6
IX. Reaction Rates	9
X. Equilibrium	10
XI. Thermodynamics, including energy changes in chemical reactions, randomness, and criteria for spontaneity	8
XII. Descriptive Chemistry: physical and chemical properties of elements and their familiar compounds; organic chemistry; periodic properties	1, 2, 13, 14
XIII. Laboratory: equipment, procedures, observations, safety, calculations, and interpretation of results.	All lab diagrams, 16

Answer Sheet for Practice Test 3

Determine the correct answer for each question. Then blacken the oval containing the letter of your choice.

1. Ⓐ Ⓑ Ⓒ Ⓓ Ⓔ
2. Ⓐ Ⓑ Ⓒ Ⓓ Ⓔ
3. Ⓐ Ⓑ Ⓒ Ⓓ Ⓔ
4. Ⓐ Ⓑ Ⓒ Ⓓ Ⓔ
5. Ⓐ Ⓑ Ⓒ Ⓓ Ⓔ
6. Ⓐ Ⓑ Ⓒ Ⓓ Ⓔ
7. Ⓐ Ⓑ Ⓒ Ⓓ Ⓔ
8. Ⓐ Ⓑ Ⓒ Ⓓ Ⓔ
9. Ⓐ Ⓑ Ⓒ Ⓓ Ⓔ
10. Ⓐ Ⓑ Ⓒ Ⓓ Ⓔ
11. Ⓐ Ⓑ Ⓒ Ⓓ Ⓔ
12. Ⓐ Ⓑ Ⓒ Ⓓ Ⓔ
13. Ⓐ Ⓑ Ⓒ Ⓓ Ⓔ
14. Ⓐ Ⓑ Ⓒ Ⓓ Ⓔ
15. Ⓐ Ⓑ Ⓒ Ⓓ Ⓔ
16. Ⓐ Ⓑ Ⓒ Ⓓ Ⓔ
17. Ⓐ Ⓑ Ⓒ Ⓓ Ⓔ
18. Ⓐ Ⓑ Ⓒ Ⓓ Ⓔ
19. Ⓐ Ⓑ Ⓒ Ⓓ Ⓔ
20. Ⓐ Ⓑ Ⓒ Ⓓ Ⓔ
21. Ⓐ Ⓑ Ⓒ Ⓓ Ⓔ
22. Ⓐ Ⓑ Ⓒ Ⓓ Ⓔ
23. Ⓐ Ⓑ Ⓒ Ⓓ Ⓔ
24. Ⓐ Ⓑ Ⓒ Ⓓ Ⓔ
25. Ⓐ Ⓑ Ⓒ Ⓓ Ⓔ
26. Ⓐ Ⓑ Ⓒ Ⓓ Ⓔ
27. Ⓐ Ⓑ Ⓒ Ⓓ Ⓔ
28. Ⓐ Ⓑ Ⓒ Ⓓ Ⓔ
29. Ⓐ Ⓑ Ⓒ Ⓓ Ⓔ

30. Ⓐ Ⓑ Ⓒ Ⓓ Ⓔ
31. Ⓐ Ⓑ Ⓒ Ⓓ Ⓔ
32. Ⓐ Ⓑ Ⓒ Ⓓ Ⓔ
33. Ⓐ Ⓑ Ⓒ Ⓓ Ⓔ
34. Ⓐ Ⓑ Ⓒ Ⓓ Ⓔ
35. Ⓐ Ⓑ Ⓒ Ⓓ Ⓔ
36. Ⓐ Ⓑ Ⓒ Ⓓ Ⓔ
37. Ⓐ Ⓑ Ⓒ Ⓓ Ⓔ
38. Ⓐ Ⓑ Ⓒ Ⓓ Ⓔ
39. Ⓐ Ⓑ Ⓒ Ⓓ Ⓔ
40. Ⓐ Ⓑ Ⓒ Ⓓ Ⓔ
41. Ⓐ Ⓑ Ⓒ Ⓓ Ⓔ
42. Ⓐ Ⓑ Ⓒ Ⓓ Ⓔ
43. Ⓐ Ⓑ Ⓒ Ⓓ Ⓔ
44. Ⓐ Ⓑ Ⓒ Ⓓ Ⓔ
45. Ⓐ Ⓑ Ⓒ Ⓓ Ⓔ
46. Ⓐ Ⓑ Ⓒ Ⓓ Ⓔ
47. Ⓐ Ⓑ Ⓒ Ⓓ Ⓔ
48. Ⓐ Ⓑ Ⓒ Ⓓ Ⓔ
49. Ⓐ Ⓑ Ⓒ Ⓓ Ⓔ
50. Ⓐ Ⓑ Ⓒ Ⓓ Ⓔ
51. Ⓐ Ⓑ Ⓒ Ⓓ Ⓔ
52. Ⓐ Ⓑ Ⓒ Ⓓ Ⓔ
53. Ⓐ Ⓑ Ⓒ Ⓓ Ⓔ
54. Ⓐ Ⓑ Ⓒ Ⓓ Ⓔ
55. Ⓐ Ⓑ Ⓒ Ⓓ Ⓔ
56. Ⓐ Ⓑ Ⓒ Ⓓ Ⓔ
57. Ⓐ Ⓑ Ⓒ Ⓓ Ⓔ
58. Ⓐ Ⓑ Ⓒ Ⓓ Ⓔ

59. Ⓐ Ⓑ Ⓒ Ⓓ Ⓔ
60. Ⓐ Ⓑ Ⓒ Ⓓ Ⓔ
61. Ⓐ Ⓑ Ⓒ Ⓓ Ⓔ
62. Ⓐ Ⓑ Ⓒ Ⓓ Ⓔ
63. Ⓐ Ⓑ Ⓒ Ⓓ Ⓔ
64. Ⓐ Ⓑ Ⓒ Ⓓ Ⓔ
65. Ⓐ Ⓑ Ⓒ Ⓓ Ⓔ
66. Ⓐ Ⓑ Ⓒ Ⓓ Ⓔ
67. Ⓐ Ⓑ Ⓒ Ⓓ Ⓔ
68. Ⓐ Ⓑ Ⓒ Ⓓ Ⓔ
69. Ⓐ Ⓑ Ⓒ Ⓓ Ⓔ
70. Ⓐ Ⓑ Ⓒ Ⓓ Ⓔ
71. Ⓐ Ⓑ Ⓒ Ⓓ Ⓔ
72. Ⓐ Ⓑ Ⓒ Ⓓ Ⓔ
73. Ⓐ Ⓑ Ⓒ Ⓓ Ⓔ
74. Ⓐ Ⓑ Ⓒ Ⓓ Ⓔ
75. Ⓐ Ⓑ Ⓒ Ⓓ Ⓔ
76. Ⓐ Ⓑ Ⓒ Ⓓ Ⓔ
77. Ⓐ Ⓑ Ⓒ Ⓓ Ⓔ
78. Ⓐ Ⓑ Ⓒ Ⓓ Ⓔ
79. Ⓐ Ⓑ Ⓒ Ⓓ Ⓔ
80. Ⓐ Ⓑ Ⓒ Ⓓ Ⓔ
81. Ⓐ Ⓑ Ⓒ Ⓓ Ⓔ
82. Ⓐ Ⓑ Ⓒ Ⓓ Ⓔ
83. Ⓐ Ⓑ Ⓒ Ⓓ Ⓔ
84. Ⓐ Ⓑ Ⓒ Ⓓ Ⓔ
85. Ⓐ Ⓑ Ⓒ Ⓓ Ⓔ

Practice Test 3

Questions 1–4 refer to the following diagram:

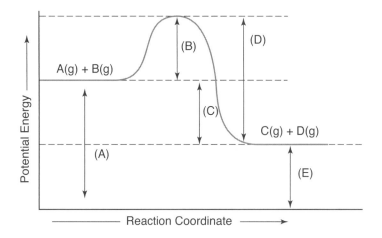

1. The activation energy of the forward reaction

2. The activation energy of the reverse reaction

3. The heat of the reaction for the forward reaction

4. The potential energy of the reactants

    (A) A
    (B) B
    (C) C
    (D) D
    (E) E

Questions 5–7 *refer to the following diagram.*

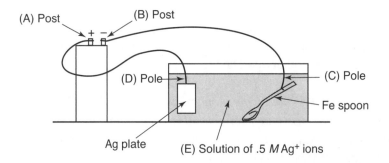

5. To plate silver on the spoon, the position to which the spoon must be connected

6. The position of the anode

7. The position from which silver that is plated out emerges

Questions 8–11 *refer to the following diagram:*

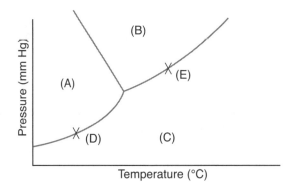

8. The solid area

9. The liquid area

10. The gaseous area

11. The area that can have both liquid and gas

Questions 12–14

    (A) 1
    (B) 2
    (C) 3
    (D) 4
    (E) 5

12. When the equation $Cu + HNO_3 \rightarrow Cu(NO_3)_2 + H_2O + NO$ is balanced, the coefficient, in the lowest whole number, of NO is _____.

13. If 6 moles of Cu react according to the above *balanced* equation, how many moles of NO will be formed?

14. If $Cu(NO_3)_2$ goes into solution as ions, it will dissociate into how many ions?

Questions 15–18

    (A) Ionic substance
    (B) Polar covalent substance
    (C) Nonpolar covalent substance
    (D) Amorphous substance
    (E) Metallic network

15. $KCl(s)$

16. $HCl(g)$

17. $CH_4(g)$

18. $Li(s)$

Questions 19–23

    (A) Brownian movement
    (B) Litmus paper reaction
    (C) Phenolphthalein reaction
    (D) Dehydration
    (E) Deliquescent

19. The reason why a blue crystal of $CuSO_4 \cdot 5H_2O$ turns white when heated

20. The zigzag path of colloidal particles in light

21. The pink color in a basic solution

22. The pink color in an acid solution

23. The adsorption of water to the surface of a crystal

24. Which of the following is the correct structural representation of sodium?

(A) Nucleus of 11 protons and 11 neutrons, electron configuration of:
$1s^2\ 2s^2\ 2p^6\ 3s^2\ 3p^6\ 4s^2\ 4p^3$

(B) Nucleus of 11 protons and 12 neutrons, electron configuration of:
$1s^2\ 2s^2\ 2p^6\ 3s^2\ 3p^6\ 4s^2\ 3d^2\ 4p^2$

(C) Nucleus of 23 protons and 23 neutrons, electron configuration of:
$1s^2\ 2s^2\ 2p^6\ 3s^2$

(D) Nucleus of 23 protons and 23 neutrons, electron configuration of:
$1s^2\ 2s^2\ 2p^6\ 3s^2\ 3p^6\ 3d^3\ 4s^2$

(E) Nucleus of 11 protons and 22 neutrons, electron configuration of:
$1s^2\ 2s^2\ 2p^6\ 3s^1$

25. Which of the following statements is true?

(A) A catalyst cannot lower the activation energy.
(B) A catalyst can lower only the forward activation energy.
(C) A catalyst affects only the product of the forward reaction.
(D) A catalyst affects only the activation energy of the reverse reaction.
(E) A catalyst is not permanently changed in the reaction.

26. Which bond is the most ionic?

(A) $H-Cl(g)$
(B) $S-Cl(g)$
(C) $Cs-F(s)$
(D) $Li-I(g)$
(E) $H-Br(g)$

27. The correct Lewis electron-dot diagram for the polyatomic ion of NH_4^+ is

(A)
$$\left[\begin{array}{c} H \\ \cdot\cdot \\ H:N:H \\ \cdot\cdot \\ H \end{array} \right]^+$$

(B)
$$\left[\begin{array}{c} H \\ \cdot\cdot \\ \cdot\cdot \\ H:\ N:H \\ \cdot\cdot \\ H \end{array} \right]^+$$

(C)
$$\left[\begin{array}{c} H \\ \cdot\cdot \\ \cdot\cdot \\ H:\ N::H \\ \cdot\cdot \\ H \end{array} \right]^+$$

(D)
$$\left[\begin{array}{c} \cdot \\ H \\ \cdot\cdot \\ \cdot H:N:H\cdot \\ \cdot\cdot \\ H \\ \cdot \end{array} \right]^+$$

(E)
$$\left[\begin{array}{c} \cdot\cdot \\ H \\ \cdot\cdot \\ \cdot H:N:H\cdot \\ \cdot\cdot \\ H \\ \cdot \end{array} \right]^+$$

28. If you want to make a dilute solution of 0.500 M NaCl and you have 250 mL of 1.00 M salt solution, how much water must be added to this solution to achieve the 0.500 M concentration?

(A) 125 mL
(B) 250 mL
(C) 375 mL
(D) 500 mL
(E) 735 mL

29. Of the following, which is an example of an acid that has two stages of ionization in a dilute solution?

(A) $HC_2H_3O_2$
(B) H_3PO_4
(C) $NaHCO_3$
(D) HCl
(E) H_2SO_4

30. In a Brønsted-Lowry acid-base reaction, which of the following statements is true?

(A) Water is always a product.
(B) The acid is the proton donor and the base the proton acceptor.
(C) The acid becomes the conjugate acid, and the base becomes the conjugate base.
(D) The acid becomes the conjugate base, and the base becomes the conjugate acid.
(E) Neutralization occurs between the hydronium ion and the hydroxide ion to form water.

31. The reaction $H_2(g) + I_2(g) + heat \leftrightarrow 2 HI(g)$ at 763 K has a $K_{eq} = 45.9$. Which of the following statements about the reaction above is true?

(A) Adding more H_2 and I_2 will increase the K_{eq} at 763 K.
(B) Increasing the pressure on this system will increase the production of HI(g).
(C) Increasing the amount of H_2 will decrease the production of HI(g).
(D) 2 moles of H_2 with sufficient I_2 will form 2 moles of HI in the forward reaction.
(E) At a given temperature, the equilibrium K_{eq} will not change with the addition of reactants to the system.

32. A small value for a K_{sp}, a solubility constant, indicates that which of these statements is true?

 (A) The concentration of the nonionized molecule must be relatively small compared with the ion concentrations.
 (B) The concentration of the ions of the molecule must be relatively small compared with the solute molecular concentration.
 (C) The substance ionizes to a large degree.
 (D) The concentration of the nonionized molecules must be about the same as the ion concentrations.
 (E) The concentrations of the ions appear in the denominator of the expression.

33. What is the K_w of distilled water ?

 (A) 1×10^{-7}
 (B) 1×10^{-0}
 (C) 1×10^{-14}
 (D) 1×10^{-1}
 (E) 0

34. What is the ΔH° value for the decomposition of sodium chlorate, given the following information?

$$NaClO_3(s) \rightarrow NaCl(s) + \frac{3}{2} O_2(g)$$

$$(\Delta H_f^\circ \text{ values}): NaClO_3(s) = -358 \text{ kJ/mol},$$
$$NaCl(s) = -410 \text{ kJ/mol}$$
$$O_2(g) = 0 \text{ kcal/mol}$$

 (A) 52 kJ/mol
 (B) −52 kJ/mol
 (C) 768 kJ/mol
 (D) −768 kJ/mol
 (E) $\frac{3}{2}$ (768 kJ/mol)

Questions 35–39 use the following choices:

 (A) $H_2(g)$
 (B) $NH_3(g)$
 (C) $CO_2(g)$
 (D) $HCl(g)$
 (E) $O_2(g)$

35. A gas produced by the reaction of zinc with hydrochloric acid

36. A gas combustion product that is heavier than air

37. A gas produced by the heating of potassium chlorate

38. A gas that is slightly soluble in water and gives a weakly acid solution

39. A gas that is slightly soluble in water and gives a slightly basic solution.

40. What are the simplest whole-number coefficients that balance this equation?

$$\underline{\quad}C_4H_{10} + \underline{\quad}O_2 \rightarrow \underline{\quad}CO_2 + \underline{\quad}H_2O$$

   (A) 1, 6, 4, 2
   (B) 2, 13, 8, 10
   (C) 1, 6, 1, 5
   (D) 3, 10, 16, 20
   (E) 4, 26, 16, 20

41. How many atoms are present in the formula $KAl(SO_4)_2$?

   (A) 7
   (B) 9
   (C) 11
   (D) 12
   (E) 13

42. All of the following are compounds EXCEPT

   (A) copper sulfate
   (B) carbon dioxide
   (C) washing soda
   (D) air
   (E) lime

43. What volume of gas, in liters, would 1.5 mol of hydrogen occupy at STP?

   (A) 11.2
   (B) 22.4
   (C) 33.6
   (D) 44.8
   (E) 67.2

44. What is the maximum number of electrons held in the d orbitals?

   (A) 2
   (B) 6
   (C) 8
   (D) 10
   (E) 14

45. If an element has an atomic number of 11, it will combine most readily with an element that has an atomic configuration of

(A) $1s^2 2s^2 2p^6 3s^2 3p^1$
(B) $1s^2 2s^2 2p^6 3s^2 3p^2$
(C) $1s^2 2s^2 2p^6 3s^2 3p^3$
(D) $1s^2 2s^2 2p^6 3s^2 3p^4$
(E) $1s^2 2s^2 2p^6 3s^2 3p^5$

46. An example of a physical property is

(A) rusting
(B) decay
(C) souring
(D) low melting point
(E) high heat of formation

47. A gas at STP which contains 6.02×10^{23} atoms and forms diatomic molecules will occupy

(A) 11.2 L
(B) 22.4 L
(C) 33.6 L
(D) 67.2 L
(E) 1.06 qt

48. When excited electrons cascade to lower energy levels in an atom,

(A) visible light is always emitted
(B) the potential energy of the atom increases
(C) the electrons always fall back to the first energy level
(D) the electrons fall indiscriminately to all levels
(E) the electrons fall back to a lower unfilled energy level

49. Mass spectroscopy uses the concept that

(A) charged particles are evenly deflected in a magnetic field
(B) charged particles are deflected in a magnetic field inversely to the mass of the particles
(C) particles of heavier mass are deflected in a magnetic field to a greater degree than lighter particles
(D) particles are evenly deflected in a magnetic field

50. The bond that includes an upper and a lower sharing of electron orbitals is called

   (A) a pi bond
   (B) a sigma bond
   (C) a hydrogen bond
   (D) a covalent bond
   (E) an ionic bond

51. What is the boiling point of water at the top of Pike's Peak (about 14,000 feet altitude)?

   (A) It is 100°C.
   (B) It is >100°C since the pressure is less than at ground level.
   (C) It is <100°C since the pressure is less than at ground level.
   (D) It is >100°C since the pressure is greater than at ground level.
   (E) It is <100°C since the pressure is greater than at ground level.

52. The atomic structure of the alkane series contains the hybrid orbitals designated as

   (A) sp
   (B) sp^2
   (C) sp^3
   (D) sp^3d^2
   (E) sp^4d^3

53. Which of the following is(are) true for this reaction?

$$Cu + 4HNO_3 \rightarrow Cu(NO_3)_2 + 2H_2O + 2NO_2(g)$$

   I. It is an oxidation-reduction reaction.
   II. Copper is oxidized.
   III. The oxidation number of nitrogen goes from +5 to +4.

   (A) I, II, and III
   (B) I and II only
   (C) II and III only
   (D) I only
   (E) III only

54. Which of the following properties can be attributed to water?

   I. It has a permanent dipole moment attributed to its molecular structure.
   II. It is a very good conductor of electricity.
   III. It has its polar covalent bonds with hydrogen on opposite sides of the oxygen atom, and thus the molecule is linear.

   (A) I, II, and III
   (B) I and II only
   (C) II and III only
   (D) I only
   (E) III only

55. Which of the following define(s) an acid according to conventional acid theories?

   I. It is a good proton donor.
   II. It is a good electron-pair acceptor.
   III. It has an excess of H_3O^+ in solution.

   (A) I, II, and III
   (B) I and II only
   (C) II and III only
   (D) I only
   (E) III only

56. A nuclear reactor must include which of the following parts?

   I. Electric generator
   II. Fissionable fuel elements
   III. Moderator

   (A) I, II, and III
   (B) I and II only
   (C) II and III only
   (D) I only
   (E) III only

57. Which of the following salts will hydrolyze in water to form basic solutions?

   I. NaCl
   II. $CuSO_4$
   III. K_3PO_4

   (A) I, II, and III
   (B) I and II only
   (C) II and III only
   (D) I only
   (E) III only

58. The dissolving of 1 mol of NaCl in 1000 g of water will change the boiling point of the water to

(A) 100.51°C
(B) 101.02°C
(C) 101.53°C
(D) 101.86°C
(E) 103.62°C

59. What is the structure associated with the BF_3 molecule?

(A) rectangle
(B) trigonal planar
(C) tetrahedron
(D) octahedron
(E) square

60. Which of these statements is the best expression for the sp^3 hybridization of carbon electrons?

(A) The new orbitals are one s orbital and three p orbitals.
(B) The s electron is promoted to the p orbitals.
(C) The s orbital is deformed into a p orbital.
(D) Four new and equivalent orbitals are formed.
(E) The s orbital electron loses energy to fall back into a partially filled p orbital.

61. The bonding that explains the variation of the boiling point of water from the boiling points of similarly structured molecules is

(A) hydrogen bonding
(B) van der Waals forces
(C) covalent bonding
(D) ionic bonding
(E) coordinate covalent bonding

62. If K for the reaction $H_2 + I_2 \rightleftharpoons 2HI$ is equal to 45.9 at 450°C, and 1 mol of H_2 and 1 mol of I_2 are introduced into a 1-L box at that temperature, what will be the concentration of H_2 at equilibrium?

(A) 0.114 mol/L
(B) 0.228 mol/L
(C) 0.456 mol/L
(D) 0.516 mol/L
(E) 0.772 mol/L

63. What is the gram-molecular mass of a nonionizing solid if 10 g of this solid, dissolved in 100 g of water, formed a solution that froze at −1.22°C?

(A) 0.65 g
(B) 6.5 g
(C) 130 g
(D) 154 g
(E) 265 g

64. What is the pH of a solution with a hydroxide ion concentration of 0.00001 mol/L?

(A) −5
(B) −1
(C) 5
(D) 9
(E) 14

65. Electrolysis of a dilute solution of sodium chloride results in the cathode product

(A) sodium
(B) hydrogen
(C) chlorine
(D) oxygen
(E) peroxide

66. _____$C_2H_4(g)$ + _____$O_2(g)$ → _____$CO_2(g)$ + _____$H_2O(\ell)$

If the equation for the above reaction is balanced with the smallest whole numbers, what is the appropriate coefficient for oxygen gas?

(A) 1
(B) 2
(C) 3
(D) 4
(E) 5

67. Five liters of gas at STP have a mass of 12.5 g. What is the molar mass of the gas?

(A) 12.5 g/mol
(B) 25.0 g/mol
(C) 47.5 g/mol
(D) 56.0 g/mol
(E) 125 g/mol

68. A compound whose molecular mass is 90 g contains 40.0% carbon, 6.67% hydrogen, and 53.33% oxygen. What is the true formula of the compound?

(A) $C_2H_2O_4$
(B) CH_2O_4
(C) C_3H_6O
(D) C_3HO_3
(E) $C_3H_6O_3$

69. How many moles of CaO are needed to react with an excess of water to form 370 g of calcium hydroxide?

(A) 1
(B) 2
(C) 3
(D) 4
(E) 5

70. To what volume, in milliliters, must 50.0 mL of 3.50 M H_2SO_4 be diluted in order to make 2 M H_2SO_4?

(A) 25
(B) 60.1
(C) 87.5
(D) 93.2
(E) 101

71. If K_{eq} is small, it indicates that equilibrium occurs

(A) at a low product concentration
(B) at a high product concentration
(C) after considerable time
(D) with the help of a catalyst
(E) with no forward reaction

72. A student measured 10.0 mL of an HCl solution into a beaker and titrated it with a standard NaOH solution that was 0.09 M. The initial NaOH buret reading was 34.7 mL while the final reading showed 49.2 mL. What is the molarity of the HCl solution?

(A) 0.13
(B) 0.47
(C) 0.52
(D) 1.32
(E) 2.43

73. A student made the following observations in the laboratory:

(a) Sodium metal reacted vigorously with water, while a strip of magnesium did not seem to react at all.

(b) The magnesium strip reacted with dilute hydrochloric acid faster than an iron strip.

(c) A copper rivet suspended in silver nitrate solution was covered with silver-colored stalactites in several days, and the resulting solution had a blue color.

(d) Iron filings dropped into the blue solution were coated with an orange color.

The order of *decreasing* strength-reducing agents is

(A) Na, Mg, Fe, Ag, Cu
(B) Mg, Na, Fe, Cu, Ag
(C) Ag, Cu, Fe, Mg, Na
(D) Na, Fe, Mg, Cu, Ag
(E) Na, Mg, Fe, Cu, Ag

74. A student placed water, sodium chloride, potassium dichromate, sand, chalk, and hydrogen sulfide into a distilling flask and proceeded to distill. What ingredient besides water would be found in the distillate?

(A) sodium chloride
(B) chalk
(C) sand
(D) hydrogen sulfide
(E) chrome sulfate

75. Which test would you use to prove the presence of the other substance in the distillate flask in question 74?

(A) flame test for sodium
(B) barium chloride test for sulfate
(C) silver nitrate test for chloride
(D) acid test for carbonate in chalk
(E) lead acetate test for hydrogen sulfide

76. A student used a steam-jacketed eudiometer and filled it with 32 mL of oxygen and 4 mL of hydrogen over mercury. How much of which gas would be left uncombined after sparking?

(A) none of either
(B) 3 mL H_2
(C) 24 mL O_2
(D) 28 mL O_2
(E) 30 mL O_2

77. What would be the *total* volume, in milliliters, of the gases in question 76 after sparking?

   (A) 16
   (B) 24
   (C) 34
   (D) 36
   (E) 40

78. A student mixed a small amount of amyl alcohol, glacial acetic acid, and a few drops of concentrated H_2SO_4. She then gently heated the mixture until she observed an odor. What odor should she expect to be given off?

   (A) banana oil
   (B) oil of wintergreen
   (C) pineapple
   (D) apple
   (E) formaldehyde

79. In which period of the periodic table is the most electronegative element found?

   (A) 1
   (B) 2
   (C) 3
   (D) 4
   (E) 5

80. What could be the equilibrium constant for the reaction $aA + bB \rightleftharpoons cC + dD$ if A and D are solids?

   (A) $\dfrac{[C]^c [D]^d}{[A]^a [B]^b}$

   (B) $\dfrac{[A]^a [B]^b}{[C]^c [D]^d}$

   (C) $\dfrac{[C]^c}{[B]^b}$

   (D) $\dfrac{[C]^c [D]^d}{[A]^a}$

   (E) $[A]^a[B]^b[C]^c[D]^d$

81. Which of the following is NOT a commonly occurring sulfur compound?

(A) H_2S
(B) H_2SO_4
(C) SO_2
(D) Ag_2S
(E) SO_3

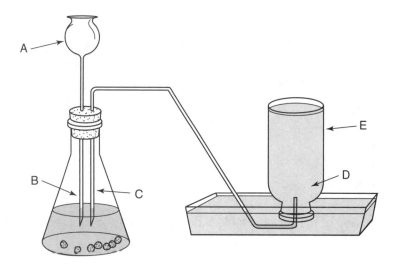

82. What letter designates an error in the above laboratory setup?

(A) A
(B) B
(C) C
(D) D
(E) E

83. The bonding that is most significant in explaining the variation in the boiling point of water from the boiling points of similarly structured molecules is

(A) hydrogen bonding
(B) van der Waals forces
(C) convalent bonding
(D) ionic bonding
(E) coordinate covalent bonding

84. If K for the reaction $H_2 + I_2 \rightleftharpoons 2HI$ is equal to 45.9 at 450°C and 1 mole of H_2 and 1 mole of I_2 are introduced into a 1-liter box at that temperature, what will be the expression of K at equilibrium?

(A) $\dfrac{\left[x^2\right]^2}{[1-x][1-x]}$

(B) $\dfrac{\left[2x\right]^2}{[1-x][1-x]}$

(C) $\dfrac{\left[2x^2\right]^2}{[x][x]}$

(D) $\dfrac{[1-x][1-x]}{\left[2x^2\right]^2}$

(E) $\dfrac{[1-x][1-x]}{\left[x^2\right]^2}$

85. What is the molar mass of a nonionizing solid if 10 grams of this solid, dissolved in 200 grams of water, forms a solution that freezes at −3.72°C?

(A) 25 g/mol
(B) 50 g/mol
(C) 100 g/mol
(D) 150 g/mol
(E) 1000 g/mol

STOP
IF YOU FINISH BEFORE ONE HOUR IS UP, YOU MAY GO BACK
TO CHECK YOUR WORK OR COMPLETE UNANSWERED QUESTIONS.

Answers and Explanations for Test 3

1. **(B)** The activation energy is the energy needed to begin the reaction.

2. **(D)** For the reverse reaction to occur, the sum B + C is needed. This is shown by D.

3. **(C)** The heat of the reaction is the heat liberated between the level of potential energy of the reactants and that of the products. This is the quantity C on the diagram.

4. **(A)** The potential energy of the activated complex is the total of the original potential energy (A) and the activation energy (B).

5. **(B)** The spoon must be made the cathode to attract the Ag^+ ions.

6. **(D)** The silver plate is the anode.

7. **(E)** The solution of Ag^+ provides the silver for plating.

8. **(A)**

9. **(B)** } In a phase diagram, the zones are as shown below.

10. **(C)**

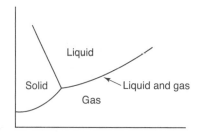

11. **(E)** The boundary of liquid and gas is shown on the diagram.

12–14. The balanced equation with half-reactions is:

12. **(B)** $3Cu^0 \rightarrow 3Cu^{2+} + 3(2e^-)$

$$\frac{2NO_3^- + 8H^+ + 2(3e^-) \rightarrow 2NO + 4H_2O}{3Cu + 8HNO_3 \rightarrow 3Cu(NO_3)_2}$$
$$+ 4H_2O + 2NO$$

13. **(D)** The coefficients show that 3 moles of Cu produces 2 moles of NO; so 6 moles of Cu produces 4 moles of NO.

14. **(C)** $Cu(NO_3)_2 \rightarrow Cu^{2+} + 2NO_3^-$ shows that the dissociation yields three ions.

15. **(A)** KCl(s) is ionic because it is the product of a very active metal combining with a very active nonmetal.

16. **(B)** The electronegativity difference between H and Cl is between 0.5 and 1.7. This indicates an unequal sharing of electrons, which results in a polar covalent bond.

17. **(C)** Because these polar bonds are symmetrically arranged in the methane molecule, the molecule is nonpolar covalent.

18. **(E)** Lithium, Li(s) is a metal.

19. **(D)** When hydrated copper sulfate is heated, the crystal crumples as the water is forced out of the structure, and a white powder is the result.

20. **(A)** Brownian movement is due to molecular collisions with colloidal particles, which knock the particles about in a zigzag path noted by the reflected light from these particles.

21. **(C)** The indicator phenolphathalein turns pink in a basic solution.

22. **(B)** Litmus paper turns pink in an acid solution.

23. **(E)** A substance that is deliquescent draws water to its surface. At times it can draw enough water to form a water solution.

24. **(A)** The atomic number gives the number of protons in the nucleus and the total number of electrons. The mass number indicates the total number of protons and neutrons in the nucleus—for Na, 23 (11 protons + 12 neutrons).

25. **(E)** A catalyst can lower the activation energy, and that is its purpose. However, it lowers the activation energy for both the forward and reverse reactions. So the only correct answer is (E), the catalyst itself is not changed.

26. **(C)** Cesium (Cs) and fluorine (F) are elements from the extreme sides of the periodic table. Therefore, they will form the strongest ionic bond.

27. **(A)** The nitrogen atom has five outer electrons and the three hydrogens have one electron each. The one remaining hydrogen is using the electron pair of the nitrogen but lacks one electron. This accounts for the +1 charge on the ammonium ion.

$$\left[\begin{array}{c} \text{H} \\ \cdot\cdot \\ \text{H} : \text{N} : \text{H} \\ \cdot\cdot \\ \text{H} \end{array} \right]^{+}$$

28. **(B)** Use the dilution equation:

$$M_{\text{orginal}} \times V_{\text{original}} = M_{\text{final}} \times V_{\text{final}}$$

By inserting the original values, you have

$$250 \text{ mL} \times 1.0 \, M = 0.50 \, M \times V$$

Solving for volume gives 500 mL total solution. Since you already had 250 mL of solution, you have to add 250 mL to make 500 mL.

29. **(E)** The only acid that shows two stages of ionization in a dilute solution is H_2SO_4.

1st stage: $H_2SO_4 + H_2O \rightarrow H_3O^+ + HSO_4^-$

2nd stage: $HSO_4^- + H_2O \rightarrow H_3O^+ + SO_4^{2-}$

30. **(B)** Only statement (B) is true. Water is not always a product. The acid becomes the conjugate base, and the base becomes the conjugate acid in a Brønsted-Lowry acid-base reaction.

31. **(E)** At a given temperature, the K_{eq} will not change with the addition of reactants to the system. The equilibrium will shift in the direction that reestablishes the equilibrium. Because the reaction has two gas volumes on the left and two on the right, pressure will not affect the production.

32. **(B)** Since the product of the ionic concentrations are the numerator of the solubility constant, K_{sp}, the value will be small if the numerator is small.

33. **(C)** The equilibrium constant for water has the designation K_w and is
$$K_w = [H_3O^+][OH^-] = 1 \times 10^{-14}$$

34. **(B)** $\Delta H^\circ_{reaction} = \Delta H^\circ_f(\text{products}) - \Delta H^\circ_f(\text{reactants})$
$\Delta H^\circ_{reaction} = -410 \text{ kJ/mol} - (-358 \text{ kJ/mol})$
$\Delta H^\circ_{reaction} = -52 \text{ kJ/mol}$

35. **(A)** $Zn(s) + 2HCl(aq) \rightarrow ZnCl_2(aq) + H_2(g)$ is the reaction that occurs.

36. **(C)** Only CO_2, with a molecular mass of 44, is denser than air, which has an assigned molecular mass of 29.

37. **(E)** $2KClO_3 \rightarrow 2KCl + 3O_2(g)$ is the reaction that occurs.

38. **(C)** CO_2 is slightly soluble in water, forming carbonic acid, H_2CO_3, which is a weak acid.

39. **(B)** NH_3 is very soluble in water and forms a solution of the weak ammonium hydroxide base.

40. **(B)** The correct coefficients are 2, 13, 8, and 10.

41. **(D)** $1K + 1Al + 2S + 8O = 12$ total.

42. **(D)** Air is a mixture; all others are compounds. Washing soda (C) is sodium carbonate, and lime (E) is calcium oxide.

43. **(C)** One mol of a gas at STP occupies 22.4 L, and so 1.5 mol $\times$ 22.4 L = 33.6 L.

44. **(D)** The maximum number of electrons in each kind of orbital is

s = 2 in one orbital
p = 6 in three orbitals
d = 10 in five orbitals
f = 14 in seven orbitals

45. **(E)** The element with atomic number 11 is Na with 1 electron in the $3s$ orbital. It would readily combine with the element that has $3p^5$ as the outer orbital since it needs only 1 more electron to be filled.

46. **(D)** The only physical property named in the list is low melting point.

47. **(A)** If the gas is diatomic, then 6.02×10^{23} atoms will form $6.02 \times 10^{23}/2$ molecules. At STP, 6.02×10^{23} molecules occupy 22.4 L. So half that amount would occupy 11.2 L.

48. **(E)** Cascading excited electrons can fall only to the lowest energy level that is unfilled.

49. **(B)** The spectroscope uses a magnetic field to separate isotopes by bending their path. The lighter ones are bent further than the heavier ones.

50. **(A)** The pi bond is a bond between two p orbitals, like this:

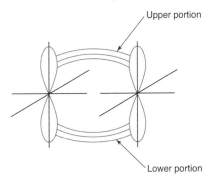

51. **(C)** At Pike's Peak (alt. approx. 14,000 ft) the pressure is lower than at ground level; therefore the vapor pressure at a lower temperature will equal the outside pressure and boiling will occur.

52. **(C)** The alkanes use the sp^3 hybrid orbitals.

53. **(A)** I, II, and III are correct.

54. **(D)** Only I is correct.

55. **(A)** I, II, and III are acid definitions.

56. **(C)** I is not necessary for the reactor, but nuclear energy is often used to operate an electric generator. The others, II and III, are necessary for fuel and neutron-speed control, respectively.

57. **(E)** III is a salt from a strong base and a weak acid, which hydrolyzes to form a basic solution with water.

58. **(B)** Since the boiling point is increased by 0.51°C for each mole of particles, 1 mol of $NaCl \rightarrow Na^+ + Cl^-$ gives 2 mol of particles. Therefore, the boiling point will be 1.02° higher or 101.02°C.

59. **(B)** The sp^2 hybridized orbital in BF_3 is trigonal, planar and so is related to the triangle.

60. **(D)** When hybridization forms the sp^3 orbitals, four entirely new orbitals different from the former s and p orbitals result.

61. **(A)** Hydrogen bonding between water molecules causes the boiling point to be higher than would be expected.

62. **(B)** At the beginning of the reaction

$$[H_2] = 1 \text{ mol/L}$$
$$[I_2] = 1 \text{ mol L}$$
$$[HI] = 0$$

At equilibrium,

$$(H_2 + I_2 \rightleftharpoons 2HI)$$

(Let x = moles/liter of H_2 and I, in HI form)

$$[H_2] = (1 - x) \text{ mole/liter}$$
$$[I_2] = (1 - x) \text{ mole/liter}$$
$$[HI] = 2x \text{ mole/liter}$$

$$K = \frac{[HI]^2}{[H_2][I_2]} = \frac{(2x)^2}{(1-x)(1-x)} = 45.9$$

Taking the square root of both sides gives

$$\frac{2x}{1-x} = 6.8$$
$$x = 0.772 \text{ mol/L}$$

Therefore $[H_2] = 1 - x = 1 - 0.0772 = 0.228$ mol/L.

63. **(D)** 10 g/100 g water = 100 g/1000 g water. The freezing point depression, 1.22, is divided by 1.86, which is the depression caused by 1 mol in 1000 g of water, to find how many moles are dissolved.

1.22/1.86 = 0.65 mol

So 100 g = 0.65 mol. Then 1 mol = $\dfrac{100 \text{ g}}{0.65 \text{ mol}}$ = 153.8 or 154 g/mol

64. **(D)** The K_w of water = $[H^+] [OH^-] = 10^{-14}$. If $[OH^-] = 10^{-5}$ mol/L, then $[H^+] = 10^{-14}/10^{-5} = 10^{-9}$

$$pH = -\log [H^+] \text{ (by definition)}$$
$$pH = -[-9]$$
$$pH = 9$$

65. **(B)** When a dilute NaCl solution is electrolyzed, hydrogen is given off at the cathode, chlorine is given off at the anode, and sodium hydroxide is left in the container.

66. **(C)** The correctly balanced equation is

$$C_2H_4(g) + 3O_2(g) \rightarrow 2CO_2(g) + 2H_2O(\ell)$$

67. **(D)** One mol of a gas at STP occupies 22.4 L.

If $\dfrac{5 \text{ L}}{12.5 \text{ g}}$, then $\dfrac{22.4 \text{ L}}{\text{gram-molecular mass}}$

$$\frac{5}{12.5} = \frac{22.4}{x}$$
$$x = 56 \text{ g}$$

68. **(E)** To find the simple or empirical formula, divide each percent by the element's atomic mass.

Carbon	Hydrogen	Oxygen
$40 \div 12 = 3.333$	$6.67 \div 1 = 6.67$	$53.33 \div 16 = 3.33$

Next, divide each quotient by the smallest quotient in an attempt to get small whole numbers.

$$3.33 \div 3.33 = 1 \text{ carbon}$$
$$6.67 \div 3.33 = 2 \text{ hydrogen}$$
$$3.33 \div 3.33 = 1 \text{ oxygen}$$

The simplest formula is CH_2O, which has a molecular mass of 30.

The true molecular mass is given as 90, which is three times the simplest. Therefore, the true formula is $C_3H_6O_3$.

69. **(E)** The reaction is

$$\overset{370 \text{ g}}{CaO} + H_2O \rightarrow Ca(OH)_2$$

$Ca(OH)_2$ molecular wt. $= 74$

So $370 \text{ g} \div 74 = 5$ mol of $Ca(OH)_2$ wanted. The reaction shows 1 mol of CaO produces 1 mol of $Ca(OH)_2$, and so the answer is 5 mols.

70. **(C)** In dilution problems, this formula can be used.

$$M_{before} \times V_{before} = M_{after} \times V_{after}$$

Substituting gives

$$3.5 \times 50 = 2 \times (?x)$$
$$x = 87.5 \text{ mL, new volume after dilution}$$

71. **(A)** For the K_{eq} to be small, the numerator, which is made up of the concentration(s) of the product(s) at equilibrium, must be smaller than the denominator.

72. **(A)** The amount of NaOH used is

$$49.2 - 34.7 = 14.5 \text{ mL}$$

Using $M_1 \times V_1 \times M_2 \times V_2$ gives

$$0.09 \, M \times 14.5 \text{ mL} = M_2 \times 10 \text{ mL}$$
$$M_2 = 0.13 \, M$$

73. **(E)** The reactions recorded indicated that the ease of losing electrons is greater in sodium than in magnesium, greater in magnesium than in iron, greater in iron than in copper, and finally greater in copper than in silver.

74. **(D)** Distillation removes only dissolved solids from the distillate. The volatile gases, like H_2S, will be carried into the distillate.

75. **(E)** The test for H_2S is to moisten lead acetate paper and look for brown-black precipitate to form, which is a positive result.

76. **(E)** H_2 to O_2 ratio by volume is 2 : 1 in the formation of water. Therefore, 4 mL H_2 will react with 2 mL of O_2 to make 4 mL of steam.

$$2H_2(g) + O_2(g) \rightarrow 2H_2O(g)$$

This leaves 30 mL of O_2 uncombined.

77. **(C)** There will be 30 mL of O_2 + 4 mL of steam = 34 mL total.

78. **(A)** She would make a small amount of amyl acetate which is an ester with a bananalike odor.

79. **(B)** The most electronegative element is fluorine (F), found in period 2.

80. **(C)** Solids are incorporated into the K value and therefore do not appear on the right side of the equation.

81. **(E)** SO_3 is not easily formed. The commercial process uses a catalyst.

82. **(B)** The thistle tube is below the fluid level in the flask and will cause liquid to be forced up the tube when gas is evolved in the reaction.

83. **(A)** Hydrogen bonding between water molecules causes the boiling point to be higher than would be expected.

84. **(B)** At the beginning of the reaction,

$$[H_2] = 1 \text{ mol/L}$$
$$[I_2] = 1 \text{ mol/L}$$
$$[HI] = 0$$

At equilibrium,

$$H_2 + I_2 \rightleftharpoons 2HI$$

(Let x = moles/liter of H_2 and I_2 that now exist in HI form.)

Then, at equilibrium,

$$[H_2] = (1 - x) \text{ mol/L}$$
$$[I_2] = (1 - x) \text{ mol/L}$$
$$[HI] = 2x \text{ mol/L}$$

Then, substituting the above values into the equation gives

$$K = \frac{[HI]^2}{[H_2][I_2]} = \frac{(2x)^2}{(1 - x)(1 - x)} = 45.9$$

85. **(A)** 10 g/200 g of water = 50 g/1000 g of water (5 times as much). The freezing point depression, 3.72°C, is divided by 1.86°C, which is the depression caused by 1 mol in 1000 g of water, to find how many moles are dissolved:

$$3.72°C/1.86°C/\text{mol} = 2 \text{ mol}$$

So if there were 50 g causing this depression and equal to 2 moles, 1 mol would be one half of 50 g, or 25 g.

Diagnosing Your Needs

After taking Practice Test 3, check your answers against the correct ones. Then fill in the chart below.

In the space under each equation number, place a check if you answered that question correctly.

Example

If your answer to question 5 was correct, place a check in the appropriate box.

Next, total the check marks for each section and insert the number in the designated block. Now do the arithmetic indicated and insert your percent for each area.

SUBJECT AREA

(✓) QUESTIONS ANSWERED CORRECTLY

I.	**Atomic Theory and Structure**, including periodic relationships	14	24	25	26	44	48	50	59	60

☐ No. of checks ÷ 9 × 100 = _____%

II.	**Nuclear Reactions**								49	56

☐ No. of checks ÷ 2 × 100 = _____%

III.	**Chemical Bonding and Molecular Structure**	15	16	17	18	27	41	45	54	61	84

☐ No. of checks ÷ 10 × 100 = _____%

IV.	**States of Matter and Kinetic Molecular Theory of Gases**	8	9	10	11	43	47	61	67

☐ No. of checks ÷ 8 × 100 = _____%

V.	**Solutions**, including concentration units, solubility, and colligative properties	19	20	28	58	70

☐ No. of checks ÷ 5 × 100 = _____%

VI. **Acids and Bases**	21	22	29	30	55	57	64

☐ No. of checks ÷ 7 × 100 = _____%

VII. **Oxidation-Reduction and** **Electrochemistry**	5	6	7	53	65

☐ No. of checks ÷ 5 × 100 = _____%

VIII. **Stoichiometry**	12	13	40	63	66	68	69	76	77

☐ No. of checks ÷ 9 × 100 = _____%

IX. **Reaction Rates**	1	31

☐ No. of checks ÷ 2 × 100 = _____%

X. **Equilibrium**	32	33	62	71	80	85

☐ No. of checks ÷ 6 × 100 = _____%

XI. **Thermodynamics:** energy changes in chemical reactions, randomness, and criteria for spontaneity	2	3	4	34

☐ No. of checks ÷ 4 × 100 = _____%

XII. **Descriptive Chemistry:** physical and chemical properties of elements and their familiar compounds; organic chemistry; periodic properties	23	35	36	37	38	39	42
	46	52	78	79	81

☐ No. of checks ÷ 12 × 100 = _____%

XIII. **Laboratory:** equipment, procedures, observations, safety, calculations, and interpretation of results	72	73	74	75	82	83

☐ No. of checks ÷ 6 × 100 = _____ %

Planning Your Study

The percentages give you an idea of how you have done on the various major areas of the test. Because of the limited number of questions on some parts, these percentages may not be as reliable as the percentages for parts with larger numbers of questions. However, you should now have at least a rough idea of the areas in which you have done well and those in which you need more study.

Start your study with the areas in which you are weakest. The corresponding chapters are indicated below.

Subject Area	Chapters to Review
I. Atomic Theory and Structure, including periodic relationships	2
II. Nuclear Reactions	15
III. Chemical Bonding and Molecular Structure	3, 4
IV. States of Matter and Kinetic Molecular Theory of Gases	1
V. Solutions, including concentration units, solubility, and colligative properties	7
VI. Acids and Bases	11
VII. Oxidation-Reduction and Electrochemistry	12
VIII. Stoichiometry	5, 6
IX. Reaction Rates	9
X. Equilibrium	10
XI. Thermodynamics, including energy changes in chemical reactions, randomness, and criteria for spontaneity	8
XII. Descriptive Chemistry: physical and chemical properties of elements and their familiar compounds; organic chemistry; periodic properties	1, 2, 13, 14
XIII. Laboratory: equipment, procedures, observations, safety, calculations, and interpretation of results.	All lab diagrams, 16

EQUATIONS AND TABLES FOR REFERENCE

Some Important Formulas and Equations

Density (*d*)

$$\text{Density} = \frac{\text{mass}}{\text{volume}} \quad d = m/v$$

Percent Error

Percent error =
$$\frac{\text{measured value} - \text{accepted value}}{\text{accepted value}} \times 100\%$$

Percent Yield

$$\text{Percent yield} = \frac{\text{actual yield}}{\text{expected yield}} \times 100\%$$

Percentage Composition

Percentage composition by mass
$$= \frac{\text{mass of element}}{\text{mass of compound}} \times 100\%$$

Boyle's Law

$$P_1 V_1 = P_2 V_2$$

Charles's Law

$$V_1 T_2 = V_2 T_1$$

Dalton's Law of Partial Pressures

$$P_T = p_a + p_b + p_c + \cdots$$

Ideal Gas Law

$$PV = nRT$$

Molarity (*M*)

$$\text{Molarity} = \frac{\text{moles of solute}}{\text{liters of solution}}$$

Molality (*m*)

$$\text{molality} = \frac{\text{moles of solute}}{\text{kilograms of solvent}}$$

Boiling Point Elevation

$$\Delta T_b = K_b m$$

where K_b is the molal boiling point elevation constant

Freezing Point Depression

$$\Delta T_f = K_f m$$

where K_f is the molal boiling point elevation constant

Rate of Reaction

$$\text{Rate} = k[A]^x[B]^y$$

where [A] and [B] are molar concentrations of reactants and k is a rate constant

Hess's Law

$$\Delta H_{\text{net}} = \Delta H_1 + \Delta H_2$$

Entropy Change

$$\Delta S = S_{\text{products}} - S_{\text{reactants}}$$

Gibbs Free Energy

$$\Delta G - \Delta H - T\Delta S$$

Tables for Reference

DENSITY AND BOILING POINTS OF SOME COMMON GASES			
Name		*Density* grams/liter at STP*	*Boiling Point* (at 1 atm) K
Air	—	1.29	—
Ammonia	NH_3	0.771	240
Carbon dioxide	CO_2	1.98	195
Carbon monoxide	CO	1.25	82
Chlorine	Cl_2	3.21	238
Hydrogen	H_2	0.0899	20
Hydrogen chloride	HCl	1.64	188
Hydrogen sulfide	H_2S	1.54	212
Methane	CH_4	0.716	109
Nitrogen	N_2	1.25	77
Nitrogen (II) oxide	NO	1.34	121
Oxygen	O_2	1.43	90
Sulfur dioxide	SO_2	2.92	263

*STP is defined as 273 K and 1 atm.

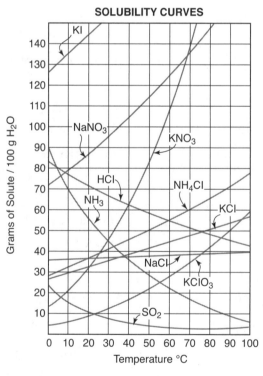

SOLUBILITY CURVES

TABLE OF SOLUBILITIES IN WATER

i—nearly insoluble ss—slightly soluble s—soluble d—decomposes n—not isolated	acetate	bromide	carbonate	chloride	chromate	hydroxide	iodide	nitrate	phosphate	sulfate	sulfide
Aluminum	ss	s	n	s	n	i	s	s	i	s	d
Ammonium	s	s	s	s	s	s	s	s	s	s	s
Barium	s	s	i	s	i	s	s	s	i	i	d
Calcium	s	s	i	s	s	ss	s	s	i	ss	d
Copper II	s	s	i	s	i	i	n	s	i	s	i
Iron II	s	s	i	s	n	s	s	s	i	s	i
Iron III	s	s	n	s	i	i	n	s	i	ss	d
Lead	s	ss	i	ss	i	i	ss	s	i	i	i
Magnesium	s	s	i	s	s	i	s	s	i	s	d
Mercury I	ss	i	i	i	ss	n	i	s	i	ss	i
Mercury II	s	ss	i	s	ss	i	i	s	i	d	i
Potassium	s	s	s	s	s	s	s	s	s	s	s
Silver	ss	i	i	i	ss	n	i	s	i	ss	i
Sodium	s	s	s	s	s	s	s	s	s	s	s
Zinc	s	s	i	s	s	i	s	s	i	s	i

SELECTED POLYATOMIC IONS

Hg_2^{2+}	dimercury (I)	CrO_4^{2-}	chromate
NH_4^+	ammonium	$Cr_2O_7^{2-}$	dichromate
$C_2H_3O_2^-$	acetate	MnO_4^-	permanganate
CH_3COO^-		MnO_4^{2-}	manganate
CN^-	cyanide	NO_2^-	nitrite
CO_3^{2-}	carbonate	NO_3^-	nitrate
HCO_3^-	hydrogen carbonate	OH^-	hydroxide
		PO_4^{3-}	phosphate
$C_2O_4^{2-}$	oxalate	SCN^-	thiocyanate
ClO^-	hypochlorite	SO_3^{2-}	sulfite
ClO_2^-	chlorite	SO_4^{2-}	sulfate
ClO_3^-	chlorate	HSO_4^-	hydrogen sulfate
ClO_4^-	perchlorate	$S_2O_3^{2-}$	thiosulfate

Ⓔ

STANDARD ENERGIES OF FORMATION OF COMPOUNDS AT 1 atm AND 298 K				
Compound	Heat (Enthalpy) of Formation* ($\Delta H_f°$)		Free Energy of Formation ($\Delta G_f°$)	
	kJ/mol	kcal/mol	kJ/mol	kcal/mol
Aluminum oxide $Al_2O_3(s)$	21676.5	2400.5	21583.1	2378.2
Ammonia $NH_3(g)$	246.0	211.0	216.3	23.9
Barium sulfate $BaSO_4(s)$	21473.9	2352.1	21363.0	2325.6
Calcium hydroxide $Ca(OH)_2(s)$	2986.6	2235.7	2899.2	2214.8
Carbon dioxide $CO_2(g)$	2393.9	294.1	2394.7	294.3
Carbon monoxide $CO(g)$	2110.5	226.4	2137.3	232.8
Copper(II) sulfate $CuSO_4(s)$	2771.9	2184.4	2662.2	2158.2
Ethane $C_2H_6(g)$	284.6	220.2	233.1	27.9
Ethene (ethylene) $C_2H_4(g)$	52.3	12.5	68.2	16.3
Ethyne (acetylene) $C_2H_2(g)$	226.9	54.2	209.3	50.0
Hydrogen fluoride $HF(g)$	2271.3	264.8	2273.3	265.3
Hydrogen iodide $HI(g)$	26.4	6.3	1.7	0.4
Iodine chloride $ICl(g)$	18.0	4.3	25.4	21.3
Lead(II) oxide $PbO(s)$	2215.6	251.5	2188.4	245.0
Methane $CH_4(g)$	274.9	217.9	250.7	212.1
Magnesium oxide $MgO(s)$	2601.9	2143.8	2569.7	2136.1
Nitrogen(II) oxide $NO(g)$	90.4	21.6	86.7	20.7
Nitrogen(IV) oxide $NO_2(g)$	33.1	7.9	51.5	12.3
Potassium chloride $KCl(s)$	2437.0	2104.4	2409.4	297.8
Sodium chloride $NaCl(s)$	2411.5	298.3	2384.3	291.8
Sulfur dioxide $SO_2(g)$	2296.8	270.9	2300.1	271.7
Water $H_2O(g)$	2242.0	257.8	2228.6	254.6
Water $H_2O(\ell)$	2285.9	268.3	2237.3	256.7

*Minus sign indicates an exothermic reaction.
Sample equations:

$$2Al(s)\ 1\ \frac{3}{2}O_2(g) \rightarrow Al_2O_3(s)\ 1\ 400.5\ kcal$$

$$2Al(s)\ 1\ \frac{3}{2}O_2(g) \rightarrow Al_2O_3(s)\quad \Delta H\ 5\ 2400.5\ kcal/mol$$

Ⓕ

SELECTED RADIOISOTOPES		
Nuclide	Half-Life	Decay Mode
^{198}Au	2.69 d	β^-
^{14}C	5730 y	β^-
^{60}Co	5.26 y	β^-
^{137}Cs	30.23 y	β^-
^{220}Fr	27.5 s	α
^{3}H	12.26 y	β^-
^{131}I	8.07 d	β^-
^{37}K	1.23 s	β^+
^{42}K	12.4 h	β^-
^{85}Kr	10.76 y	β^-
^{85m}Kr*	4.39 h	γ
^{16}N	7.2 s	β^-
^{32}P	14.3 d	β^-
^{239}Pu	2.44×10^4 y	α
^{226}Ra	1600 y	α
^{222}Rn	3.82 d	α
^{90}Sr	28.1 y	β^-
^{99}Tc	2.13×10^5 y	β^-
^{99m}Tc*	6.01 h	γ
^{232}Th	1.4×10^{10} y	α
^{233}U	1.62×10^5 y	α
^{235}U	7.1×10^8 y	α
^{238}U	4.51×10^9 y	α

y = years; d = days; h = hours;
s = seconds
*m = meta stable or excited state of the same nucleus. Gamma decay from such a state is called an isomeric transition (IT).

Nuclear isomers are different energy states of the same nucleus, each having a different measurable lifetime.

HEATS OF REACTION AT 1 atm AND 298 K

Reaction	ΔH	
	kJ	kcal
$CH_4(g) + 2O_2(g) \rightarrow CO_2(g) + 2H_2O(\ell)$	−890.8	−212.8
$C_3H_8(g) + 5O_2(g) \rightarrow 3CO_2(g) + 4H_2O(\ell)$	−2221.1	−530.6
$CH_3OH(\ell) + \frac{3}{2}O_2(g) \rightarrow CO_2(g) + 2H_2O(\ell)$	−726.7	−173.6
$C_6H_{12}O_6(s) + 6O_2(g) \rightarrow 6CO_2(g) + 6H_2O(\ell)$	−2804.2	−669.9
$CO(g) + \frac{1}{2}O_2(g) \rightarrow CO_2(g)$	−283.4	−67.7
$C_8H_{18}(\ell) + \frac{25}{2}O_2(g) \rightarrow 8CO_2(g) + 9H_2O(\ell)$	−5453.1	−1302.7
$KNO_3(s) \xrightarrow{H_2O} K^1(aq) + NO_3{}^2(aq)$	34.7	8.3
$NaOH(s) \xrightarrow{H_2O} Na^1(aq) + OH^2(aq)$	−44.4	−10.6
$NH_4Cl(s) \xrightarrow{H_2O} NH_4{}^1(aq) + Cl^2(aq)$	14.7	3.5
$NH_4NO_3(s) \xrightarrow{H_2O} NH_4{}^1(aq) + NO_3{}^2(aq)$	25.5	6.1
$NaCl(s) \xrightarrow{H_2O} Na^1(aq) + Cl^2(aq)$	3.8	0.9
$KClO_3(s) \xrightarrow{H_2O} K^1(aq) + ClO_3{}^2(aq)$	41.4	9.9
$LiBr(s) \xrightarrow{H_2O} Li^1(aq) + Br^2(aq)$	−49.0	−11.7
$H^1(aq) + OH^2(aq) \rightarrow H_2O(\ell)$	−57.8	−13.8

SYMBOLS USED IN NUCLEAR CHEMISTRY

alpha particle	^4_2He	α
beta particle (electron)	$^0_{-1}e$	β^-
gamma radiation		γ
neutron	^1_0n	n
proton	^1_1H	p
deuteron	^2_1H	
triton	^3_1H	
positron	$^0_{+1}e$	β^+

RELATIVE STRENGTHS OF ACIDS IN AQUEOUS SOLUTION AT 1 ATM AND 298 K

Conjugate Pairs ACID BASE	K_a	Conjugate Pairs ACID BASE	K_a
$HI = H^+ + I^-$	very large	$Al(H_2O)_6{}^{3+} = H^+ + Al(H_2O)_5(OH)^{2+}$	1.1×10^{-5}
$HBr = H^+ + Br^-$	very large	$H_2O + CO_2 = H^+ + HCO_3{}^-$	4.3×10^{-7}
$HCl = H^+ + Cl^-$	very large	$HSO_3{}^- = H^+ + SO_3{}^{2-}$	1.1×10^{-7}
$HNO_3 = H^+ + NO_3{}^-$	very large	$H_2S = H^+ + HS^-$	9.5×10^{-8}
$H_2SO_4 = H^+ + HSO_4{}^-$	large	$H_2PO_4{}^- = H^+ + HPO_4{}^{2-}$	6.2×10^{-8}
$H_2O + SO_2 = H^+ + HSO_3{}^-$	1.5×10^{-2}	$NH_4{}^+ = H^+ + NH_3$	5.7×10^{-10}
$HSO_4{}^- = H^+ + SO_4{}^{2-}$	1.2×10^{-2}	$HCO_3{}^- = H^+ + CO_3{}^{2-}$	5.6×10^{-11}
$H_3PO_4 = H^+ + H_2PO_4{}^-$	7.5×10^{-3}	$HPO_4{}^{2-} = H^+ + PO_4{}^{3-}$	2.2×10^{-13}
$Fe(H_2O)_6{}^{3+} = H^+ + Fe(H_2O)_5(OH)^{2+}$	8.9×10^{-4}	$HS^- = H^+ + S^{2-}$	1.3×10^{-14}
$HNO_2 = H^+ + NO_2{}^-$	4.6×10^{-4}	$H_2O = H^+ + OH^-$	1.0×10^{-14}
$HF = H^+ + F^-$	3.5×10^{-4}	$OH^- = H^+ + O^{2-}$	$< 10^{-36}$
$Cr(H_2O)_6{}^{3+} = H^+ + Cr(H_2O)_5(OH)^{2+}$	1.0×10^{-4}	$NH_3 = H^+ + NH_2{}^-$	very small
$CH_3COOH = H^+ + CH_3COO^-$	1.8×10^{-5}		

Note: $H^+(aq) = H_3O^+$

Sample equation: $HI + H_2O = H_3O^+ + I^-$

CONSTANTS FOR VARIOUS EQUILIBRIA AT 1 ATM AND 298 K

$H_2O(\ell) = H^+(aq) + OH^-(aq)$	$K_w = 1.0 \times 10^{-14}$
$H_2O(\ell) + H_2O(\ell) = H_3O^+(aq) + OH^-(aq)$	$K_w = 1.0 \times 10^{-14}$
$CH_3COO^-(aq) + H_2O(\ell) = CH_3COOH(aq) + OH^-(aq)$	$K_b = 5.6 \times 10^{-10}$
$Na^+F^-(aq) + H_2O(\ell) = Na^+(OH)^- + HF(aq)$	$K_b = 1.5 \times 10^{-11}$
$NH_3(aq) + H_2O(\ell) = NH_4^+(aq) + OH^-(aq)$	$K_b = 1.8 \times 10^{-5}$
$CO_3^{2+}(aq) + H_2O(\ell) = HCO_3^-(aq) + OH^-(aq)$	$K_b = 1.8 \times 10^{-4}$
$Ag(NH_3)_2^-(aq) = Ag^+(ag) + 2NH_3(aq)$	$K_{eq} = 8.9 \times 10^{-8}$
$N_2(g) + 3H_2(g) = 2NH_3(g)$	$K_{eq} = 6.7 \times 10^{5}$
$H_2(g) + I_2(g) = 2HI(g)$	$K_{eq} = 3.5 \times 10^{-1}$

Compound	K_{sp}	Compound	K_{sp}
AgBr	5.0×10^{-13}	Li_2CO_3	2.5×10^{-2}
AgCl	1.8×10^{-10}	$PbCl_2$	1.6×10^{-5}
Ag_2CrO_4	1.1×10^{-12}	$PbCO_3$	7.4×10^{-14}
AgI	8.3×10^{-17}	$PbCrO_4$	2.8×10^{-13}
$BaSO_4$	1.1×10^{-10}	PbI_2	7.1×10^{-9}
$CaSO_4$	9.1×10^{-6}	$ZnCO_3$	1.4×10^{-11}

STANDARD ELECTRODE POTENTIALS
Ionic Concentrations 1 M Water At 298 K, 1 atm

Half-Reaction	E^0 (volts)	Half-Reaction	E^0 (volts)
$F_2(g) + 2e^- \rightarrow 2F^-$	+2.87	$Sn^{2+} + 2e^- \rightarrow Sn(s)$	−0.14
$8H^+ + MnO_4^- + 5e^- \rightarrow Mn^{2+} + 4H_2O$	+1.51	$Ni^{2+} + 2e^- \rightarrow Ni(s)$	−0.26
$Au^{3+} + 3e^- \rightarrow Au(s)$	+1.50	$Co^{2+} + 2e^- \rightarrow Co(s)$	−0.28
$Cl_2(g) + 2e^- \rightarrow 2Cl^-$	+1.36	$Fe^{2+} + 2e^- \rightarrow Fe(s)$	−0.45
$14H^+ + Cr_2O_7^{2-} + 6e^- \rightarrow 2Cr^{3+} + 7H_2O$	+1.23	$Cr^{3+} + 3e^- \rightarrow Cr(s)$	−0.74
$4H^+ + O_2(g) + 4e^- \rightarrow 2H_2O$	+1.23	$Zn^{2+} + 2e^- \rightarrow Zn(s)$	−0.76
$4H^+ + MnO_2(s) + 2e^- \rightarrow Mn^{2+} + 2H_2O$	+1.22	$2H_2O + 2e^- \rightarrow 2OH^- + H_2(g)$	−0.83
$Br_2(\ell) + 2e^- \rightarrow 2Br^-$	+1.09	$Mn^{2+} + 2e^- \rightarrow Mn(s)$	−1.19
$Hg^{2+} + 2e^- \rightarrow Hg(\ell)$	+0.85	$Al^{3+} + 3e^- \rightarrow Al(s)$	−1.66
$Ag^+ + e^- \rightarrow Ag(s)$	+0.80	$Mg^{2+} + 2e^- \rightarrow Mg(s)$	−2.37
$Hg_2^{2+} + 2e^- \rightarrow 2Hg(\ell)$	+0.80	$Na^+ + e^- \rightarrow Na(s)$	−2.71
$Fe^{3+} + e^- \rightarrow Fe^{2+}$	+0.77	$Ca^{2+} + 2e^- \rightarrow Ca(s)$	−2.87
$I_2(s) + 2e^- \rightarrow 2I^-$	+0.54	$Sr^{2+} + 2e^- \rightarrow Sr(s)$	−2.89
$Cu^+ + e^- \rightarrow Cu(s)$	+0.52	$Ba^{2+} + 2e^- \rightarrow Ba(s)$	−2.91
$Cu^{2+} + 2e^- \rightarrow Cu(s)$	+0.34	$Cs^+ + e^- \rightarrow Cs(s)$	−2.92
$4H^+ + SO_4^{2-} + 2e^- \rightarrow SO_2(aq) + 2H_2O$	+0.17	$K^+ + e^- \rightarrow K(s)$	−2.93
$Sn^{4+} + 2e^- \rightarrow Sn^{2+}$	+0.15	$Rb^+ + e^- \rightarrow Rb(s)$	−2.98
$2H^+ + 2e^- \rightarrow H_2(g)$	0.00	$Li^+ + e^- \rightarrow Li(s)$	−3.04
$Pb^{2+} + 2e^- \rightarrow Pb(s)$	−0.13		

PHYSICAL CONSTANTS AND CONVERSION FACTORS

Name	Symbol	Value(s)	Units
Angstrom unit	Å	1×10^{-10} m	meter
Avogadro number	N_A	6.02×10^{23} per mol	
Charge of electron	e	1.60×10^{-19} C	coulomb
electron volt	eV	1.60×10^{-19} J	joule
Speed of light	c	3.00×10^{8} m/s	meters/second
Planck's constant	h	6.63×10^{-34} J·s	joule-second
		1.58×10^{-37} kcal·s	kilocalorie-second
Universal gas constant	R	0.0821 L·atm/mol·K	liter-atmosphere/ mole-kelvin
		1.98 cal/mol·K	calories/mole-kelvin
		8.31 J/mol·K	joules/mole-kelvin
Atomic mass unit	μ(amu)	1.66×10^{-24} g	gram
Volume standard, liter	L	1×10^{3} cm^3 = 1 dm^3	cubic centimeters, cubic decimeter
Standard pressure, atmosphere	atm	101.3 kPa	kilopascals
		760 mm Hg	millimeters of mercury
		760 torr	torr
Heat equivalent, kilocalorie	kcal	4.18×10^{3} J	joules

Physical Constants for H$_2$O

Molal freezing point depression . . . . .1.86°C
Molal boiling point elevation . . . . .0.52°C
Heat of fusion . . . . .79.72 cal/g
Heat of vaporization . . . . .539.4 cal/g

VAPOR PRESSURE OF WATER

°C	torr (mm Hg)	°C	torr (mm Hg)
0	4.6	26	25.2
5	6.5	27	26.7
10	9.2	28	28.3
15	12.8	29	30.0
16	13.6	30	31.8
17	14.5	40	55.3
18	15.5	50	92.5
19	16.5	60	149.4
20	17.5	70	233.7
21	18.7	80	355.1
22	19.8	90	525.8
23	21.1	100	760.0
24	22.4	105	906.1
25	23.8	110	1074.6

STANDARD UNITS

Symbol	Name	Quantity
m	meter	length
kg	kilogram	mass
Pa	pascal	pressure
K	kelvin	thermodynamic temperature
mol	mole	amount of substance
J	joule	energy, work, quantity of heat
s	second	time
C	coulomb	quantity of electricity
V	volt	electric potential, potential difference
L	liter	volume

PREFIXES USED WITH SI UNITS
See Table on page 11, "Prefixes Used with SI Units."

Radii of Some Atoms and Ions

	2 IIA	13 IIIA	14 IVA	15 VA	16 VIA	

1

0.037 H

? 1+

0.037 H

0.208

2

Li 0.152 — 1+ 0.060

Be 0.111 — 2+ 0.031

B 0.088 — 3+ 0.020

C 0.077 — 4+ 0.015

N 0.070 — 3− 0.071

O 0.066 — 2− 0.140

F 0.064 — 1− 0.136

3

Na 0.154 — 1+ 0.095

Mg 0.160 — 2+ 0.065

Al 0.143 — 3+ 0.050

Si 0.117 — 4+ 0.041

P 0.110 — 3− 0.212

S 0.104 — 2− 0.184

Cl 0.099 — 1− 0.181

4

K 0.227 — 1+ 0.133

Ca 0.197 — 2+ 0.099

Ga 0.122 — 3+ 0.062

Ge 0.122 — 4+ 0.053

As 0.121 — 3− 0.222

Se 0.116 — 2− 0.198

Br 0.110 — 1− 0.195

5

Rb 0.244 — 1+ 0.148

Sr 0.215 — 2+ 0.113

In 0.163 — 3+ 0.081

Sn 0.141 — 4+ 0.071

Sb 0.141 — 5+ 0.062

Te 0.137 — 2− 0.221

I 0.133 — 1− 0.216

6

Cs 0.265 — 1+ 0.169

Ba 0.217 — 2+ 0.135

Tl 0.170 — 3+ 0.095

Pb 0.175 — 4+ 0.084

Bi 0.155 — 5+ 0.074

Po 0.167

At 0.140

7

Fr 0.270

Ra 0.220 — 2+ 0.152

*Preferred IUPAC designation

Note: The atomic radius is usually given for metal atoms, which are shown in gray, and the covalent radius is usually given for atoms of nonmetals, which are shown in black.

Periodic Table of the Elements

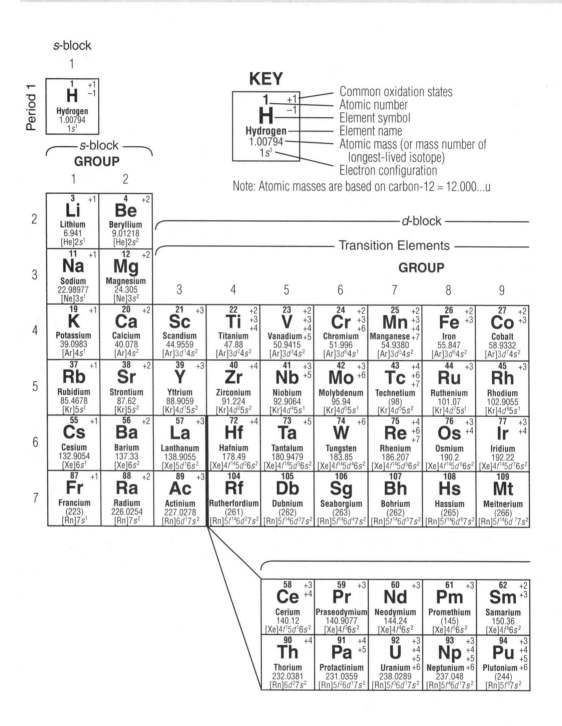

s-block

1

Period 1

1	+1 −1
H	
Hydrogen	
1.00794	
1*s*¹	

KEY

1	+1 −1
H	
Hydrogen	
1.00794	
1*s*¹	

Common oxidation states
Atomic number
Element symbol
Element name
Atomic mass (or mass number of longest-lived isotope)
Electron configuration

Note: Atomic masses are based on carbon-12 = 12.000...u

┌─ *s*-block ─┐
GROUP
1   2

d-block

Transition Elements

GROUP
3   4   5   6   7   8   9

Period	1	2	3	4	5	6	7	8	9
2	3 +1 **Li** Lithium 6.941 [He]2*s*¹	4 +2 **Be** Beryllium 9.01218 [He]2*s*²							
3	11 +1 **Na** Sodium 22.98977 [Ne]3*s*¹	12 +2 **Mg** Magnesium 24.305 [Ne]3*s*²							
4	19 +1 **K** Potassium 39.0983 [Ar]4*s*¹	20 +2 **Ca** Calcium 40.078 [Ar]4*s*²	21 +3 **Sc** Scandium 44.9559 [Ar]3*d*¹4*s*²	22 +2 +3 +4 **Ti** Titanium 47.88 [Ar]3*d*²4*s*²	23 +2 +3 +4 +5 **V** Vanadium 50.9415 [Ar]3*d*³4*s*²	24 +2 +3 +6 **Cr** Chromium 51.996 [Ar]3*d*⁵4*s*¹	25 +2 +3 +4 +7 **Mn** Manganese 54.9380 [Ar]3*d*⁵4*s*²	26 +2 +3 **Fe** Iron 55.847 [Ar]3*d*⁶4*s*²	27 +2 +3 **Co** Cobalt 58.9332 [Ar]3*d*⁷4*s*²
5	37 +1 **Rb** Rubidium 85.4678 [Kr]5*s*¹	38 +2 **Sr** Strontium 87.62 [Kr]5*s*²	39 +3 **Y** Yttrium 88.9059 [Kr]4*d*¹5*s*²	40 +4 **Zr** Zirconium 91.224 [Kr]4*d*²5*s*²	41 +3 +5 **Nb** Niobium 92.9064 [Kr]4*d*⁴5*s*¹	42 +3 +6 **Mo** Molybdenum 95.94 [Kr]4*d*⁵5*s*¹	43 +6 +7 **Tc** Technetium (98) [Kr]4*d*⁵5*s*²	44 +3 **Ru** Ruthenium 101.07 [Kr]4*d*⁷5*s*¹	45 +3 **Rh** Rhodium 102.9055 [Kr]4*d*⁸5*s*¹
6	55 +1 **Cs** Cesium 132.9054 [Xe]6*s*¹	56 +2 **Ba** Barium 137.33 [Xe]6*s*²	57 +3 **La** Lanthanum 138.9055 [Xe]5*d*¹6*s*²	72 +4 **Hf** Hafnium 178.49 [Xe]4*f*¹⁴5*d*²6*s*²	73 +5 **Ta** Tantalum 180.9479 [Xe]4*f*¹⁴5*d*³6*s*²	74 +6 **W** Tungsten 183.85 [Xe]4*f*¹⁴5*d*⁴6*s*²	75 +6 +7 **Re** Rhenium 186.207 [Xe]4*f*¹⁴5*d*⁵6*s*²	76 +3 +4 **Os** Osmium 190.2 [Xe]4*f*¹⁴5*d*⁶6*s*²	77 +3 +4 **Ir** Iridium 192.22 [Xe]4*f*¹⁴5*d*⁷6*s*²
7	87 +1 **Fr** Francium (223) [Rn]7*s*¹	88 +2 **Ra** Radium 226.0254 [Rn]7*s*²	89 +3 **Ac** Actinium 227.0278 [Rn]6*d*¹7*s*²	104 **Rf** Rutherfordium (261) [Rn]5*f*¹⁴6*d*²7*s*²	105 **Db** Dubnium (262) [Rn]5*f*¹⁴6*d*³7*s*²	106 **Sg** Seaborgium (263) [Rn]5*f*¹⁴6*d*⁴7*s*²	107 **Bh** Bohrium (262) [Rn]5*f*¹⁴6*d*⁵7*s*²	108 **Hs** Hassium (265) [Rn]5*f*¹⁴6*d*⁶7*s*²	109 **Mt** Meitnerium (266) [Rn]5*f*¹⁴6*d* ⁷7*s*²

58 +3 +4 **Ce** Cerium 140.12 [Xe]4*f*¹5*d*¹6*s*²	59 +3 **Pr** Praseodymium 140.9077 [Xe]4*f*³6*s*²	60 +3 **Nd** Neodymium 144.24 [Xe]4*f*⁴6*s*²	61 +3 **Pm** Promethium (145) [Xe]4*f*⁵6*s*²	62 +2 +3 **Sm** Samarium 150.36 [Xe]4*f*⁶6*s*²
90 +4 **Th** Thorium 232.0381 [Rn]6*d*²7*s*²	91 +4 +5 **Pa** Protactinium 231.0359 [Rn]5*f*²6*d*¹7*s*²	92 +3 +4 +5 +6 **U** Uranium 238.0289 [Rn]5*f*³6*d*¹7*s*²	93 +3 +4 +5 +6 **Np** Neptunium 237.048 [Rn]5*f*⁴6*d*¹7*s*²	94 +3 +4 +5 +6 **Pu** Plutonium (244) [Rn]5*f*⁶7*s*²

s-block

18

2	0
He	
Helium	
4.00260	
$1s^2$	

p-block
GROUP

10	11	12	13	14	15	16	17	18

5 +3	6 −4, +2, +4	7 −3, −2, −1, +1, +2, +3, +4, +5	8 −2	9 −1	10 0
B	**C**	**N**	**O**	**F**	**Ne**
Boron	Carbon	Nitrogen	Oxygen	Fluorine	Neon
10.81	12.011	14.0067	15.9994	18.998403	20.1797
$[He]2s^22p^1$	$[He]2s^22p^2$	$[He]2s^22p^3$	$[He]2s^22p^4$	$[He]2s^22p^5$	$[He]2s^22p^6$

13 +3	14 −4, +2, +4	15 −3, +3, +5	16 −2, +4, +6	17 −1, +1, +3, +5, +7	18 0
Al	**Si**	**P**	**S**	**Cl**	**Ar**
Aluminum	Silicon	Phosphorus	Sulfur	Chlorine	Argon
26.98154	28.0855	30.97376	32.066	35.453	39.948
$[Ne]3s^23p^1$	$[Ne]3s^23p^2$	$[Ne]3s^23p^3$	$[Ne]3s^23p^4$	$[Ne]3s^23p^5$	$[Ne]3s^23p^6$

28 +2, +3	29 +1, +2	30 +2	31 +3	32 −4, +2, +4	33 −3, +3, +5	34 −2, +4, +6	35 −1, +1, +5	36 0, +2
Ni	**Cu**	**Zn**	**Ga**	**Ge**	**As**	**Se**	**Br**	**Kr**
Nickel	Copper	Zinc	Gallium	Germanium	Arsenic	Selenium	Bromine	Krypton
58.69	63.546	65.39	69.72	72.61	74.9216	78.96	79.904	83.80
$[Ar]3d^84s^2$	$[Ar]3d^{10}4s^1$	$[Ar]3d^{10}4s^2$	$[Ar]3d^{10}4s^24p^1$	$[Ar]3d^{10}4s^24p^2$	$[Ar]3d^{10}4s^24p^3$	$[Ar]3d^{10}4s^24p^4$	$[Ar]3d^{10}4s^24p^5$	$[Ar]3d^{10}4s^24p^6$

46 +2, +4	47 +1	48 +2	49 +3	50 +2, +4	51 −3, +3, +5	52 −2, +4, +6	53 −1, +1, +5, +7	54 0, +2, +4, +6
Pd	**Ag**	**Cd**	**In**	**Sn**	**Sb**	**Te**	**I**	**Xe**
Palladium	Silver	Cadmium	Indium	Tin	Antimony	Tellurium	Iodine	Xenon
106.42	107.8682	112.41	114.82	118.710	121.757	127.60	126.9045	131.29
$[Kr]4d^{10}$	$[Kr]4d^{10}5s^1$	$[Kr]4d^{10}5s^2$	$[Kr]4d^{10}5s^25p^1$	$[Kr]4d^{10}5s^25p^2$	$[Kr]4d^{10}5s^25p^3$	$[Kr]4d^{10}5s^25p^4$	$[Kr]4d^{10}5s^25p^5$	$[Kr]4d^{10}5s^25p^6$

78 +2, +4	79 +1, +3	80 +1, +2	81 +1, +3	82 +2, +4	83 +3, +5	84 +2, +4	85	86 0
Pt	**Au**	**Hg**	**Tl**	**Pb**	**Bi**	**Po**	**At**	**Rn**
Platinum	Gold	Mercury	Thallium	Lead	Bismuth	Polonium	Astatine	Radon
195.08	196.9665	200.59	204.383	207.2	208.9804	(209)	(210)	(222)
$[Xe]4f^{14}5d^96s^1$	$[Xe]4f^{14}5d^{10}6s^1$	$[Xe]4f^{14}5d^{10}6s^2$	$[Xe]4f^{14}5d^{10}6s^26p^1$	$[Xe]4f^{14}5d^{10}6s^26p^2$	$[Xe]4f^{14}5d^{10}6s^26p^3$	$[Xe]4f^{14}5d^{10}6s^26p^4$	$[Xe]4f^{14}5d^{10}6s^26p^5$	$[Xe]4f^{14}5d^{10}6s^26p^6$

110	111	112
Ds	**Rg**	**Uub**
Darmstadtium	Roentgenium	Ununbium
(271)	(272)	(277)
$[Rn]5f^{14}6d^97s^2$	$[Rn]5f^{14}6d^87s^2$	$[Rn]5f^{14}6d^{10}7s^2$

f-block

63 +2, +3	64 +3	65 +3	66 +3	67 +3	68 +3	69 +3	70 +2, +3	71 +3
Eu	**Gd**	**Tb**	**Dy**	**Ho**	**Er**	**Tm**	**Yb**	**Lu**
Europium	Gadolinium	Terbium	Dysprosium	Holmium	Erbium	Thulium	Ytterbium	Lutetium
151.96	157.25	158.9254	162.50	164.9304	167.26	168.9342	173.04	174.967
$[Xe]4f^76s^2$	$[Xe]4f^75d^16s^2$	$[Xe]4f^96s^2$	$[Xe]4f^{10}6s^2$	$[Xe]4f^{11}6s^2$	$[Xe]4f^{12}6s^2$	$[Xe]4f^{13}6s^2$	$[Xe]4f^{14}6s^2$	$[Xe]4f^{14}5d^16s^2$

95 +3, +4, +5, +6	96 +3	97 +3, +4	98 +3	99	100	101	102	103
Am	**Cm**	**Bk**	**Cf**	**Es**	**Fm**	**Md**	**No**	**Lr**
Americium	Curium	Berkelium	Californium	Einsteinium	Fermium	Mendelevium	Nobelium	Lawrencium
(243)	(247)	(247)	(251)	(252)	(257)	(258)	(259)	(262)
$[Rn]5f^77s^2$	$[Rn]5f^76d^17s^2$	$[Rn]5f^97s^2$	$[Rn]5f^{10}7s^2$	$[Rn]5f^{11}7s^2$	$[Rn]5f^{12}7s^2$	$[Rn]5f^{13}7s^2$	$[Rn]5f^{14}7s^2$	$[Rn]5f^{14}6d^17s^2$

*Due to disputes over the discovery of some of the heavier elements, the International Union for Pure and Applied Chemistry (IUPAC) has devised a systematic naming scheme, based on Greek and Latin roots to temporarily name element 112.

The Chemical Elements

(Atomic masses in this table are based on the atomic mass of carbon-12 being exactly 12.)

Name	Symbol	Atomic Number	Atomic Mass
Actinium	Ac	89	(227)
Aluminum	Al	13	26.98
Americium	Am	95	(243)
Antimony	Sb	51	121.75
Argon	Ar	18	39.95
Arsenic	As	33	74.92
Astatine	At	85	(210)
Barium	Ba	56	137.34
Berkelium	Bk	97	(247)
Beryllium	Be	4	9.01
Bismuth	Bi	83	208.98
Bohrium	Bh	107	(262)
Boron	B	5	10.81
Bromine	Br	35	79.90
Cadmium	Cd	48	112.40
Cesium	Cs	55	132.91
Calcium	Ca	20	40.08
Californium	Cf	98	(251)
Carbon	C	6	12.01
Cerium	Ce	58	140.12
Chlorine	Cl	17	35.45
Chromium	Cr	24	52.00
Cobalt	Co	27	58.93
Copper	Cu	29	63.55
Curium	Cm	96	(247)
Darmstadtium	Ds	110	(271)
Dubnium	Db	105	(262)
Dysprosium	Dy	66	162.50
Einsteinium	Es	99	(254)
Erbium	Er	68	167.26
Europium	Eu	63	151.96
Fermium	Fm	100	(257)
Fluorine	F	9	19.00
Francium	Fr	87	(223)
Gadolinium	Gd	64	157.25
Gallium	Ga	31	69.72
Germanium	Ge	32	72.59
Gold	Au	79	196.97
Hafnium	Hf	72	178.49
Hassium	Hs	108	(265)
Helium	He	2	4.00
Holmium	Ho	67	164.93
Hydrogen	H	1	1.008
Indium	In	49	114.82
Iodine	I 5	3	126.90
Iridium	Ir	77	192.2
Iron	Fe	26	55.85
Krypton	Kr	36	83.80
Lanthanum	La	57	138.91
Lawrencium	Lr	103	(261)
Lead	Pb	82	207.19
Lithium	Li	3	6.94
Lutetium	Lu	71	174.97
Magnesium	Mg	12	24.31
Manganese	Mn	25	54.94
Meitnerium	Mt	109	(266)
Mendelevium	Md	101	(258)
Mercury	Hg	80	200.59
Molybdenum	Mo	42	95.94
Neodymium	Nd	60	144.24
Neon	Ne	10	20.18
Neptunium	Np	93	237.05
Nickel	Ni	28	58.71
Niobium	Nb	41	92.90
Nitrogen	N	7	14.01
Nobelium	No	102	(259)
Osmium	Os	76	190.2
Oxygen	O	8	16.00
Palladium	Pd	46	106.4
Phosphorus	P	15	30.97
Platinum	Pt	78	195.09
Plutonium	Pu	94	(244)
Polonium	Po	84	(210)
Potassium	K	19	39.10
Praseodymium	Pr	59	140.90
Promethium	Pm	61	(145)
Protactinium	Pa	91	231.04
Radium	Ra	88	(226)
Radon	Rn	86	(222)
Rhenium	Re	75	186.2
Rhodium	Rh	45	102.91
Roentgenium	Rg	111	(272)
Rubidium	Rb	37	85.47
Ruthenium	Ru	44	101.07
Rutherfordium	Rf	104	(261)
Samarium	Sm	62	150.35
Scandium	Sc	21	44.95
Seaborgium	Sg	106	(263)
Selenium	Se	34	78.96
Silicon	Si	14	28.09
Silver	Ag	47	107.89
Sodium	Na	11	22.99
Strontium	Sr	38	87.62
Sulfur	S	16	32.06
Tantalum	Ta	73	180.95
Technetium	Tc	43	(99)
Tellurium	Te	52	127.60
Terbium	Tb	65	158.92
Thallium	Tl	81	204.37
Thorium	Th	90	232.03
Thulium	Tm	69	168.93
Tin	Sn	50	118.69
Titanium	Ti	22	47.90
Tungsten	W	74	183.85
Uranium	U	92	238.03
Vanadium	V	23	50.94
Xenon	Xe	54	131.30
Ytterbium	Yb	70	173.04
Yttrium	Y	39	88.91
Zinc	Zn	30	65.37
Zirconium	Zr	40	91.22

*A number in parentheses is the mass number of the most stable isotope.

Glossary of Common Terms

(See Index for additional references.)

absolute temperature Temperature measured on the absolute scale, which has its origin at absolute zero. *See also* **Kelvin scale**.

absorption The process of taking up by capillary, osmotic, chemical, or solvent action, as a sponge absorbs water.

acid A water solution that has an excess of hydrogen ions; an acid turns litmus paper pink or red, has a sour taste, and neutralizes bases to form salts.

acid anhydride A nonmetallic oxide that, when placed in water, reacts to form an acid solution.

acid salt A salt formed by replacing part of the hydrogen ions of a dibasic or tribasic acid with metallic ions.

Examples: $NaHSO_4$, NaH_2PO_4.

actinide series The series of radioactive elements starting with actinium, no. 89, and ending with lawrencium, no. 103.

activated charcoal A specially treated and finely divided form of carbon, which possesses a high degree of adsorption.

activation energy The minimum energy necessary to start a reaction.

adsorption The adhesion (in an extremely thin layer) of the molecules of gases, of dissolved substances, or of liquids to the surfaces of solid or liquid bodies with which they come into contact.

alcohol An organic hydroxyl compound formed by replacing one or more hydrogen atoms of a hydrocarbon with an equal number of hydroxyl (OH) groups.

aldehyde An organic compound formed by dehydrating oxidized alcohol; contains the characteristic –CHO group.

alkali Usually, a strong base, such as sodium hydroxide or potassium hydroxide.

alkaline Referring to any substance that has basic properties.

alkyl A substituent obtained from a saturated hydrocarbon by removing one hydrogen atom.

Examples: methyl (CH_3^-, ethyl $C_2H_5^-$).

alkylation The combining of a saturated hydrocarbon with an unsaturated one.

allotropic forms Forms of the same element that differ in their crystalline structures.

alloy A substance composed of two or more metals, which are intimately mixed; usually made by melting the metals together.

alpha particles Positively charged helium nuclei.

alum A double sulfate of a monovalent metal and a trivalent metal, such as $K_2SO_4 \ Al_2(SO_4)_3 \ 24H_2O$; also, a common name for commercial aluminum sulfate.

amalgam An alloy of mercury and another metal.

amine A compound such as CH_3NH_2, derived from ammonia by substituting one or more hydrocarbon substituents for hydrogen atoms.

amino acid One of the "building blocks" of proteins; contains one or more NH_2^- groups that have replaced the same number of hydrogen atoms in an organic acid.

amorphous Having no definite crystalline structure.

amphoteric Referring to a hydroxide that may have either acidic or basic properties, depending on the substance with which it reacts.

analysis The breaking down of a compound into two or more simpler substances.

anhydride A compound derived from another compound by the removal of water; it will combine with water to form an acid (acid anhydride) or a base (basic anhydride).

anhydrous containing no water.

anion An ion or particle that has a negative charge and thus is attracted to a positively charged anode.

anode The electrode in an electrolytic cell that has a positive charge and attracts negative ions.

aromatic compounds A compound whose basic structure contains the benzene ring; it usually has an odor.

aryl A radical obtained from a ring hydrocarbon by removing one hydrogen atom from the ring.

atmosphere The layer of gases surrounding the earth; also, a unit of pressure (1 atm = approx. 760 mm Hg or 1 torr).

atom   The smallest particle of an element that retains the properties of that element and can enter into a chemical reaction.

atomic energy   *See* **nuclear energy**, a more accurate term.

atomic mass   The average mean value of the isotopic masses of the atoms of an element. It indicates the relative mass of the element as compared with that of carbon-12, which is assigned a mass of exactly 12 atomic mass units.

atomic mass unit   One twelfth of the mass of a carbon-12 atom; equivalent to 1.660531×10^{-27} kg (abbreviation: amu or μ).

atomic number   The number that indicates the order of an element in the periodic system; numerically equal to the number of protons in the nucleus of the atom, or the number of negative electrons located outside the nucleus of the atom.

atomic radius   One-half the distance between adjacent nuclei in the crystalline or solid phase of an element; the distance from the atomic nucleus to the valence electrons.

Aufbau Principle   A principle stating that an electron occupies the lowest energy orbital that can receive it.

Avogadro's hypothesis   *See under* **laws**.

Avogadro's number   The number of molecules in 1 gram-molecular volume of a substance, or the number of atoms in 1 gram-atomic mass of an element; equal to 6.022169×10^{23}. *See also* **mole**.

barometer   An instrument, invented by Torricelli in 1643, used for measuring atmospheric pressure.

base   A water solution that contains an excess of hydroxide ions; a proton acceptor; a base turns litmus paper blue and neutralizes acids to form salts.

basic anhydride   A metallic oxide that forms a base when placed in water.

beta particles   High-speed, negatively charged electrons emitted in radiation.

binary   Referring to a compound composed of two elements, such as H_2O.

boiling point   The temperature at which the vapor pressure of a liquid equals the atmospheric pressure.

bond energy   The energy needed to break a chemical bond and form a neutral atom.

bonding   The union of atoms to form compounds or molecules by filling their outer shells of electrons. This can be done through giving and taking electrons (ionic) or by sharing electrons (covalent).

Boyle's Law   *See under* **laws**.

brass   An alloy of copper and zinc.

breeder reactor   A nuclear reactor in which more fissionable material is produced than is used up during operation.

Brownian movement   Continuous zigzagging movement of colloidal particles in a dispersing medium, as viewed through an ultramicroscope.

buffer   A substance that when added to a solution, makes changing the pH of the solution more difficult.

calorie   A unit of heat; the amount of heat needed to raise the temperature of 1 gram of water 1 degree on the Celsius scale.

calorimeter   An instrument used to measure the amount of heat liberated or absorbed during a change.

carbohydrate   A compound of carbon, hydrogen, and oxygen, with hydrogen and oxygen usually present in the ratio of 2 : 1, as in H_2O.

carbonated water   Water containing dissolved carbon dioxide.

carbon dating   The use of radioactive carbon-14 to estimate the age of ancient materials, such as archeological or paleontological specimens.

catalyst   A substance that speeds up a reaction without being permanently changed itself.

cathode   The electrode in an electrolytic cell that is negatively charged and attracts positive ions.

cathode rays   Streams of electrons given off by the cathode of a vacuum tube.

cation   An ion that has a positive charge.

Celsius scale   A temperature scale divided into 100 equal divisions and based on water freezing at 0° and boiling at 100°. Synonymous with centigrade.

chain reaction   A reaction produced during nuclear fission when at least one neutron from each fission produces another fission, so that the process becomes self-sustaining without additional external energy.

Charles's Law   *See under* **laws**.

chemical change   A change that alters the atomic structures of the substances involved and results in different properties.

chemical property   A property that determines how a substance will behave in a chemical reaction.

chemistry   The science concerned with the compositions of substances and the changes that they undergo.

colligative properties   Properties of a solution that depend primarily on the concentration, not the type, of particles present.

colloids   Particles larger than those found in a solution but smaller than those in a suspension.

Combining volumes   *See Gay-Lussac's, under* **laws**.

combustion   A chemical action in which both heat and light are given off.

compound   A substance composed of elements chemically united in definite proportions by weight.

condensation   (a) A change from gaseous to liquid state; (b) the union of like or unlike molecules with the elimination of water, hydrogen chloride, or alcohol.

Conservation of energy   *See under* **laws**.

Conservation of matter   *See under* **laws**.

control rod   In a nuclear reactor, a rod of a certain metal such as cadmium, which controls the speed of the chain reaction by absorbing neutrons.

coordinate covalence   Covalence in which both electrons in a pair come from the same atom.

covalent bonding   Bonding accomplished through the sharing of electrons so that atoms can fill their outer shells.

critical mass   The smallest amount of fissionable material that will sustain a chain reaction.

critical temperature   The temperature above which no gas can be liquefied, regardless of the pressure applied.

crystalline   Having a definite molecular or ionic structure.

crystallization   The process of forming definitely shaped crystals when water is evaporated from a solution of the substance.

cyclotron   A device used to accelerate charged particles to high energies for bombarding the nuclei of atoms.

Dalton's law of partial pressures   *See under* **laws**.

decomposition   The breaking down of a compound into simpler substances or into its constituent elements.

definite composition   *See under* **laws**.

dehydrate   To take water from a substance.

dehydrating agent   A substance able to withdraw water from another substance, thereby drying it.

deliquescence   The absorption by a substance of water from the air so that the substance becomes wet.

denatured alcohol   Ethyl alcohol that has been "poisoned" in order to produce (by avoiding federal tax) a cheaper alcohol for industrial purposes.

density   The mass per unit volume of a substance; the mathematical formula is $D = m/V$, where D = density, m = mass, and V = volume.

destructive distillation   The process of heating an organic substance, such as coal, in the absence of air to break it down into solid and volatile products.

deuterium   An isotope of hydrogen, sometimes called heavy hydrogen, with an atomic mass of 2.

dew point   The highest temperature at which water vapor condenses out of the air.

dialysis   The process of separation of a solution by diffusion through a semipermeable membrane.

diffusion   The process whereby gases or liquids intermingle freely of their own accord.

dipole-dipole attraction   A relatively weak force of attraction between polar molecules; a component of van der Waals forces.

disaccharide   A sugar, like $C_{12}H_{22}O_{11}$, that hydrolyzes to form two molecules of simpler sugars.

displacement   A change by which an element takes the place of another element in a compound.

dissociation (ionic)   The separation of the ions of an electrovalent compound as a result of the action of a solvent.

distillation   The process of first vaporizing a liquid and then condensing the vapor back into a liquid, leaving behind the nonvolatile impurities.

double bond   A bond between atoms involving two electron pairs. In organic chemistry: unsaturated.

double displacement   A reaction in which two chemical substances exchange ions with the formation of two new compounds.

dry ice   Solid carbon dioxide.

ductile   Capable of being drawn into thin wire.

effervescence   The rapid escape of excess gas that has been dissolved in a liquid.

efflorescence   The loss by a substance of its water of hydration on exposure to air at ordinary temperatures.

effusion   The flow of a gas through a small aperture.

Einstein equation   The equation $E = mc^2$, which relates mass to energy; E is energy in ergs, m is mass in grams, and c is velocity of light, 3×10^{10} cm/s.

electrode   A terminal of an electrolytic cell.

electrode potential   The difference in potential between an electrode and the solution in which it is immersed.

electrolysis   The process of separating the ions in a compound by means of electrically charged poles.

electrolysis (electrolytic) cell   A cell in which electrolysis is carried out.

electrolyte   A liquid that will conduct an electric current.

electron   A negatively charged particle found outside the nucleus of the atom; it has a mass of 9.109×10^{-28} g.

electron dot symbol   *See* **Lewis dot symbol**.

electronegativity   The numerical expression of the relative strength with which the atoms of an element attract valence electrons to themselves; the higher the number, the greater the attraction.

electron volt   A unit for expressing the kinetic energy of subatomic particles; the energy acquired by an electron when it is accelerated by a potential difference of 1 V; equals 1.6×10^{-12} erg or 23.1 kcal/mol (abbreviation: eV).

electroplating   Depositing a thin layer of (usually) a metallic element on the surface of another metal by electrolysis.

electrovalence   The bond of attraction between negatively and positively charged ions.

element   One of the more than 100 "building blocks" of which all matter is composed. An element consists of atoms of only one kind and cannot be decomposed further by ordinary chemical means.

empirical formula   A formula that shows only the simplest ratio of the numbers and kinds of atoms, such as CH_4.

emulsifying agent   A colloidal substance that forms a film about the particles of two immiscible liquids so that one liquid remains suspended in the other.

emulsion   A suspension of fine particles or droplets of one liquid in another, the two liquids being immiscible in each other; the droplets are surrounded by a colloidal (emulsifying) agent.

endothermic   Referring to a chemical reaction that results in an overall absorption of heat from its surroundings.

energy   The capacity to do work. In every chemical change energy is either given off or taken in. Forms of energy are heat, light, motion, sound, and electrical, chemical, and nuclear energy.

enthalpy   The heat content of a chemical system.

entropy   The measure of the randomness that exists in a system.

equation   A shorthand method of showing the changes that take place in a chemical reaction.

equilibrium   The point in a reversible reaction at which the forward reaction is occurring at the same rate as the opposing reaction.

erg   A unit of energy or work done by a force of 1 dyne (1/980 g of force) acting through a distance of 1 cm; equals 2.4×10^{-11} kcal.

ester   An organic salt formed by the reaction of an alcohol with an organic (or inorganic) acid.

esterification   A chemical reaction between an alcohol and an acid in which an ester is formed.

ether   An organic compound containing the —O— group.

eudiometer   A graduated glass tube into which gases are placed and subjected to an electric spark; used to measure the individual volumes of combining gases.

evaporation   The process in which molecules of a liquid (or a solid) leave the surface in the form of vapor.

exothermic   Referring to a chemical reaction that results in the giving off of heat to its surroundings.

Fahrenheit scale   The temperature scale that has 32° as the freezing point of water and 212° as the boiling point.

fallout   The residual radioactivity from an atmospheric nuclear test, which eventually settles on the surface of the earth.

Faraday's Law   *See under* **laws**.

filtration   The process by which suspended matter is removed from a liquid by passing the liquid through a porous material.

First Law of Thermodynamics   *See under* **laws**.

fission   A nuclear reaction that releases energy as a result of the splitting of large nuclei into smaller ones.

fixation of nitrogen   Any process for converting atmospheric nitrogen into compounds such as ammonia and nitric acid.

flame   The glowing mass of gas and luminous particles produced by the burning of a gaseous substance.

flammable   Capable of being easily set on fire; combustible (same as *inflammable*).

fluorescence   Emission by a substance of electromagnetic radiation, usually visible, as the immediate result of (and only during) absorption of energy from another source.

fluoridation   Addition of small amounts of fluoride (usually NaF) to drinking water to help prevent tooth decay.

flux   In metallurgy: a substance that helps to melt and remove the solid impurities as slag. In soldering: a substance that cleans the surface of the metal to be soldered. In nucleonics: the concentration of nuclear particles or rays.

formula   An expression that uses the symbols for elements and subscripts to show the basic makeup of a substance.

formula mass   The sum of the atomic masses of all the atoms (or ions) contained in a formula.

fractional crystallization   The separation of the components in a mixture of dissolved solids by evaporation according to individual solubilities.

fractional distillation   The separation of the components in a mixture of liquids having different boiling points by vaporization.

free energy   *See* **Gibbs free energy**.

freezing point   The specific temperature at which a given liquid and its solid form are in equilibrium.

fuel   Any substance used to furnish heat by combustion. *See also* **nuclear fuel**.

fuel cell   A device for converting an ordinary fuel such as hydrogen or methane directly into electricity.

functional group   A group of atoms that characterizes certain types of organic compounds, such as —OH for alcohols, and that reacts more or less independently.

fusion   A nuclear reaction that releases energy as a result of the union of smaller nuclei to form larger ones.

fusion melting   Changing a solid to the liquid state by heating.

galvanizing   Applying a coating of zinc to iron or steel to protect the latter from rusting.

gamma rays   A type of radiation consisting of high-energy waves that can pass through most materials.

gas   A phase of matter that has neither definite shape nor definite volume.

Gay-Lussac's Law   *See under* **laws**.

Gibbs free energy   Changes in Gibbs free energy, ΔG, are useful in indicating the conditions under which a chemical reaction will occur. The equation is $\Delta G = \Delta H - T\Delta S$, where ΔH is the change in enthalpy and ΔS is the change in entropy. If ΔG is negative, the reaction will proceed spontaneously to equilibrium.

glass   An amorphous, usually translucent substance consisting of a mixture of silicates. Ordinary glass is made by fusing together silica and sodium carbonate and lime; the various forms of glass contain many other silicates.

Graham's Law   *See under* **laws**.

gram   A unit of weight in the metric system; the weight of 1 mL of water at 4°C (abbreviation: g).

gram-atomic mass   The atomic mass, in grams, of an element.

gram-molecular mass   The molecular mass, in grams, of a substance.

gram-molecular volume   The volume occupied by 1 gram-molecular mass of a gaseous substance at standard conditions; equals 22.4 L.

group   A vertical column of elements in the periodic table that generally have similar properties.

Haber process   A catalytic method for the union of atmospheric nitrogen with hydrogen to form ammonia.

half-life   The time required for half of the mass of a radioactive substance to disintegrate.

half-reaction   One of the two parts, either the reduction part or the oxidation part, of a redox reaction.

halogen   Any of the five nonmetallic elements (fluorine, chlorine, bromine, iodine, astatine) that form part of Group 17 (Group VII) of the periodic table.

heat   A form of molecular energy; it passes from a warmer body to a cooler one.

heat capacity (specific heat)   The quantity of heat, in calories, needed to raise the temperature of 1 g of a substance 1° on the Celsius scale.

heat of formation   The quantity of heat either given off or absorbed in the formation of 1 mol of a substance from its elements.

heat of fusion   The amount of heat, in calories, required to melt 1 g of a solid; for water, 80 cal.

heat of vaporization   The quantity of heat needed to vaporize 1 g of a liquid at constant temperature and pressure; for water at 100°C, 540 cal.

heavy water (deuterium oxide, D_2O)   Water in which the hydrogen atoms are replaced by atoms of the isotope of hydrogen, deuterium.

Henry's Law   *See under* **laws**.

homogeneous   Uniform; having every portion exactly like every other portion.

homologous   Alike in structure; referring to series of organic compounds, such as hydrocarbons, in which each member differs from the next by the addition of the same group.

humidity   The amount of moisture in the air.

hybridization   The combination of two or more orbitals to form new orbitals.

hydrate   A compound that has water molecules included in its crystalline makeup.

hydride   Any binary compound containing hydrogen, such as HCl.

hydrogenation   A process in which hydrogen is made to combine with another substance, usually organic, in the presence of a catalyst.

hydrogen bond   A weak chemical linkage between the hydrogen of one polar molecule and the oppositely charged portion of a closely adjacent molecule.

hydrolysis   Of carbohydrates: the action of water in the presence of a catalyst upon one carbohydrate to form simpler carbohydrates. Of salts: a reaction involving the splitting of water into its ions by the formation of a weak acid, a weak base, or both.

hydronium ion   A hydrated ion, $H_2O \cdot H^+$ or H_3O^+.

hydroponics   Growing plants without the use of soil, as in nutrient solution or in sand irrigated with nutrient solution.

hydroxyl   Referring to the —OH radical.

hygroscopic   Referring to the ability of a substance to draw water vapor from the atmosphere to itself and become wet.

hypothesis   A possible explanation of the nature of an action or phenomenon; a hypothesis is not as completely developed as a theory.

Ideal Gas Law   *See under* **laws**.

immiscible   Referring to the inability of two liquids to mix.

indicator   A dye that shows one color in the presence of the hydrogen ion (acid) and a different color in the presence of the hydroxyl ion (base).

inertia   The property of matter whereby it remains at rest or, if in motion, remains in motion in a straight line unless acted upon by an outside force.

ion   An atom or a group of combined atoms that carries one or more electric charges.
Examples: NH_4^+, OH^-.

ionic bonding   The bonding of ions as a result of their opposite charges.

ionic equation   An equation showing a reaction among ions.

ionization   The process in which ions are formed from neutral atoms.

ionization equation   An equation showing the ions set free from an electrolyte.

isomerization   The rearrangement of atoms in a molecule to form isomers.

isomers   Two or more compounds having the same percentage composition but different arrangements of atoms in their molecules and hence different properties.

isotopes   Two or more forms of an element that differ only in the number of neutrons in the nucleus and hence in their mass numbers.

IUPAC   International Union of Pure and Applied Chemistry, an organization that establishes standard rules for naming compounds.

joule   The SI unit of work or of energy equal to work done; 1 joule = 0.2388 calories.

Kelvin scale   A temperature scale based on water freezing at 273 and boiling at 373 Kelvin units; its origin is absolute zero. Synonymous with *absolute scale*.

kernel (atomic)   The nucleus and all the electron shells of an atom except the outer one; usually designated by the symbol for the atom.

ketone   An organic compound containing the —CO— group.

kilocalorie   A unit of heat; the amount of heat needed to raise the temperature of 1 kg of water 1° on the Celsius scale.

kindling temperature   The temperature to which a given substance must be raised before it ignites.

Kinetic Molecular Theory   The theory that all molecules are in motion; this motion is most rapid in gases, less rapid in liquids, and very slow in solids.

lanthanide series   The "rare earth" series of elements starting with lanthanum, no. 57, and ending with lutetium, no. 71.

law (in science)   A generalized statement about the uniform behavior in natural processes.

laws

Avogadro's   Equal volumes of gases under identical conditions of temperature and pressure contain equal numbers of particles (atoms, molecules, ions, or electrons).

Boyle's   The volume of a confined gas is inversely proportional to the pressure to which it is subjected, provided that the temperature remains the same.

Charles's   The volume of a confined gas is directly proportional to the absolute temperature, provided that the pressure remains the same.

Combining Volumes   See *Gay-Lussac's* under **laws**.

Conservation of Energy   Energy can be neither created nor destroyed, so that the energy of the universe is constant.

Conservation of Matter   Matter can be neither created nor destroyed (or weight remains constant in an ordinary chemical change).

Dalton's   When a gas is made up of a mixture of different gases, the pressure of the mixture is equal to the sum of the partial pressures of the components.

Definite Composition   A compound is composed of two or more elements chemically combined in a definite ratio by weight.

Faraday's   During electrolysis, the weight of any element liberated is proportional (1) to the quantity of electricity passing through the cell, and (2) to the equivalent weight of the element.

First Law of Thermodynamics   The total energy of the universe is constant and cannot be created or destroyed.

Gay-Lussac's   The ratio between the combining volumes of gases and the product, if gaseous, can be expressed in small whole numbers.

Graham's   The rate of diffusion (or effusion) of a gas is inversely proportional to the square root of its molecular mass.

Henry's   The solubility of a gas (unless the gas is very soluble) is directly proportional to the pressure applied to the gas.

Hess's Law   If a series of reactions are added together, the enthalpy change for the net reaction will be the sum of the enthalpy changes for the individual steps.

Ideal Gas   Any gas that obeys the gas laws perfectly. No such gas actually exists.

Multiple Proportions   When any two elements, A and B, combine to form more than one compound, the different weights of B that unite with a fixed weight of A bear a small whole-number ratio to each other.

Periodic   The chemical properties of elements vary periodically with their atomic numbers.

Second Law of Thermodynamics   Heat cannot, of itself, pass from a cold body to a hot body.

Le Châtelier's Principle   If a stress is placed on a system in equilibrium, the system will react in the direction that relieves the stress.

leptons   Elementary particles that are believed to make up the electron and neutrino.

Lewis dot symbol   The chemical symbol (kernel) for an atom, surrounded by dots to represent its outer level electrons.

Examples: K·, Sr:.

liquid   A phase of matter that has a definite volume but takes the shape of the container.

liquid air   Air that has been cooled and compressed until it liquefies.

litmus   An organic substance obtained from the lichen plant and used as an indicator; it turns red in acidic solution and blue in basic solution.

London dispersion force   The weakest of the van der Waal forces between molecules. These weak, attractive forces become apparent only when the molecules approach one another closely (usually at low temperatures and high pressure). They are due to the way the positive charges of one molecule attract the negative charges of another molecule because of the charge distribution at any one instant.

luminous   Emitting a steady, suffused light.

malleable   Capable of being hammered or pounded into thin sheets.

manometer   A U-tube (containing mercury or some other liquid) used to measure the pressure of a confined gas.

mass   The quantity of matter that a substance possesses; it can be measured by its resistance to a change in position or motion and is not related to the force of gravity.

mass number   The nearest whole number to the combined atomic mass of the individual atoms of an isotope when that mass is expressed in atomic mass units.

mass spectograph   A device for determining the masses of electrically charged particles by separating them into distinct streams by means of magnetic deflection.

matter   A substance that occupies space, has mass, and cannot be created or destroyed easily.

melting   The change in phase of a substance from solid to liquid.

melting point   The specific temperature at which a given solid changes to a liquid.

meson   Any unstable, elementary nuclear particle having a mass between that of an electron and that of a proton.

metal   (a) An element whose oxide combines with water to form a base; (b) an element that readily loses electrons and acquires a positive valence.

metallurgy   The process involved in obtaining a metal from its ores.

meter   The basic unit of length in the metric system; defined as 1,650,763.73 times the wavelength of krypton-86 when excited to give off an orange-red spectral line.

MeV   A unit for expressing the kinetic energy of subatomic particles; equals 10^6 V.

micron   One thousandth of a millimeter (abbreviation: μ).

millibar   A unit of pressure used in meteorology; equal 1000 dynes/cm^2.

millimicron   One millionth of a millimeter (abbreviation: mμ).

mineral   An inorganic substance of definite composition found in nature.

miscible   Referring to the ability of two fluids to mix with one another in all portions.

mixture   A substance composed of two or more components, each of which retains its own properties.

moderator   A substance such as graphite, paraffin, or heavy water used in a nuclear reactor to slow down neutrons.

molal solution   A solution containing 1 mol of solute in 1000 g of solvent (indicated by m).

molar mass   The mass arrived at by the addition of the atomic masses of the units that make up a molecule of an element or compound. Expressed in grams, the molar mass of a gaseous substance at STP occupies a molar volume equal to 22.4 L.

molar solution   A solution containing 1 mol of solute in 1000 mL of solution (indicated by M).

mole   A unit of quantity that consists of 6.02×10^{23} particles. Abbreviation is mol.

mole fraction   Moles of one substance divided by the total moles of solution parts.

molecular mass   The sum of the atomic masses of all the atoms in a molecule of a substance.

molecular theory   *See* **Kinetic Molecular Theory**.

molecule   The smallest particle of a substance that retains the physical and chemical properties of that substance.

monobasic acid   An acid having only one hydrogen atom which can be replaced by a metal or a positive radical.

monosaccharide   A simple sugar, such as $C_6H_{12}O_6$.

mordant   A chemical, such as aluminum sulfate, used for fixing colors on textiles.

multiple proportions   *See under* **laws**.

nascent (atomic)   Referring to an element in the atomic form as it has just been liberated in a chemical reaction.

neutralization   The union of the hydrogen ion of an acid and the hydroxyl ion of a base to form water.

neutron   A subatomic particle found in the nucleus of the atom; it has no charge and has the same mass as the proton.

neutron capture   A nuclear reaction in which a neutron attaches itself to a nucleus; a gamma ray is usually emitted simultaneously.

nitriding   A process in which ammonia or a cyanide is used to produce case-hardened steel; a nitride is formed instead of a carbide.

nitrogen fixation   Any process by which atmospheric nitrogen is converted into a compound such as ammonia or nitric acid.

noble gas   A gaseous element that has a complete outer level of electrons; any of a group of rare gases (helium, neon, argon, krypton, etc.) that exhibit great stability and very low reaction rates.

nonelectrolyte   A substance whose solution does not conduct a current of electricity.

nonmetal   (a) An element whose oxide reacts with water to form an acid; (b) an element that takes on electrons and acquires a negative valence.

nonpolar compound   A compound in whose molecules the atoms are arranged symmetrically so that the electric charges are uniformly distributed.

normal salt   A salt in which all the hydrogen of the acid has been displaced by a metal.

normal solution   A solution that contains 1 gram of H^+ (or its equivalent: 17 g of OH^-, 23 g of Na^+, 20 g of Ca^{2+}, etc.) in 1 L of solution (indicated by N).

nuclear energy   The energy released by spontaneously or artificially produced fission, fusion, or disintegration of the nuclei of atoms.

nuclear fuel   A substance that is consumed during nuclear fission or fusion.

nuclear reaction   Any reaction involving a change in nuclear structure.

nuclear reactor   A device in which a controlled chain reaction of fissionable material can be produced.

nucleonics   The science that deals with the constituents and all the changes in the atomic nucleus.

nucleus   The center of the atom, which contains protons and neutrons.

nuclide   A species of atom characterized by the constitution of its nucleus.

octane number   A conventional rating for gasoline based on its behavior as compared with that of isooctane as 100.

orbital   A subdivision of a nuclear shell; it may contain none, one, or two electrons.

ore   A natural mineral substance from which an element, usually a metal, may be obtained with profit.

organic acid   An organic compound that contains the —COOH group.

organic chemistry   The branch of chemistry dealing with carbon compounds, usually those found in nature.

oxidation   The chemical process by which oxygen is attached to a substance; the process of losing electrons.

oxidation number (state)   A positive or negative number representing the charge that an ion has or an atom appears to have when its electrons are counted according to arbitrarily accepted rules: (1) electrons shared by two unlike atoms are counted with the more electronegative atom; (2) electrons shared by two like atoms are divided equally between the atoms.

oxidation potential   An electrode potential associated with the oxidation half-reaction.

oxidizing agent   A substance that (a) gives up its oxygen readily, (b) removes hydrogen from a compound, (c) takes electrons from an element.

ozone   An allotropic and very active form of oxygen, has the formula O_3.

paraffin series   The methane series of hydrocarbons.

pascal   The SI unit of pressure equal to 1 newton per square meter.

pasteurization   Partial sterilization of a substance, such as milk, by heating to approximately 65°C for $\frac{1}{2}$ hour.

Pauli Exclusion Principle   Each electron orbital of an atom can be filled by only two electrons, each with an opposite spin.

period   A horizontal row of elements in the periodic table.

periodic law   *See under* **laws**.

petroleum (meaning "oil from stone")   A complex mixture of gaseous, liquid, and solid hydrocarbons obtained from the earth.

pH   A numerical expression of the hydrogen or hydronium ion concentration in a solution; defined as $-\log[H^+]$, where $[H^+]$ is the concentration of hydrogen ions, in moles per liter.

phenolphthalein   An organic indicator; it is colorless in acid solution and red in the presence of OH^- ions.

photosynthesis   The reaction taking place in all green plants that produces glucose from carbon dioxide and water under the catalytic action of chlorophyll in the presence of light.

physical change   A change that does not involve any alteration in chemical composition.

physical property   A property of a substance arrived at through observation of its smell, taste, color, density, and so on, which does not relate to chemical activity.

pi bond   A bond between p orbitals.

pile   A general term for a nuclear reactor; specifically, a graphite-moderated reactor in which uranium fuel is distributed throughout a "pile" of graphite blocks.

pitchblende   A massive variety of uraninite that contains a small amount of radium.

plasma   Very hot ionized gases.

polar covalent bond   A bond in which electrons are closer to one atom than to another. *See also* **polar molecule**.

polar dot structure   Representation of the arrangements of electrons around the atoms of a molecule in which the polar characteristics are shown by placing the electrons closer to the more electronegative atom.

polar molecule   A molecule that has differently charged areas because of unequal sharing of electrons.

polyatomic ion   A group of chemically united atoms that react as a unit and have an electric charge.

polymerization   The process of combining several molecules to form one large molecule (polymer). (a) Additional polymerization: The addition of unsaturated molecules to each other. (b) Condensation polymerization: The reaction of two molecules by loss of a molecule of water.

polysaccharide   A large complex molecule with the general formula $(C_6H_{10}O_5)_n$.

positron   A positively charged particle of electricity with about the same weight as the electron.

potential energy   Energy due to the position of a body or to the configuration of its particles.

precipitate   An insoluble compound formed in the chemical reaction between two or more substances in solution.

proteins   Large, complex organic molecules, with nitrogen an essential part, found in plants and animals.

proton   A subatomic particle found in the nucleus that has a positive charge.

qualitative analysis   A term applied to the methods and procedures used to determine any or all of the constituent parts of a substance.

quantitative analysis   A term applied to the methods and procedures used to determine the definite quantity or percentage of any or all of the constituent parts of a substance.

quark   A subatomic particle with a fractional charge of one third or two thirds the charge of an electron. Six quarks have been described.

quenching   Cooling a hot piece of metal rapidly, as in water or oil.

radiation   The emission of particles and rays from a radioactive source; usually alpha and beta particles and gamma rays.

radioactive   Referring to substances that have the ability to emit radiations (alpha or beta particles or gamma rays).

radioisotopes   Isotopes that are radioactive, such as uranium-235.

reactant   A substance involved in a reaction.

reaction   A chemical transformation or change. The four basic types are combination (synthesis), decomposition (analysis), single replacement or single displacement, and double replacement or double displacement.

reaction potential   The sum of the oxidation potential and reduction potential for a particular reaction.

reagent   Any chemical taking part in a reaction.

recrystallization   A series of crystallizations, repeated for the purpose of greater purification.

redox   A shortened name for a reaction that involves reduction and oxidation.

reducing agent   From an electron standpoint, a substance that loses its valence electrons to another element; a substance that is readily oxidized.

reduction   A chemical reaction that removes oxygen from a substance; a gain of electrons.

reduction potential   An electrode potential associated with a reduction half-reaction.

refraction (of light)   The bending of light rays as they pass from one material into another.

relative humidity   The ratio, expressed in percent, between the amount of water vapor in a given volume of air and the amount the same volume can hold when saturated at the same temperature.

resonance   The phenomenon in a molecular structure that exhibits properties between those of a single bond and those of a double bond and thus possesses two or more alternate structures.

reversible reaction   Any reaction that reaches an equilibrium or that can be made to proceed from right to left as well as from left to right.

roasting   Heating an ore (usually a sulfide) in an excess of air to convert the ore to an oxide, which can then be reduced.

saccharin $C_6H_4COSO_2NH$, a white, crystalline, organic compound derived from coal tar, and 400 times sweeter than cane sugar.

salt   A compound, such as NaCl, made up of a positive metallic ion and a negative nonmetallic ion or radical.

saponification The reaction taking place when an alkali reacts with a fat or vegetable oil; used to make soap.

saturated solution A solution that contains the maximum amount of solute under the existing temperature and pressure.

secondary cell A cell (battery) that can be regenerated or "charged."

Second Law of Thermodynamics *See under* **laws**.

shell A level outside the nucleus of an atom that electrons can inhabit without expending energy.

sigma bond A bond between *s* orbitals or between an *s* orbital and another kind of orbital.

significant figures All the certain digits recorded in a measurement plus one uncertain digit.

slag The product formed when the flux reacts with the impurities of an ore in a metallurgical process.

solid A phase of matter that has a definite size and shape.

solubility A measure of the amount of solute that will dissolve in a given quantity of solvent at a given temperature.

solute The material that is dissolved to make a solution.

solution A uniform mixture of a solute in a solvent.

solvent The dispersing substance that allows the solute to go into solution.

specific gravity (mass) The ratio between the mass of a certain volume of a substance and the mass of an equal volume of water (or, in the case of gases, an equal volume of air); expressed as a single number.

specific heat The ratio between the number of calories needed to raise the temperature of a certain mass of a substance 1° on the Celsius scale and the number of calories needed to raise the temperature of the same mass of water 1° on the Celsius scale.

spectroscope An instrument used to analyze light by separating it into its component wavelengths.

spectrum The image formed when radiant energy is dispersed by a prism or grating into its various wavelengths.

spinthariscope A device for viewing through a microscope the flashes of light made by particles from radioactive materials against a sensitized screen.

spontaneous combustion (ignition) The process in which slow oxidation produces enough heat to raise the temperature of a substance to its kindling temperature.

stable Referring to a substance not easily decomposed or dissociated.

standard conditions An atmospheric pressure of 760 mm or torr or 1 atm (mercury pressure) and a temperature of 0°C (273 K) (abbreviation: STP).

stratosphere The upper portion of the atmosphere, in which the temperature changes little with altitude and clouds of water never form.

strong acid (or base) An acid (or a base) capable of a high degree of ionization in water solution. Example: sulfuric acid, sodium hydroxide.

structural (graphic) formula A pictorial representation of the atomic arrangement of a molecule.

sublime To vaporize directly from the solid to the gaseous state and then condense back to the solid.

substance A single kind of matter, element, or compound.

substitution products Products formed by the substitution of other elements or radicals for hydrogen atoms in hydrocarbons.

sulfation An accumulation of lead sulfate on the plates and at the bottom of a (lead) storage cell.

supersaturated solution A solution that contains a greater quantity of solute than is normally possible at a given temperature.

suspension A mixture of finely divided solid material in a liquid, from which the solid settles on standing.

symbol A letter or letters representing an element of the periodic table. Examples: O, Mn.

synthesis The chemical process of forming a substance from its individual parts.

Systéme International d'Unitès (SI) The modernized metric system of measurements universally used by scientists. There are seven base units: kilogram, meter, second, ampere, kelvin, mole, and candela. Called SI units.

temperature The intensity or the degree of heat of a body, measured by a thermometer.

tempering The heating and then rapid cooling of a metal to increase its hardness.

ternary Referring to a compound composed of three different elements, such as H_2SO_4.

theory An explanation used to interpret the "mechanics" of nature's actions; a theory is more fully developed than a hypothesis.

thermochemical equation An equation that includes values for the calories absorbed or evolved.

thermoplastic Capable of being softened by heat; may be remolded.

thermosetting Capable of being permanently hardened by heat and pressure; resistant to the further effects of heat.

tincture An alcoholic solution of a substance, such as a tincture of iodine.

torr A unit of pressure defined as 1 mm Hg; 1 torr equals 133.32 Pa.

tracer A minute quantity of radioactive isotope used in medicine and biology to study chemical changes within living tissues.

transmutation Conversion of one element into another, either by bombardment or by radioactive disintegration.

tribasic acid An acid that contains three replaceable hydrogen atoms in its molecule, such as H_3PO_4.

tritium A very rare, unstable, "triple-weight" hydrogen isotope (H^3) that can be made synthetically.

ultraviolet light The portion of the spectrum that lies just beyond the violet; therefore of short wavelength.

USP (United States Pharmacopeia) chemicals Chemicals certified as having a standard of purity that demonstrates their fitness for use in medicine.

valence The combining power of an element; the number of electrons gained, lost, or borrowed in a chemical reaction.

valence electrons The electrons in the outermost level or levels of an atom that determine its chemical properites.

van der Waals forces Weak attractive forces existing between molecules.

vapor The gaseous phase of a substance that normally exists as a solid or liquid at ordinary temperatures.

vapor density The density of a gas in relation to another gas taken as a standard under identical conditions of temperature and pressure.

vapor pressure The pressure exerted by a vapor given off by a confined liquid or solid when the vapor is in equilibrium with its liquid or solid form.

volatile Easily changed to a gas or a vapor at relatively low pressure.

volt A unit of electrical potential or voltage, equal to the difference of potential between two points in a conducting wire carrying a constant current of 1 ampere when the power dissipated between these two points is equal to 1 watt (abbreviation: V).

volume The amount of three-dimensional space occupied by a substance.

VSEPR Valence shell electron pair repulsion. This model explains the non-90° variations of bond angles for *p* orbitals in the outer energy levels of atoms in molecules because of electron repulsions.

vulcanization The process of combining sulfur or other additives, using heat, with rubber to improve its properties, particularly to give it added hardness.

water of hydration Water that is held in chemical combination in a hydrate and can be removed without essentially altering the composition of the substance. *See also* **hydrate**.

weak acid (or base) An acid (or base) capable of being only slightly ionized in an aqueous solution. Examples: Acetic acid, ammonium hydroxide.

weak electrolyte A substance that, when dissolved in water, ionizes only slightly and hence is a poor conductor of electricity.

weight The measure of the force with which a body is attracted toward the earth by gravity.

work The product of the force exerted on a body and the distance through which the force acts; expressed mathematically by the equation $W = Fs$, where W = work, F = force, and s = distance.

X-rays Penetrating radiations, of extremely short wavelength, emitted when a stream of electrons strikes a solid target in a vacuum tube.

zeolite A natural or synthesized silicate used to soften water.

Index

history of, 31–32
hydrogen, 36
radii of, 54–55
structure of, 35–37
Aufbau Principle, 46, 48
Automobile lead storage battery, 273
Average atomic mass, 37
Avogadro, Amedeo, 136
Avogadro's Hypothesis, 136
Avogadro's Law, 136, 142
Avogadro's Number, 134

B

Balanced reaction equations, 139
Balancing equations, 87–90, 274
Balmer series, 41
Barometer, 115–116
Barometric pressure, 115
Bases
    Arrhenius Theory of, 235–236
    concentration of, 238–239
    conjugate, 236–238
    definition of, 235
    degree of ionization, 235
    Lewis Theory of, 238
Battery, 273
Becquerel, Henri, 32, 347–348
Benzene, 323
Benzoic acid, 330
Beta decay, 353
Beta particles
    characteristics of, 349
    detection methods for, 350–351
    nature of, 348–349
Binary compounds
    naming of, 90–91
    writing of, 87
Bismuth, 294
Bohr, Niels, 32, 34
Bohr model, 34–40, 42
Boiling point, 154
Bond
    covalent, 65–67

double, 70
electrovalence, 64–65
hydrogen, 69–70
ionic, 64–65
metallic, 68
pi, 78–79, 319–320
sigma, 78–79
triple, 70
Bond dissociation energy, 199
Bond energies, 199
Bone black, 309
Boundary surface diagram, 44
Boyle, Robert, 120
Boyle's Law, 120–121
Brass, 297–298
Broglie, Louis de, 32, 43
Bromide, 374
Bromine, 289
Brønsted-Lowry Theory, 236
Bronze, 297–298
Brownian movement, 152, 169
Bubble chamber, 351
Buffer solutions, 244
Bureau of Reclamation, 159
Buret setup for titration, 241
Burning, 261
n-Butane, 314, 324

C

Calories, 15
Carbohydrates, 336
Carbolic acid, 328
Carbon
    allotropes of, 309
    forms of, 307–309
Carbon dioxide, 86, 234, 310–312, 373
Carbon monoxide, 86, 373
Carbon-14, 353
Carbonate, 310, 374
Carboxylic acid, 329–330, 335
Catalysts, 110, 207–208
Cathode, 263, 273
Cations, 253

Cavendish, Henry, 112, 156
Cell potential, 264
Cellulose, 337
Celsius, 13
Chadwick, James, 34, 35
Chain reaction of fission, 354
Change of phase, 153
Charcoal, 309
Charles, Jacques, 119
Charles's Law, 119–120
Chemical bond, 63
Chemical changes, 6
Chemical formulas
  balancing, 88–90
  definition of, 86–87, 95
  flowchart for, 94
  naming compounds from, 90–91
  writing of, 87
Chemical properties, 6
Chemical reactions. *See* Reactions
Chloride, 290, 374
Chromatography, 369
Citric acid, 331
Clean Air Act of 1967, 246
Cloud chamber, 350
Coal tars, 308
Cockcroft, John Sir, 354
Coke, 309
Colligative properties, 174–177
Collision theory of reaction rates, 207
Colloids, 169
Combination reaction, 187–189
Combined gas law, 121–122
Combustion, 256
Combustion reactions, 261–262
Commercial galvanic cells, 272–273
Common ion effect, 225–226
Complex organic acids, 331
Compound
  binary, 87, 94
  definition of, 4
  formula mass of, 95
  molecular mass of, 95

naming of, 90–91
other than binary, 87, 94
writing of, 86–87, 94
Concentrated, 169
Concentration, expressions of, 169–170
Condensation polymerization, 337–338
Conjugate acids, 236–238
Conjugate base, 236–238
Conservation of energy, 9
Conservation of mass, 7
Coordinate covalent bonds, 66–67
Copper, 190, 297
Cosmic-ray levels, 357
Covalent bonds, 65–67
Covalent radius, 54
Cracking, 324
Crisscross and reduce method, 88–90
Critical mass, 355
Critical pressure, 155
Critical temperature, 155
Crystal, 177
Crystal structure, 298
Crystalline, allotropes of carbon, 309
Crystallization, 177–178
Curie, Marie, 32, 348
Curie, Pierre, 32, 348
Current, 271
Cycloalkanes, 317

D

Dalton, John, 31, 85
Dalton's Law of Partial Pressures,
  122–123
Dead Sea Scrolls, 353
n-Decane, 314
Decay series, 351–353
Decomposition reaction, 188, 190
Deflection of radioactive emissions,
  348
Degree of ionization
  common acids, 233–234
  common bases, 235
Dehydrogenation, 326

Deliquescent, 179
Density, 3, 138
Derived quantity, 12
Desalinization, 159
Destructive distillation, 308
Deuterium, 36, 157
Dextrin, 337
Diamond, 307–309
Diffusion, 118
Dilute, 169
Dilution, 174
Dimensional analysis, 17, 133
Dimethylbenzene, 324
Dipole-dipole attraction, 68
Dipoles, 66, 68, 168
Direct combination of elements, 245
Disaccharides, 336
Dissociation, 252
Distillation of water, 158
Distinct substances, 4–5
Double bond, 70
Double displacement reaction, 188
Double replacement reaction, 188,
    191–192, 245
Dow process, 296
Dry cell, 272
Dynamic equilibrium, 153–154

E

Efflorescent, 179
Effusion, 118
Einstein, Albert, 7, 9, 43, 354
Electrochemical cell
    applications of, 269, 272–273
    battery, 273
Electrochemistry, 253–256
Electrode potentials, 264–268
Electrolysis, 269
Electrolytes, 251
Electrolytic cells
    description of, 269–271
    quantitative aspects of, 271–272
Electrolytic reactions, 269

Electromagnetic process, 356
Electromotive force (EMF), 264
Electron capture, 353
Electron dot notation, 50
Electronegativity, 56
Electron(s)
    configuration of, 47–50
    definition of, 32–33
    transfer of, 255
Electroplating, 269
Electrostatic repulsion, 71–74
Electrovalence bonds, 64–65
Element, 4, 490
Empirical formula, 95, 312
Endothermic reaction, 6, 8–9, 168
Endpoint, 240
Energy
    activation, 208
    bond dissociation, 199
    conservation of, 9
    conversion of, 7
    definition of, 7
    forms of, 7–8
    free, 228
    nuclear, 354–356
Energy levels, 34
Enthalpy
    changes in, 194–195
    from bond energies, 199
    k magnitude, 226–227
Entropy, 193–194, 226–227
Equations
    ionic, 101
    list of, 485
    reaction, 98–101
    redox, 274
Equilibrium
    acid ionization constants, 221
    common ion effect, 225–226
    constant, 215–216
    description of, 213–218
    ionization constant of, 222–223
    pressure effects, 219–220

schematic diagram of, 5
states of, 4
Measurements
    accuracy of, 17–19
    heat, 15
    observations, 10
    precision in, 17–19
    reaction rate, 205
    scientific method, 9–10
    scientific notation, 15–16
    SI system, 10–13
    significant figures, 19–21
    temperature, 13–14
    uncertainty of, 17–18
Melting point, 155, 298
Mendeleev, Dimitry, 51
Mendeleev's table, 52
Meniscus, 18
Mercury barometer, 115
Metallic bond, 68
Metalloids, 54, 298–299
Metal(s)
    acids on, 234
    definition of, 38
    extracting, 295
    periodic table location of, 53–54
    properties of, 294–296
    qualitative tests for, 375
    reduction methods for, 296–298
Methane, 313–314
Methanoic acid, 330
Methanol, 327
Methyl anthranilate, 334
Methyl salicylate, 334
Methylbenzene, 323
Meyer, Lothar, 51
Microcircuitry, 299
Millikan, Robert, 33
Miscible, 168
Mixtures
    composition of, 4–5
    water, 168–169
Modern atomic theory, 32
Molality, 172, 175–176

Molar heat of reaction, 195
Molar mass, 134–136, 138
Molar volume, 136–137
Molarity, 171–172
Mole
    definition of, 134
    fraction, 173
    rations, 139
    relationships, 139–140
Molecular crystals, 80
Molecular geometry, 73–74
Molecular liquids, 80
Molecular mass, 95
Molecule
    definition of, 4, 63, 134
    forces of attraction, 68
    polar, 66
    types of, 74
Mole-mole relationships, 139
Monomers, 337
Monosaccharides, 336
Moseley, Henry, 52
Multiple-effect evaporation, 159

N
Natural gas, 261
Negative ions, 374
Neutralization, 234, 245
Neutron, 34
Nitrate, 374
Nitric acid, 291–292
Nitric oxide, 293, 373
Nitrite, 374
Nitrogen family
    compounds of, 293
    description of, 291
    nitric acid, 291–292
    properties of, 294
Nitrous oxide, 293, 373
Noble gases
    description of, 38, 53
    electron distribution of, 63–64
n-Nonane, 314
Nonelectrolytes, 252